STATISTICS WITH VAGUE DATA

THEORY AND DECISION LIBRARY

General Editors: W. Leinfellner and G. Eberlein

Series A: Philosophy and Methodology of the Social Sciences
Editors: W. Leinfellner (Technical University of Vienna)
G. Eberlein (Technical University of Munich)

Series B: Mathematical and Statistical Methods
Editor: H. Skala (University of Paderborn)

Series C: Game Theory, Mathematical Programming and Mathematical Economics
Editor: S. Tijs (University of Nijmegen)

Series D: System Theory, Knowledge Engineering and Problem Solving
Editor: W. Janko (University of Vienna)

SERIES B: MATHEMATICAL AND STATISTICAL METHODS

Editor: H. Skala (Paderborn)

Scope

The series focuses on the application of methods and ideas of logic, mathematics and statistics to the social sciences. In particular, formal treatment of social phenomena, the analysis of decision making, information theory and problems ofinference will be central themes of this part of the library. Besides theoretical results, empirical investigations and the testing of theoretical models of real world problems will be subjects of interest. In addition to emphasizing interdisciplinary communication, the series will seek to support the rapid dissemination of recent results.

STATISTICS WITH VAGUE DATA

by

RUDOLF KRUSE
Institut für Betriebssysteme und Rechnerverbund,
Technische Universität Braunschweig, F.R.G.

and

KLAUS DIETER MEYER
Institut für Mathematische Stochastik,
Technische Universität Braunschweig, F.R.G.

D. REIDEL PUBLISHING COMPANY

A MEMBER OF THE KLUWER  ACADEMIC PUBLISHERS GROUP

DORDRECHT / BOSTON / LANCASTER / TOKYO

Library of Congress Cataloging in Publication Data

Kruse, Rudolf.
Statistics with vague data.

(Theory and decision library. Series B, Mathematical and statistical methods)
Bibliography: p.
Includes index.
1. Fuzzy sets. 2. Mathematical statistics.
I. Meyer, Klaus Dieter, 1960– II. Title. III. Series.
QA248.K744 1987 519.5 87–16449
ISBN 90–277–2562–4

Published by D. Reidel Publishing Company,
P.O. Box 17, 3300 AA Dordrecht, Holland.

Sold and distributed in the U.S.A. and Canada
by Kluwer Academic Publishers,
101 Philip Drive, Assinippi Park, Norwell, MA 02061, U.S.A.

In all other countries, sold and distributed
by Kluwer Academic Publishers Group,
P.O. Box 322, 3300 AH Dordrecht, Holland.

Printed in The Netherlands

TABLE OF CONTENTS

PREFACE

This monograph is an attempt to unify existing works in the field of random sets, random variables, and linguistic random variables with respect to statistical analysis. It is intended to be a tutorial research compendium.

The material of the work is mainly based on the postdoctoral thesis (Habilitationsschrift) of the first author and on several papers recently published by both authors. The methods form the basis of a user-friendly software tool which supports the statistical inference in the presence of vague data.

Parts of the manuscript have been used in courses for graduate level students of mathematics and computer sciences held by the first author at the Technical University of Braunschweig. The textbook is designed for readers with an advanced knowledge of mathematics.

The idea of writing this book came from Professor Dr. H. Skala. Several of our students have significantly contributed to its preparation. We would like to express our gratitude to Reinhard Elsner for his support in typesetting the book, Jörg Gebhardt and Jörg Knop for preparing the drawings, Michael Eike and Jürgen Freckmann for implementing the programming system and Günter Lehmann and Winfried Böer for proofreading the manuscript. This work was partially supported by the Fraunhofer-Gesellschaft.

We are indebted to D. Reidel Publishing Company for making the publication of this book possible and would especially like to acknowledge the support which we received from our families on this project.

Braunschweig
May 1987

R. Kruse
K.D. Meyer

1 Introduction

In this monograph, problems resulting from two types of uncertainty – randomness and vagueness – are treated. Randomness involves only uncertainties in the outcomes of an experiment; vagueness, on the other hand, involves uncertainties in the meaning of the data. Examples of randomness can be found in any well-defined random experiment such as tossing a coin, observing queues, and recording observed signals. Examples of vagueness include experiments involving linguistic data, which for the purpose of information processing have to be modeled with greater care. A typical example of the occurrence of vague data is to be seen in knowledge-based systems, in which the combined knowledge of a group of experts is often vague.

Ideally, such an approach must be both rigorous and heuristic in the sense that the mathematical aspects of the model contain all necessary expert knowledge appropriately encoded.

The work is composed of twelve chapters. With respect to the mathematics involved, a standard knowledge of mathematical logic, measure theory, probability theory, and mathematical statistics is required, and graduate students in applied mathematics, engineering sciences, and mathematical statistics should find little difficulty in reading the treatise.

Chapter 2 deals with the mathematical concepts of vague data, L-sets, and fuzzy sets. Chapter 3 treats useful classes of vague data of the real line, while in Chapter 4 several operations with fuzzy sets are discussed. Chapter 5 contains a detailed application of the techniques developed in the preceding chapters, in order to represent vague data in a digital computer and to provide efficient algorithms for the operations with those data. Chapter 6 discusses topological properties of vague data. Chapter 7 gives an introduction to the well-established (but relatively little known) theory of random sets and fuzzy random variables. In Chapter 8 the descriptive statistics with vague data are developed. Based on the notions distribution function and i.i.d. sequences described in Chapter 9, limit theorems are proved in Chapter 10. In Chapter 11 an application of these results to statistical inference can be found. Finally Chapter 12 presents a software tool by which the methods described in this monograph can be supported. Furthermore it contains a sample session which may clarify these methods.

2 Vague data

This chapter deals with the problems of obtaining vague data and their mathematical description. Normally we can obtain data material e.g. by interviewing people or by using measuring methods. The persons or objects, resp., which are the subjects of the measurements are called the elements of a population. The quantities which are related to the questions or the measurements are called characteristics. By observing the characteristics of the elements of a population we obtain the value of a characteristic. Taken together, these values constitute the data.

Example 2.1 Suppose we seek information about the sex, the age, and the size of a number of mayflies. The mayflies then represent the population and the characteristics are their sex, age, and size. Sex is a qualitative characteristic with only two different possible values, whereas age and size are quantitative characteristics with an infinite number of different possible values.

Before we set out to measure or to investigate, resp., the values of a characteristic, we have to define a scale which contains all possible values of the characteristic. Several different scales are possible. The nominal scale represents the lowest level; it is not possible to give a ranking list or to compare values of a nominal scale. Typical examples of nominally scaled characteristics are colour, profession, sex, or race. The ordinal scale has a higher level. Here it is possible to distinguish the values by their intensities and to rank these intensities. An example of an ordinal scaled characteristic is a school report with the possible values of very good, good, ..., unsatisfactory. Here, we are not able to give a reasonable interpretation of the distance between two different values. This, however, is possible on a scale of the highest level, the ratio scale. Typical examples are ages, lifetimes, weights, sizes, and intensities of current. If there is a value of 10 ampères and a value of 1 ampère, the difference of these values can clearly be interpreted.

In the following pages we will restrict ourselves to characteristics with a ratio scale, where the values are numbers of the real line $\mathbb{R}$, and our data will be real numbers. A further classification of characteristics has to be made, namely the distinction between continuous and discrete characteristics. A discrete characteristic has a finite or at most countable number of different values, whereas a characteristic is called continuous if we have a non-denumerable set of different (possible) values. The number of the inhabitants of the town 'Rotenburg' is a discrete characteristic ($\{0, 1, 2, 3, \ldots\}$); the characteristic "life-time" is continuous ($[0, +\infty)$).

In practice, however, it is in most cases not possible to observe the continuous characteristics directly. If, for example, we are measuring a current by a digital measuring apparatus, we receive data which are theoretically continuous but which, in reality, are discrete. The reason for this is that measurement proceedings are not precise. In this example a kind of uncertainty appears which is different from randomness, which we call vagueness We have the situation that the observed datum 20 ampères is different from the true datum 20.12 ampères, for example. If we want to work with such data, we have to take this problem into account. One way to achieve a mathematically precise result is to extend the data types. Suppose we know that the digital ammeter shows 20 ampères if and only if the true current is between 19.4 and 20.6 ampères. Then it seems to be reasonable to use the interval $[19.4, 20.6]$ as (vague) data for the information "the ammeter shows 20". Moreover, it should be stressed that the interval is not the true value itself, but that the original value would be located somewhere in this interval.

In practice, it also happens that experts have conflicting views about the description of problems. If for instance, one expert claims that the age of an insect is in $[10, 12]$, another expert may claim that it is in $[11, 13]$, and if we presume that at least one of them is right, we know that the real value is in the union of the intervals, here $[10, 13]$. But if we presume that both experts are right, the correct claim would be the intersection of the intervals, here $[11, 12]$. In order to describe these kinds of data, we use arbitrary subsets of $\mathbb{R}$, $\mathfrak{P}(\mathbb{R}) = \{A \mid A \subseteq \mathbb{R}\}$. $\mathfrak{P}(\mathbb{R})$ is a Boolean algebra with respect to the union ($\cup$), intersection ($\cap$), and complement ($\bar{\ }$).

Consider next the characteristics "Real number greater than 10" and "Large number". As the former refers to a characteristic which is unambiguously either possessed or not possessed by a real number under consideration, we have an objective attribute. "Large number" refers to a characteristic whose linguistic definition contains inherent ambiguity. Here a subjective interpretation is possible by an observer.

From the very fact that individuals may subjectively decide the amount of the attribute possessed by an object, we are faced with the problem of assigning numbers to subjective perceptions. The construction of such representations is the province of mathematical psychology.

We are primarily interested in the purely mathematical background. Our problem is to decide which is the right scale for measuring subjectivity.

A natural scale for our considerations is (at least) a bounded ordinal scale. We therefore have a set $L \neq \emptyset$, a total order $\preceq$ on L and two elements $l_{\min}$ and $l_{\max}$ such that $l_{\min} \preceq l \preceq l_{\max}$ for all $l \in L$. This is one assumption for our work.

Moreover, we are often able to evaluate statements of the kind

$$l_1 \preceq l_2 \text{ at least as much as } l_3 \preceq l_4.$$

If "at least as much as" is assumed to be a total ordering on $L \times L$, in sign $\sqsupseteq$, satisfying the five axioms of an algebraic difference structure (see [111,162],e.g.), then there is a bounded, realvalued function μ on L such that

$$l_1 \preceq l_2 \Leftrightarrow \mu(l_1) \leq \mu(l_2) \quad \text{and}$$

$$[l_1 \preceq l_2] \sqsupseteq [l_3 \preceq l_4] \Leftrightarrow \mu(l_2) - \mu(l_1) \geq \mu(l_4) - \mu(l_2).$$

$\mu(l)$ is called the value of acceptability μ is then unique up to linear transformations (see D.H. Krantz et al. [111]).

On the one hand, we see that interval scales are often appropriate for the task of measuring subjectivity. On the other hand, ratio scales for representing subjectivity are in most cases not obtainable.

Let us summarize the results of the preceding section. There are characteristics which have been described by subsets of the real line, and there is a bounded ordinal scale by which subjectivity can be measured. These notions will be described by vague data. Consider an expert's opinion "The insect is approximately 10 hours old". It is assumed that our expert is able to distinguish between two acceptability degrees maximum $\preceq$ sure. A natural description of these vague data is to assign to each acceptability the set of the real numbers for which the acceptability holds at least. We introduce the following notion:

Definition 2.2 *A flou set f is a pair $f = (E, F)$ with $E \subseteq F \subseteq \mathbb{R}$. E is called the sure region, F the maximum region, and $F \backslash E$ the flou region. The set of all flou sets is denoted by $Fl(\mathbb{R})$.*

We can describe the vague datum "approximately 10 hours old" by the flou set $f = (\{10\}, [8, 12])$ because precise boundaries are not defined. The assignment of the regions E and F is of subjective nature and, in general, reflects the context in which the problem is investigated.

The union, intersection, and complementation are defined canonically in the following way:

If $f = (E, F)$ and $f' = (E', F')$ are flou sets, then we define

$$f \cup f' \stackrel{\mathrm{d}}{=} (E \cup E', F \cup F'),$$
$$f \cap f' \stackrel{\mathrm{d}}{=} (E \cap E', F \cap F'),$$
$$\overline{f} \stackrel{\mathrm{d}}{=} (\overline{F}, \overline{E}).$$

$\left(Fl(\mathbb{R}), \cup, \cap, ^{-}\right)$ is a completely distributive lattice with the least element $0 = (\emptyset, \emptyset)$ and the greatest element $1 = (\mathbb{R}, \mathbb{R})$, but it is not a Boolean algebra, since from $f = (E, F)$, $E \subseteq F$, $E \neq F$, it follows

$$f \cup \overline{f} = (E \cup \overline{F}, F \cup \overline{E}) = (E \cup \overline{F}, \mathbb{R}) \neq 1,$$

$$f \cap \overline{f} = (E \cap \overline{F}, F \cap \overline{E}) = (\emptyset, F \backslash E) \neq 0.$$

The concept of a flou set from Y. Gentilhomme [61] and C.V. Negoita [150] quoted above may be generalized to arbitrary bounded ordinal scales.

Let $(L, \preceq)$ be a total ordered set with $l_{\min} \preceq l \preceq l_{\max}$ for all $l \in L$. We assume that L is complete, that is $\sup A$ and $\inf A$ exist for all $A \subseteq L$. We interprete each element of L as a "value of acceptability". The sets $L_1 =$ { sure , maximum } and $L_2 =$ { sure , vague , maximum } with the relation maximum $\preceq$ vague $\preceq$ sure are examples for the lattices quoted above.

Definition 2.3 *Let $(L, \preceq)$ be a total ordered set. An L-flou set is a family $f = (E_\alpha)_{\alpha \in L}$, $E_\alpha \subseteq \mathbb{R}$, such that*

$$E_{\inf\{\alpha_i \mid i \in I\}} = \bigcup_{i \in I} E_{\alpha_i} \quad \text{for all} \quad \{\alpha_i\}_{i \in I} \subseteq L.$$

The set of all L-flou sets is denoted by $Fl_L(\mathbb{R})$.

We can define a complete lattice structure in $Fl_L(\mathbb{R})$ by setting

$$f' \cup f'' = (E'_\alpha \cup E''_\alpha)_{\alpha \in L} \quad \text{and}$$
$$f' \cap f'' = \bigcup \{f \in Fl_L(\mathbb{R}) \mid f \subseteq f', f \subseteq f''\},$$

if $f' = (E'_\alpha)_{\alpha \in L}$, $f'' = (E''_\alpha)_{\alpha \in L}$. $f' \subseteq f''$ means: $E_\alpha \subseteq E''_\alpha$ for all $\alpha \in L$. Note that i.g. $(E'_\alpha \cap E''_\alpha)_{\alpha \in L} \notin Fl_L(\mathbb{R})$.

Example 2.4 Let us consider again the vague datum "The insect is approximately 10 hours old" (see Ex. 2.1). The expert may choose the chain

L_2 as a hierarchical acceptability, and he can describe the vague datum by an L-flou set represented in the following figure:

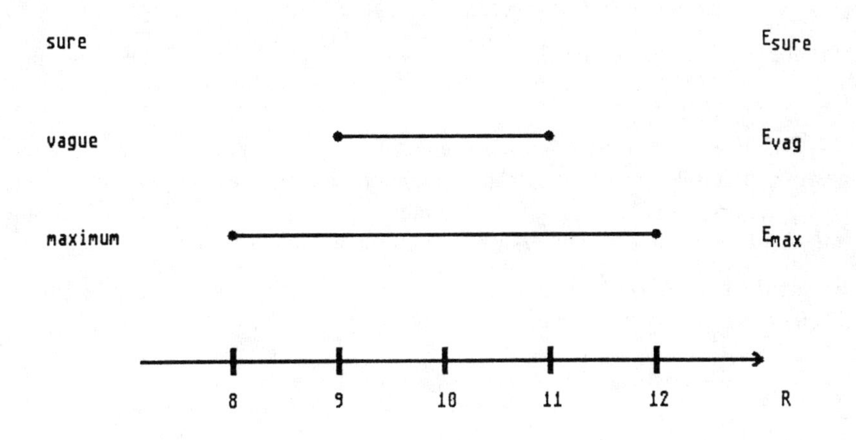

Fig. 2.1. Example for an L-flou set with $L = \{$ maximum, vague, sure $\}$ representing "approximately 10 hours".

The example motivates the following considerations.

Definition 2.5 *An L-set is a function $g : I\!R \rightarrow L$. $F_L(I\!R)$ denotes the class of all L-sets. In $F_L(I\!R)$ there are defined the following operations:*

$$(g \vee g')(x) = g(x) \vee g'(x) \quad \textit{union}$$
$$(g \wedge g')(x) = g(x) \wedge g'(x) \quad \textit{intersection}$$

where $a \vee b = \max\{a, b\}$ and $a \wedge b = \min\{a, b\}$ for $a, b \in L$.

$F_L(I\!R)$ is a completely distributive lattice. We have the following theorem, which is a special version of the so called representation theorem ([151, 178]).

Theorem 2.6 *There is a dual isomorphism of complete lattices between $F_L(I\!R)$ and $Fl_L(I\!R)$.*

Theorem 2.6 shows that $F_L(\mathbb{R})$ is our appropriate data typ. It is clear that there is an order preserving map Φ from $(L, \preceq)$ to $([0,1], \leq)$ such that $\Phi(l_{\min}) = 0$ and $\Phi(l_{\max}) = 1$. Therefore, we can restrict ourselves to consider $(L, \leq)$ with $L \subseteq [0,1]$. We define:

Functions in $F_L(\mathbb{R})$ where $L \subseteq [0,1]$ are called fuzzy sets of the real line. $\mu(x)$, $x \in \mathbb{R}$, is called the acceptability value. When A is an ordinary set, its indicator function $\mathrm{I}_A : \mathbb{R} \to [0,1]$, $x \mapsto \begin{cases} 1 & , \text{if } x \in A \\ 0 & , \text{if } x \notin A \end{cases}$, can take only two values, one or zero.

Fuzzy sets can be viewed as a generalization of ordinary sets, where we have no precisely defined criteria of membership. In these classes an object needs neither to belong or not to belong to a class, therefore it may be an intermediate grade of membership, ranging from zero to one.

Example 2.7 An expert may choose subjectively the membership function μ illustrated in Fig. 2.2 for a "class of numbers near two".

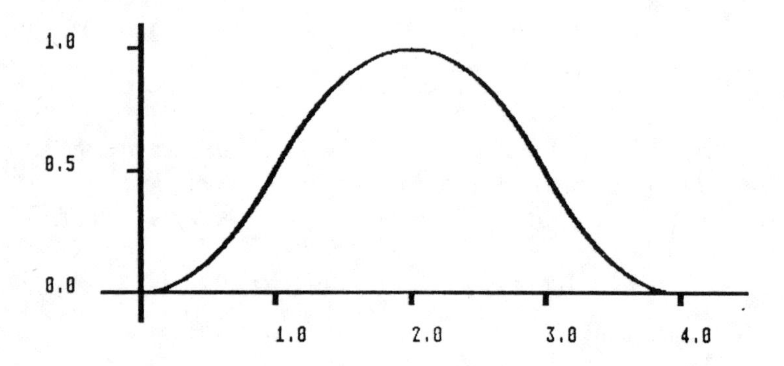

Fig. 2.2. The membership function of the "class of all numbers near two" being an L-set with $L = [0,1]$.

This membership function is defined by

$$\mu(x) = \begin{cases} 0 & \text{, if } x \leq 0 \\ 2 \cdot \left(\frac{x}{2}\right)^2 & \text{, if } 0 \leq x \leq 1 \\ 1 - 2 \cdot \left(\frac{x-2}{2}\right)^2 & \text{, if } 1 \leq x \leq 3 \\ 2 \cdot \left(\frac{x-4}{2}\right)^2 & \text{, if } 3 \leq x \leq 4 \\ 0 & \text{, if } x \geq 4 \, . \end{cases}$$

It ought to be pointed out here that the concept of a fuzzy set is completely non-probabilistic in nature, although the membership function of a fuzzy set has some resemblance to a probability density function. We use it only for a reasonable mathematical description of vague, ill defined data. Fuzzy sets generalize real numbers, intervals, subset, flou sets, and *L*-flou sets in a canonical way.

The foundation of the fuzzy set theory was given by L.A. Zadeh [212] in 1965. A basic controversy has grown up with respect to the modeling and the treatment of uncertainties. Fuzzy sets are effective tools for dealing with vagueness in our sense, but it should be clear that its use is not a panacea for the description of vagueness.

3 Fuzzy sets of the real line

In Chapter 2 we realized that the notion of a fuzzy set of the real line is an appropriate tool for the representation of vague data. In the following we introduce mathematically the concepts of a fuzzy set and of a set representation.

Definition 3.1 *A fuzzy subset μ of a set $A \neq \emptyset$ is a function from A to $[0,1]$. We denote by $E(A)$ the class of all fuzzy subsets of A. A fuzzy subset μ of A is called normal if there is a $t \in A$ such that $\mu(t) = 1$. We denote by $F(\mathbb{R})$ the class of all normal fuzzy subsets of the real line $\mathbb{R}$.*

In the following part we restrict ourselves to the class $F(\mathbb{R})$, since vague data often have at least one real number with 1 as the value of acceptability. For the representation of a fuzzy set B of $\mathbb{R}$ the notation $\int_{\mathbb{R}} \mu_B(x)dx$ is often used.

We denote by $\mathfrak{P}(\mathbb{R})$ the set of all subsets of $\mathbb{R}$, $\mathfrak{P}(\mathbb{R}) \stackrel{\mathrm{d}}{=} \{A \mid A \subseteq \mathbb{R}\}$. If $A \in \mathfrak{P}(\mathbb{R})$, then the characteristic function of A, I_A, is defined in the following way:

$$\mathrm{I}_A : \mathbb{R} \to [0,1], \quad x \mapsto \begin{cases} 1 & , \text{if } x \in A \\ 0 & , \text{if } x \in \mathbb{R} \backslash A \, . \end{cases}$$

It is easy to prove that $\mathfrak{P}(\mathbb{R})$ and the set of all characteristic functions $CH(\mathbb{R}) \stackrel{\mathrm{d}}{=} \{f \mid f : \mathbb{R} \to \{0,1\}\}$ are isomorphic (as sets) with isomorphism $\psi(A) \stackrel{\mathrm{d}}{=} \mathrm{I}_A$. In some sense fuzzy sets are generalized characteristic functions.

An important tool for dealing with fuzzy sets are level sets (C.V. Negoita and D.A. Ralescu [150]) and set representations (M. Miyakoshi and M. Shimbo [141]).

Definition 3.2 *Let $\mu \in F(\mathbb{R})$. $\{A_\alpha \mid \alpha \in (0,1)\}$ is called a set representation of μ, if and only if*
(i) $0 < \alpha \leq \beta < 1 \quad \Rightarrow \quad A_\beta \subseteq A_\alpha \subseteq \mathbb{R}$ and
(ii) $\forall t \in \mathbb{R} : \ \mu(t) = \sup \{\alpha \mathrm{I}_{A_\alpha}(t) \mid \alpha \in (0,1)\}$ *hold.*

If a system $\{A_\alpha \mid \alpha \in (0,1)\}$ is given, and if
(i) $0 < \alpha \leq \beta < 1 \quad \Rightarrow \quad A_\beta \subseteq A_\alpha \subseteq I\!R$ and
(ii) $\bigcap_{\alpha \in (0,1)} A_\alpha \neq \emptyset$ holds,
then we can define $\mu \in F(I\!R)$ by

$$\mu(t) \stackrel{d}{=} \sup\{\alpha \mathrm{I}_{A_\alpha}(t) \mid \alpha \in (0,1)\}, \quad t \in I\!R.$$

Obviously $\{A_\alpha \mid \alpha \in (0,1)\}$ is a set representation of μ. When we want to exhibit an element $x \in I\!R$ that typically belongs to a fuzzy set $\mu \in F(I\!R)$, we may demand that its membership value is greater or equal than some threshold $\alpha \in [0,1]$. The ordinary set of such elements is the α-level set $\mu_{\overline{\alpha}}$,

$$\mu_{\overline{\alpha}} \stackrel{d}{=} \{t \in I\!R \mid \mu(t) \geq \alpha\} .$$

One also defines the strong α-cut

$$\mu_\alpha \stackrel{d}{=} \{t \in I\!R \mid \mu(t) > \alpha\} .$$

We have obviously $\mu_{\overline{0}} = I\!R$ and $\mu_1 = \emptyset$. The α-level sets and the strong α-cuts will play a central role in the sequel. Therefore they have to be analysed in detail. The following theorem shows that strong α-level sets and α- cuts can be used as set representations for a given fuzzy set.

Theorem 3.3 *Let $\mu \in F(I\!R)$. Then $\{\mu_\alpha \mid \alpha \in (0,1)\}$ and $\{\mu_{\overline{\alpha}} \mid \alpha \in (0,1)\}$ are set representations of μ.*

Proof Obviously μ_α and $\mu_{\overline{\alpha}}$ are not empty for all $\alpha \in (0,1)$, and $0 < \alpha \leq \beta < 1$ implies $\mu_\beta \subseteq \mu_\alpha$ and $\mu_{\overline{\beta}} \subseteq \mu_{\overline{\alpha}}$.
Let $t \in I\!R$ arbitrary. We want to show that

$$\mu(t) \leq \sup\{\alpha \mathrm{I}_{\mu_\alpha}(t) \mid \alpha \in (0,1)\} \quad \text{and}$$
$$\mu(t) \geq \sup\{\alpha \mathrm{I}_{\mu_{\overline{\alpha}}}(t) \mid \alpha \in (0,1)\} \quad \text{is valid.}$$

For abbreviation, define $\gamma \stackrel{d}{=} \mu(t)$. If $\gamma = 0$, the first assertion is obvious. If $\gamma > 0$, then $t \in \mu_{\gamma-\epsilon}$ is valid for all $\epsilon > 0$. This implies

$$(\gamma - \epsilon)\mathrm{I}_{\mu_{\gamma-\epsilon}}(t) = \gamma - \epsilon$$

for all $\epsilon > 0$, and the first assertion follows.
For all $\alpha \in (0,1)$ we have

$$\mathrm{I}_{\mu_{\overline{\alpha}}}(t) = \begin{cases} 1 & , \text{if } \alpha \in (0,\gamma] \\ 0 & , \text{if } \alpha \in (\gamma,1). \end{cases}$$

This implies

$$\alpha \mathrm{I}_{\mu_{\overline{\alpha}}}(t) = \begin{cases} \alpha & , \text{if } \alpha \in (0, \gamma] \\ 0 & , \text{if } \alpha \in (\gamma, 1] \end{cases} \quad , \text{and}$$

$$\alpha \mathrm{I}_{\mu_{\overline{\alpha}}}(t) \leq \gamma \mathrm{I}_{\gamma_{\overline{\alpha}}}(t) = \gamma = \mu(t).$$

This demonstrates the second assertion and implies the two inequalities

$$\mu(t) \leq \sup \{\alpha \mathrm{I}_{\mu_\alpha}(t) \,|\, \alpha \in (0,1)\} \leq \sup \{\alpha \mathrm{I}_{\mu_{\overline{\alpha}}}(t) \,|\, \alpha \in (0,1)\} \leq \mu(t),$$

as $\mu_\alpha \subseteq \mu_{\overline{\alpha}}$ is valid for $\alpha \in (0,1)$. This completes the proof.

In many cases it is convenient to express the membership function of a fuzzy set in term of a standard function, whose parameters may be adjusted to fit a specified membership function in an approximate fashion. Two such functions, which may be used to describe the vague datum "approximately a", are defined by

$$g^{(a,b)}(x) \overset{\text{d}}{=} \exp\left(-\left[\frac{x-a}{b}\right]^2\right), \; b > 0, \; a \in \mathbb{R},$$

$$g^{(a,0)}(x) \overset{\text{d}}{=} \mathrm{I}_{\{a\}}(x), \; a \in \mathbb{R},$$

$$P^{(a,b)}(x) \overset{\text{d}}{=} \max\left\{1 - \frac{1}{2}b(x-a)^2, 0\right\}, \; b > 0, \; a \in \mathbb{R}.$$

It is easy to calculate that

$$\left(g^{(a,b)}\right)_{\overline{\alpha}} = \left[a - b\sqrt{\ln\frac{1}{\alpha}}, a + b\sqrt{\ln\frac{1}{\alpha}}\right], \; \alpha \in (0,1],$$

$$\left(g^{(a,b)}\right)_{\alpha} = \left(a - b\sqrt{\ln\frac{1}{\alpha}}, a + b\sqrt{\ln\frac{1}{\alpha}}\right), \; \alpha \in (0,1),$$

$$\left(g^{(a,b)}\right)_{\overline{0}} = \mathbb{R}, \; \left(g^{(a,b)}\right)_{0} = \mathbb{R}, \left(g^{(a,b)}\right)_{1} = \emptyset,$$

and

$$\left(P^{(a,b)}\right)_{\overline{\alpha}} = \left[a - \sqrt{\frac{2}{b}(1-\alpha)}, a + b\sqrt{\frac{2}{b}(1-\alpha)}\right], \; \alpha \in (0,1],$$

$$\left(P^{(a,b)}\right)_{\alpha} = \left(a - \sqrt{\frac{2}{b}(1-\alpha)}, a + b\sqrt{\frac{2}{b}(1-\alpha)}\right), \; \alpha \in [0,1),$$

$$\left(P^{(a,b)}\right)_{\overline{0}} = \mathbb{R}, \; \left(P^{(a,b)}\right)_{\overline{1}} = \emptyset.$$

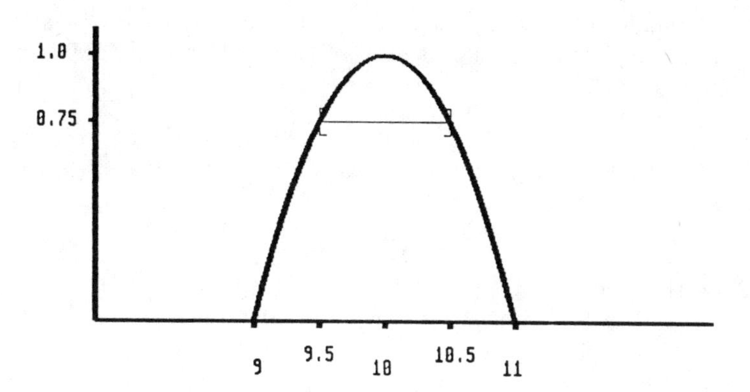

Fig. 3.1. The membership function of $P^{(10,2)}$ and its 0.75- level-set.

In the figure above we have $(P^{(10,2)})_{\overline{0.75}} = [9.5, 10.5]$.

Theorem 3.4 *Let $\mu \in F(\mathbb{R})$ and $\{A_\alpha \,|\, \alpha \in (0,1)\}$ be a set representation of μ.*

(i) Let $\alpha \in [0,1)$ and $\{\alpha_r\}_{r \in \mathbb{N}}$ be a strictly decreasing sequence such that $\lim_{r\to\infty} \alpha_r = \alpha$ is valid as well as $\alpha_r \in (\alpha, 1)$ for all $r \in \mathbb{N}$.

Then $\mu_\alpha = \bigcup_{r=1}^{\infty} A_{\alpha_r}$ is valid.

(ii) Let $\beta \in (0,1]$ and $\{\beta_r\}_{r \in \mathbb{N}}$ be a strictly increasing sequence such that $\lim_{r\to\infty} \beta_r = \beta$ is valid as well as $\beta_r \in (0, \beta)$ for all $r \in \mathbb{N}$.

Then $\mu_{\overline{\beta}} = \bigcap_{r=1}^{\infty} A_{\beta_r}$ is valid.

Proof

(i) If $t \in \mu_\alpha$, then there is a β with $\beta \cdot \mathrm{I}_{A_\beta}(t) > \alpha$. This implies $t \in A_\beta$ for a $\beta \in (\alpha, 1)$ or $t \in A_{\alpha_r}$ for an $r \in \mathbb{N}$.
If there is an $r \in \mathbb{N}$ with $t \in A_{\alpha_r}$, then $\alpha_r \mathrm{I}_{A_{\alpha_r}}(t) = \alpha_r > \alpha$ is valid, i.e. $t \in \mu_\alpha$.

(ii) Let $t \in \mu_{\overline{\beta}}$. If $r \in \mathbb{N}$ arbitrary, there is an $\epsilon > 0$ with $\beta - \epsilon > \beta_r$. As $\mu(t) \geq \beta$ we can find an $\alpha \in (0,1)$ with $\alpha \mathrm{I}_{A_\alpha}(t) \geq \beta - \epsilon$, i.e. $t \in A_\alpha$ and $\alpha \geq \beta - \epsilon$. This implies $t \in A_{\beta-\epsilon} \subseteq A_{\beta_r}$.

If $t \in A_\beta$ holds for all $r \in \mathbb{N}$, we have $\beta_r I_{A_{\beta_r}}(t) = \beta_r$ for all $r \in \mathbb{N}$, and therefore $\mu(t) \geq \sup\{\beta_r I_{A_{\beta_r}}(t) \mid r \in \mathbb{N}\} = \beta$, i.e. $t \in \mu_{\overline{\beta}}$ follows.
•

The relation between different set representations – α-level set and strong α-cuts – are clarified with the following theorem which is due to M. Miyakoshi and M. Shimbo [141].

Theorem 3.5 *Let $\mu \in F(\mathbb{R})$ and $\{A_\alpha \mid \alpha \in (0,1)\}$ be a set of subsets of $\mathbb{R}$. Then $\{A_\alpha \mid \alpha \in (0,1)\}$ is a set representation of μ if and only if $\mu_\alpha \subseteq A_\alpha \subseteq \mu_{\overline{\alpha}}$ holds for all $\alpha \in (0,1)$.*

Proof

"$\Rightarrow$" Let $\alpha \in (0,1)$ arbitrary.
If $t \in \mu_\alpha$, then Theorem 3.4(i) shows that there is a $\beta \in (\alpha, 1)$ with $t \in A_\beta$. This implies $t \in A_\alpha$ as $A_\beta \subseteq A_\alpha$ holds.
If $t \in A_\alpha$, then

$$\mu(t) = \sup\{\gamma I_{A_\gamma}(t) \mid \gamma \epsilon (0,1)\} \geq \alpha I_{A_\alpha}(t) = \alpha$$

follows, i.e. $t \in \mu_{\overline{\alpha}}$.

"$\Leftarrow$" $0 < \alpha < \beta < 1$ implies $\mu_{\overline{\beta}} \subseteq \mu_\alpha$, and therefore $A_\beta \subseteq A_\alpha$ holds for $\alpha \leq \beta$.
By Theorem 3.3 we can conclude:

$$\begin{aligned}\mu(t) = \sup\{\alpha I_{\mu_\alpha}(t) \mid \alpha \in (0,1)\} &\leq \sup\{\alpha I_{A_\alpha}(t) \mid \alpha \in (0,1)\} \leq \\ &\leq \sup\{\alpha I_{\mu_{\overline{\alpha}}}(t) \mid \alpha \in (0,1)\} = \mu(t).\end{aligned}$$

In practical application we often have to calculate $\inf \mu_\alpha$ and $\sup \mu_\alpha$. The following theorem shows a mathematical property that will be useful in the sequel.

Theorem 3.6 *Let $\mu \in F(\mathbb{R})$ and $\{A_\alpha \mid \alpha \in (0,1)\}$ be a set representation of μ. Then*

$$\lim_{r \to \infty} \inf A_{\alpha + 1/(2r)(1-\alpha)} = \inf \mu_\alpha$$

and

$$\lim_{r \to \infty} \sup A_{\alpha + 1/(2r)(1-\alpha)} = \sup \mu_\alpha$$

is valid for all $\alpha \in [0,1)$.

Proof Let $\alpha \in [0,1)$.
For abbreviation, define $\alpha_r \stackrel{\text{d}}{=} \alpha + 1/2r(1-\alpha)$ for $r \in I\!N$.
For $r, s \in I\!N$ $r \leq s$ implies $\alpha_r \geq \alpha_s$ and by this $A_{\alpha_r} \subseteq A_{\alpha_s}$ holds.
Therefore $\{\inf A_{\alpha_r}\}_{r\in I\!N}$ is monotonously decreasing and $\{\sup A_{\alpha_r}\}_{r\in I\!N}$ is monotonously increasing.
By Theorem 3.4(i) we know that $\inf \mu_\alpha \leq \inf A_{\alpha_r} \leq \sup A_{\alpha_r} \leq \sup \mu_\alpha$ holds for $r \in I\!N$.
If $\inf \mu_\alpha > -\infty$, then $\{\inf A_{\alpha_r}\}_{r\in I\!N}$ is convergent, let A denote the limit. $A \geq \inf \mu_\alpha$ is obvious.
Let $\epsilon > 0$ be arbitrary. There exists $x \in \mu_\alpha$ with $x \leq \inf \mu_\alpha + \epsilon$.
By Theorem 3.4(i) we can find an $r \in I\!N$ with $x \in A_{\alpha_r}$, i.e.

$$\inf A_{\alpha_r} \leq x \leq \inf \mu_\alpha + \epsilon.$$

Therefore $A = \lim\limits_{r\to\infty} \inf A_{\alpha_r} \leq \inf \mu_\alpha$ follows.
If $\inf \mu_\alpha = -\infty$ and $N \in I\!N$ are arbitrary, there is an $x \in \mu_\alpha$ with $x \leq -N$. We can find $r \in I\!N$ with $x \in A_{\alpha_r}$ and can follow: $\inf A_{\alpha_r} \leq -N$. As $\{\inf A_{\alpha_r}\}_{r\in I\!N}$ is decreasing, it converges against $-\infty$.
The second assertion can be shown in a similar way.

It is natural to describe vague data of the kind "approximately x" by a fuzzy set μ such that the sets $\mu_{\overline{\alpha}}$, $\alpha \in (0,1]$ are finite intervals. Consider now the vague concepts "big numbers", "long distances", etc. These concepts may be described by fuzzy sets, such that the sets $\mu_{\overline{\alpha}}$ are infinite intervals. Two such functions are defined by

$$h_{a,b}(x) \stackrel{\text{d}}{=} \begin{cases} 1-\exp\left[-b(x-a)\right] & \text{, if } x \geq a \\ 0 & \text{, if } x \leq a \end{cases}, \quad b > 0,\ a \in I\!R,$$

$$q_{a,b}(x) \stackrel{\text{d}}{=} \begin{cases} 1-a/\left[1+b(x-a)^2\right] & \text{, if } x \geq a \\ 0 & \text{, if } x \leq a \end{cases}, \quad b > 0,\ a \in I\!R.$$

We see that, for example,

$$\left\{\left(a + \frac{1}{b}\ln\frac{1}{1-\alpha}, +\infty\right) \mid \alpha \in (0,1)\right\}$$

is a set representation of $h_{a,b}$, and that

$$\left\{\left[a + \sqrt{\frac{\alpha}{b(1-\alpha)}}, +\infty\right) \mid \alpha \in (0,1)\right\}$$

is a set representation of $q_{a,b}$.

The sets in the set representation of $h_{a,b}$ and $q_{a,b}$ are convex in the following sense:

$$A \subseteq I\!R \text{ convex } \overset{\mathrm{d}}{\Leftrightarrow} \forall x, y \in A,\ \forall \lambda \in [0,1]:\ \lambda x + (1-\lambda)y \in A.$$

The fuzzy sets $h_{a,b}$ and $q_{a,b}$ satisfy the F-convexity property (L.A. Zadeh [212]):

$$\forall x, y \in I\!R\ \forall \lambda \in [0,1]:\ \mu(\lambda x + (1-\lambda)y) \geq \min\{\mu(x), \mu(y)\}.$$

We give a slightly different definition of convexity.

Definition 3.7 *A fuzzy set $\mu \in F(I\!R)$ is called convex, if there is a set representation $\{A_\alpha \mid \alpha \in (0,1)\}$ of μ such that A_α is convex and closed for all $\alpha \in (0,1)$. We denote by $U(I\!R)$ the set of all convex fuzzy sets.*

Lemma 3.8 *Let $\mu \in F(I\!R)$. Then the following three propositions are equivalent:*

(i) $\mu \in U(I\!R)$

(ii) $\mu_{\overline{\beta}}$ *is closed and convex for all* $\beta \in (0,1]$

(iii) $\begin{cases} \mu[\lambda t + (1-\lambda)v] \geq \min\{\mu(t), \mu(v)\} \text{ is valid for all} \\ t \in I\!R,\ v \in I\!R,\ \lambda \in [0,1]; \\ \inf \mu_{\overline{\beta}} > -\infty \text{ implies } \inf \mu_{\overline{\beta}} \in \mu_{\overline{\beta}} \text{ for all } \beta \in (0,1]; \\ \sup \mu_{\overline{\beta}} < +\infty \text{ implies } \sup \mu_{\overline{\beta}} \in \mu_{\overline{\beta}} \text{ for all } \beta \in (0,1]. \end{cases}$

Proof (i) $\Rightarrow$ (ii) follows from Theorem 3.4(ii) as the intersection of closed sets is closed and that one of a sequence of non-increasing convex sets is convex.

(ii) $\Rightarrow$ (iii): Let $t \in I\!R$, $v \in I\!R$ and $\lambda \in [0,1]$.
Define $\beta \overset{\mathrm{d}}{=} \min\{\mu(t), \mu(v)\}$. Then $t \in \mu_{\overline{\beta}}$ and $v \in \mu_{\overline{\beta}}$ holds. As $\mu_{\overline{\beta}}$ is convex, $\lambda t + (1-\lambda)v \in \mu_{\overline{\beta}}$ holds, i.e. $\mu[\lambda t + (1-\lambda)v] \geq \beta = \min\{\mu(t), \mu(v)\}$.

The two other assertions follow from the fact that $\mu_{\overline{\beta}}$ is closed for all $\beta \in (0,1]$.

(iii) $\Rightarrow$ (i): Theorem 3.3 shows that $\left\{\mu_{\overline{\beta}} \mid \beta \in (0,1)\right\}$ is a set representation for μ. We want to show that $\mu_{\overline{\beta}}$ is convex and closed for all $\beta \in (0,1]$.
Let $\beta \in (0,1]$ and $t \in \mu_{\overline{\beta}}$, $v \in \mu_{\overline{\beta}}$ and $\lambda \in [0,1]$.
Then $\mu[\lambda t + (1-\lambda)v] \geq \min\{\mu(t), \mu(v)\} \geq \beta$ is valid, i.e. $\lambda t + (1-\lambda)v \in \mu_{\overline{\beta}}$.
Therefore $\mu_{\overline{\beta}}$ is convex.
By this and by (iii) $\mu_{\overline{\beta}}$ is closed.

The intersection of convex subsets of $\mathbb{R}$ is also convex. We therefore can define the convex hull co A of an arbitrary subset A of $\mathbb{R}$ by setting

$$\text{co } A \stackrel{\text{d}}{=} \bigcap \{B \subseteq \mathbb{R} \mid B \text{ is convex and } A \subseteq B\} =$$
$$= \{x \in \mathbb{R} \mid \exists a \in A,\ \exists b \in A,\ \exists \lambda \in [0,1]: \quad x = \lambda a + (1-\lambda)b\}.$$

Obviously co A is convex and $A = \text{co } A$ is valid if and only if A is convex. Our purpose is to define an analogue for fuzzy subsets of $\mathbb{R}$.

Lemma 3.9 *Let $\mu \in F(\mathbb{R})$ and $\{A_\alpha \mid \alpha \in (0,1)\}$ be a set representation of μ. Define $\nu \in F(\mathbb{R})$ by*

$$\nu(t) \stackrel{\text{d}}{=} \sup\{\alpha \mathrm{I}_{\text{co } A_\alpha}(t) \mid \alpha \in (0,1)\} \quad \textit{for } t \in \mathbb{R}.$$

Then $\nu_\alpha = \text{co } \mu_\alpha$, $\inf \nu_\alpha = \inf \mu_\alpha$, and $\sup \nu_\alpha = \sup \mu_\alpha$ holds for $\alpha \in [0,1)$.

Proof Obviously $\{\text{co } A_\alpha \mid \alpha \in (0,1)\}$ is a set representation of ν.
$\nu_\alpha = \bigcup_{r=1}^{\infty} \text{co } A_{\alpha+1/2r(1-\alpha)}$ is valid for $\alpha \in [0,1)$.

Let $\alpha \in [0,1)$, define $\alpha_r \stackrel{\text{d}}{=} \alpha + 1/2r(1-\alpha)$ for $r \in \mathbb{N}$.
For all $r \in \mathbb{N}$ $A_{\alpha_r} \subseteq \mu_\alpha$ is valid, it follows: co $A_{\alpha_r} \subseteq \text{co } \mu_\alpha$.
This shows $\nu_\alpha = \bigcup_{r=1}^{\infty} \text{co } A_{\alpha_r} \subseteq \text{co } \mu_\alpha$.
If $x \in \text{co } \mu_\alpha$, there exist $u, v \in \mu_\alpha$ and $\lambda \in [0,1]$ with $x = \lambda u + (1-\lambda)v$.
There are $r \in \mathbb{N}$, $s \in \mathbb{N}$ with $u \in A_{\alpha_r}$ and $v \in A_{\alpha_s}$.
It follows: $u \in A_{\alpha_t}$ and $v \in A_{\alpha_t}$ where $t \stackrel{\text{d}}{=} \max\{r, s\}$. Therefore,
$x \in \text{co } A_{\alpha_r} \subseteq \bigcup_{r=1}^{\infty} \text{co } A_{\alpha_r} = \nu_\alpha$ follows.
The other assertions are obvious.

This lemma shows that it is possible to give a well defined notion of a convex hull of a fuzzy set, i.e. this convex hull does not depend on the choice of the set representation.

Definition 3.10 *Let $\mu \in F(\mathbb{R})$, and let $\{A_\alpha \mid \alpha \in (0,1)\}$ be a set representation of μ. Then co $\mu \in F(\mathbb{R})$, where*

$$(\text{co } \mu)(t) \stackrel{\text{d}}{=} \sup\{\alpha \mathrm{I}_{\text{co } A_\alpha}(t) \mid \alpha \in (0,1)\},\ t \in \mathbb{R},$$

is called the convex hull of μ.
Obviously $(\text{co } \mu)_\alpha = \text{co } \mu_\alpha \subseteq \text{co } A_\alpha \subseteq (\text{co } \mu)_{\overline{\alpha}}$ is valid for all $\alpha \in (0,1)$.

A (usual) set $A \subseteq \mathbb{R}$ is contained in $B \subseteq \mathbb{R}$ if and only if $I_A(t) \leq I_B(t)$ holds for all $t \in \mathbb{R}$. We transfer this notion:

Definition 3.11 *Let $\mu \in F(\mathbb{R})$ and $\nu \in F(\mathbb{R})$. Then we define*

$$\mu \subseteq \nu \overset{\text{d}}{\Leftrightarrow} \mu(t) \leq \nu(t) \ \text{ holds for } \ t \in \mathbb{R} \ .$$

We can characterize co μ as the smallest convex fuzzy set containing μ, i.e.

$$\text{co } \mu = \inf \{\nu \in U(\mathbb{R}) \mid \mu \subseteq \nu\},$$

where inf is defined pointwise.
Two fuzzy sets $\mu, \nu \in E(\mathbb{R})$ are called equal, if they coincide as functions, i.e. we define

$$\mu = \nu \overset{\text{d}}{\Leftrightarrow} \forall x \in \mathbb{R} \ : \mu(x) = \nu(x)$$
$$\Leftrightarrow \mu \subseteq \nu \ \text{ and } \ \nu \subseteq \mu.$$

If two fuzzy sets $\mu, \nu \in U(\mathbb{R})$ are given, it is easy to check their equality.

Lemma 3.12 *Let $\mu, \nu \in U(\mathbb{R})$. Then $\mu = \nu$ holds if and only if*

$$\inf \mu_\alpha = \inf \nu_\alpha \ \text{ and } \ \sup \mu_\alpha = \sup \nu_\alpha$$

is valid for all $\alpha \in [0,1)$.

Proof
"$\Rightarrow$" is obvious.
"$\Leftarrow$" Let $\{A_\alpha \mid \alpha \in (0,1)\}$ be a set representation of μ, and $\{B_\alpha \mid \alpha \in (0,1)\}$ be one of ν, such that A_α and B_α are convex and closed for $\alpha \in (0,1)$. By Theorem 3.4(i) we know that μ_α and ν_α are convex for $\alpha \in (0,1)$. Define for $\alpha \in (0,1)$:

$$C_\alpha \overset{\text{d}}{=} \begin{cases} [\inf \mu_\alpha, \sup \mu_\alpha] & , \text{if } \inf \mu_\alpha > -\infty \text{ and } \sup \mu_\alpha < +\infty \\ [\inf \mu_\alpha, +\infty) & , \text{if } \inf \mu_\alpha > -\infty \text{ and } \sup \mu_\alpha = +\infty \\ (-\infty, \sup \mu_\alpha] & , \text{if } \inf \mu_\alpha = -\infty \text{ and } \sup \mu_\alpha < +\infty \\ \mathbb{R} & , \text{if } \inf \mu_\alpha = -\infty \text{ and } \sup \mu_\alpha = +\infty \end{cases}$$

By definition, $\mu_\alpha \subseteq C_\alpha \subseteq \mu_{\overline{\alpha}}$ is valid for $\alpha \in (0,1)$; this implies

$$\text{cl } \mu_\alpha \subseteq \text{cl } A_\alpha = A_\alpha \subseteq \mu_{\overline{\alpha}}$$

for $\alpha \in (0,1)$. For $A \subseteq \mathbb{R}$ cl A denotes the closure of A.
By Lemma 3.8(ii) we know that $\mu_{\overline{\alpha}} = \text{cl } \mu_{\overline{\alpha}}$, so $\{C_\alpha \mid \alpha \in (0,1)\}$ is a set

representation of μ. By assumption, $C_\alpha = \text{cl}\ \nu_\alpha$ is valid, and we can conclude:

$$\nu_\alpha \subseteq C_\alpha = \text{cl}\ \nu_\alpha \subseteq \text{cl}\ B_\alpha = B_\alpha \subseteq \text{cl}\ \nu_{\overline{\alpha}} = \nu_{\overline{\alpha}}.$$

This implies that $\{C_\alpha \mid \alpha \in (0,1)\}$ is also a set representation of ν.

It should be clear that the set of all real numbers with a positive membership grade plays an important role. We consider all convex fuzzy sets with a bounded "support", i.e. we define

$$U_C(\mathbb{R}) \stackrel{\text{d}}{=} \{\mu \in U(\mathbb{R}) \mid \inf \mu_0 > -\infty \ \text{ and } \ \sup \mu_0 < +\infty\}.$$

Definition 3.13 *Let $(\mu,\nu) \in [U(\mathbb{R})]^2$. Define*

$$\mu \leq \nu \stackrel{\text{d}}{\Leftrightarrow} \inf \mu_\alpha \leq \inf \nu_\alpha \ \textit{ and } \ \sup \mu_\alpha \leq \sup \nu_\alpha$$

is valid for all $\alpha \in [0,1)$.

Lemma 3.12 shows that $\mu \leq \nu$ and $\nu \leq \mu$ implies $\mu = \nu$. So "$\leq$" is an order relation on $U(\mathbb{R})$. Obviously not every pair of convex fuzzy sets can be compared, as an example neither $I_{[0,1]} \leq I_{[-1,2]}$ nor $I_{[-1,2]} \leq I_{[0,1]}$ is valid.

From these considerations above, it should be clear that the fuzzy sets in $U(\mathbb{R})$ are very easy to deal with, but, from the remark in the Chapter 2, we know that convex fuzzy sets do not suffice to describe all vague data. The fuzzy set $I_{[0,1]\cup[2,3]}$, the union of two ordinary disjoint intervals, is not in $U(\mathbb{R})$.

A wider class of fuzzy sets turned out to be useful for practical calculations.

Definition 3.14 *A fuzzy set $\mu \in F(\mathbb{R})$ belongs to $N(\mathbb{R})$, if and only if the following conditions are fulfilled:*

(a) $\inf \mu_0 > -\infty$ *and* $\sup \mu_0 < +\infty$

(b) There is a set representation $\{A_\alpha \mid \alpha \in (0,1)\}$ of μ, such that for all $\alpha \in (0,1)$ there exist an integer $N_\alpha \in \mathbb{N}$ and real numbers $\{(a_j)_\alpha \mid j \in \{1,\ldots,N_\alpha\}\}$ and $\{(b_j)_\alpha \mid j \in \{1,\ldots,N_\alpha\}\}$ with

(i) $(a_j)_\alpha \leq (b_j)_\alpha$ *for* $j \in \{1,\ldots,N_\alpha\}$,

(ii) $(b_{j-1})_\alpha < (a_j)_\alpha$ *for* $j \in \{2,\ldots,N_\alpha\}$, *and*

(iii) $A_\alpha = \bigcup_{j=1}^{N_\alpha} [(a_j)_\alpha, (b_j)_\alpha]$ *for* $\alpha \in (0,1)$.

$\{A_\alpha \mid \alpha \in (0,1)\}$ is then called an interval set representation of μ.

If $\mu \in F(\mathbb{R})$ such that there is an $n \in \mathbb{N}$ and n fuzzy sets $\{\nu_1, \ldots, \nu_n\} \subseteq U_C(\mathbb{R})$ with $\mu = \bigvee_{i=1}^{n} \nu_i$, then $\mu \in N(\mathbb{R})$ follows.

E.P. Klement, M.L. Puri and D.A. Ralescu [108,109] considered the class of all fuzzy sets μ such that μ is "upper semicontinuous" and cl μ_0 is bounded. An equivalent definition can be given in the following way:

Definition 3.15 *Let $F_C(\mathbb{R})$ denote the class of all fuzzy sets $\mu \in F(\mathbb{R})$ such that*
(i) $\inf \mu_0 > -\infty$ and $\sup \mu_0 < +\infty$ is valid, and
(ii) $\mu_{\overline{\alpha}}$ is closed for all $\alpha \in [0,1)$.

The following lemma shows why $F_C(\mathbb{R})$ is called the class of all upper semicontinuous fuzzy sets.

Lemma 3.16 *Let $\mu \in F(\mathbb{R})$ such that $\inf \mu_0 > -\infty$ and $\sup \mu_0 \leq +\infty$ holds. Then the following two propositions are equivalent.*
(i) $\mu_{\overline{\beta}}$ is closed for all $\beta \in (0,1]$.
(ii) For all convergent sequences $\{u_r\}_{n\in\mathbb{N}}$ of real numbers

$$\mu\left(\lim_{r\to\infty} u_r\right) \geq \lim_{r\to\infty} \mu(u_r)$$

is valid if $\{\mu(u_r)\}_{n\in\mathbb{N}}$ is convergent.

Proof

"$\Rightarrow$" Define $\beta_r \stackrel{d}{=} \mu(u_r)$ for $r \in \mathbb{N}$ and $\beta \stackrel{d}{=} \lim_{r\to\infty} \mu(u_r)$. Define $u \stackrel{d}{=} \lim_{r\to\infty} u_r$.

Let $\epsilon > 0$ arbitrary. Choose $R \in \mathbb{N}$ such that for all $r \geq R$ $\beta_r \geq \beta - \epsilon$ is valid. Then we have for $r \geq R$ $u_r \in \mu_{\overline{\beta_r}} \subseteq \mu_{\overline{\beta-\epsilon}}$.

We conclude: $u \in$ cl $\mu_{\overline{\beta-\epsilon}} = \mu_{\overline{\beta-\epsilon}}$, i.e. $\mu(u) \geq \beta - \epsilon$. As $\epsilon > 0$ was arbitrary, we can follow: $\mu\left(\lim_{r\to\infty} u_r\right) \geq \lim_{r\to\infty} \mu(u)$.

"$\Leftarrow$" Let $\beta \in (0,1]$ and $u \in$ cl $\mu_{\overline{\beta}}$.

Then there exists a sequence $\{u_r\}_{r\in\mathbb{N}}$ being convergent against u such that $u_r \in \mu_{\overline{\beta}}$ is valid for all $r \in \mathbb{N}$.

As $[0,1]$ is compact, there exists a convergent subsequence $\{\mu(u_{r_k})\}_{k\in\mathbb{N}}$. We conclude:

$$\mu(u) = \mu\left(\lim_{k\to\infty} u_{r_k}\right) \geq \lim_{k\to\infty} \mu(u_{r_k}) \geq \beta,$$

i.e. $u \in \mu_{\overline{\beta}}$.

Lemma 3.17 *Let $\mu \in F_C(\mathbb{R})$. Let $\{A_\alpha \mid \alpha \in (0,1)\}$ be a set representation for μ. Let $\beta \in (0,1]$. Then*
(i) $\{\text{cl } A_\alpha \mid \alpha \in (0,1)\}$ is a set representation of μ,
(ii) $\inf \mu_{\overline{\beta}} = \lim_{r\to\infty} \inf A_{\beta\cdot(1-1/2r)}$ and $\sup \mu_{\overline{\beta}} = \lim_{r\to\infty} \sup A_{\beta\cdot(1-1/2r)}$ holds for $\beta \in (0,1]$
and
(iii) $(\text{co } \mu)_{\overline{\beta}} = \text{co } \mu_{\overline{\beta}}$ holds for $\beta \in (0,1]$.

Proof
(i) For all $\alpha \in (0,1]$ $\mu_{\overline{\alpha}}$ is closed. This implies

$$\mu_\alpha \subseteq A_\alpha \subseteq \text{cl } A_\alpha \subseteq \mu_{\overline{\alpha}} = \mu_{\overline{\alpha}}$$

for all $\alpha \in (0,1)$, i.e. $\{\text{cl } A_\alpha \mid \alpha \in (0,1)\}$ is a set representation for μ.

(ii.i) As $\inf \text{cl } A_\alpha = \inf A_\alpha$ and $\sup \text{cl } A_\alpha = \sup A_\alpha$ holds for $\alpha \in (0,1)$, we will only show

$$\inf \mu_{\overline{\beta}} = \lim_{r\to\infty} \inf \text{cl } A_{\beta\cdot(1-1/2r)}.$$

In a similar way it is possible to show

$$\sup \mu_{\overline{\beta}} = \lim_{r\to\infty} \sup \text{cl } A_{\beta\cdot(1-1/2r)},$$

which proves the assertion.

(ii.ii) $\left\{\inf \text{cl } A_{\beta\cdot(1-1/2r)}\right\}_{r\in\mathbb{N}}$ is monotonously non-decreasing and bounded by $\sup \mu_0 < +\infty$, so

$$A \stackrel{\text{d}}{=} \lim_{r\to\infty} \inf \text{cl } A_{\beta\cdot(1-1/2r)}$$

exists, and $A \geq \inf \text{cl } A_{\beta\cdot(1-1/2r)}$ holds for $r \in \mathbb{N}$.
As $\mu_{\overline{\beta}} \subseteq \text{cl } A_{\beta\cdot(1-1/2r)}$ holds for $r \in \mathbb{N}$, $A \leq \inf \mu_{\overline{\beta}}$ is valid.
Suppose that $A < \inf \mu_{\overline{\beta}}$. Then $\mu(A) < \beta$. There is an $R \in \mathbb{N}$ such that $A \notin \text{cl } A_{\beta\cdot(1-1/2r)}$ holds for $r \geq R$.
We can find an $\epsilon > 0$ such that $(A-\epsilon, A+\epsilon) \cap \text{cl } A_{\beta\cdot(1-1/2r)}$ is empty for $r > R$, as $\text{cl } A_\alpha$ is closed for $\alpha \in (0,1)$. This implies $A - \epsilon > \inf \text{cl } A_{\beta\cdot(1-1/2r)}$ for all $r \in \mathbb{N}$, i.e. $A - \epsilon \geq A$, which is a contradiction.

(iii) can be shown by Theorem 3.4(ii), Definition 3.10, and Lemma 3.15(ii).

This was proven in another way first by R. Lowen [131].
For $\mu \in F(\mathbb{R}) \setminus F_C(\mathbb{R})$ an analogue to Lemma 3.17 does not hold. This shows the following example. Let $\beta \in (0,1]$ arbitrary. Define $\mu \in F(\mathbb{R})$ by

$$\mu(t) \stackrel{\text{d}}{=} \begin{cases} \beta(1+t) & \text{, if } t \in (-1,0) \\ 1 & \text{, if } t \in [1,2] \\ 0 & \text{, otherwise .} \end{cases}$$

It follows

$$\mu_\alpha = \left(\frac{\alpha}{\beta} - 1, 0\right) \cup [1,2] \quad \text{for} \quad \alpha \in [0\beta)$$

and

$$\mu_{\overline{\alpha}} = \left(\frac{\alpha}{\beta} - 1, 0\right) \cup [1,2] \quad \text{for} \quad \alpha \in (0,\beta) \, .$$

For any set representation $\{A_\alpha \mid \alpha \in (0,1)\}$ of μ and for all $\alpha \in (0,\beta)$, $\inf A_\alpha = \frac{\alpha}{\beta} - 1$ is valid because of Theorem 3.5.
We have

$$\lim_{r \to \infty} \inf A_{\beta \cdot (1 - 1/2r)} = 0.$$

But $\mu_{\overline{\beta}} = [1,2]$ holds, i.e. $\inf \mu_{\overline{\beta}} = 1$.

The class $F_C(\mathbb{R})$ is rather wide but some sensible fuzzy sets do not belong to it. Let us give an example. One expert describes an observation by "approximately a_1", i.e. $g^{(a_1,1)}$ (see the definition after Theorem 3.3 on page 12) and another by "approximately a_2", i.e. $g^{(a_2,1)}$ with $a_1 \neq a_2$. We know that one of these is right, and we can summarize these results in the fuzzy set $\mu \stackrel{\text{d}}{=} \max\{g^{(a_1,1)}, g^{(a_2,1)}\}$.
Then μ neither belongs to $U(\mathbb{R})$ nor to $F_C(\mathbb{R})$ as $\mu[(a_1 + a_2)\backslash 2] < \min[\mu(a_1), \mu(a_2)] = 1$ holds (i.e. $\mu_{\overline{1}}$ is not convex) as well as $\mu_0 = \mathbb{R}$, i.e. $\inf \mu_0 = -\infty$ and $\sup \mu_0 = +\infty$.

Definition 3.18 *A fuzzy set $\mu \in F(\mathbb{R})$ belongs to $Q(\mathbb{R})$ if and only if there is a set representation $\{A_\alpha \mid \alpha \in (0,1)\}$ of μ such that*
(i) $\inf A_\alpha > -\infty$ implies $\inf A_\alpha \in A_\alpha$,
(ii) $\sup A_\alpha < +\infty$ implies $\sup A_\alpha \in A_\alpha$,
and
(iii) if $\inf A_\alpha - \infty$ or $\sup A_\alpha = +\infty$ then A_α is convex
holds for all $\alpha \in (0,1)$.

A set representation $\{A_\alpha \mid \alpha \in (0,1)\}$ of μ fulfilling (i), (ii), and (iii) is called a normal set representation.

In this chapter several classes of fuzzy sets are defined. In the following remark we shall collect some relations between these classes.

Remark

a) If $\mu \in Q(I\!R)$, then co $\mu \in U(I\!R)$.

b) If we identify a real number $x \in I\!R$ with its membership function $I_{\{x\}} \in E(I\!R)$, we have the following inclusions:

$$I\!R \subseteq U_C(I\!R) \subseteq N(I\!R) \subseteq F_C(I\!R) \subseteq Q(I\!R) \subseteq F(I\!R) \subseteq E(I\!R)$$

and

$$I\!R \subseteq U_C(I\!R) \subseteq U(I\!R) \subseteq Q(I\!R) \subseteq F(I\!R) \subseteq E(I\!R).$$

For practical application it often suffices to consider fuzzy subsets of a finite interval $K = [a, b] \subseteq I\!R$. If we consider life times of mayflies e.g. we can restrict ourselves to the interval [0 hours , 24 hours]. The data types $F(K)$, $U(K)$, etc. can be defined analogously to $F(I\!R)$, $U(I\!R)$, etc, and we only need to replace $I\!R$ by K everywhere in the chapters above. The conditions $\inf \mu_0 > -\infty$ and $\sup \mu_0 < +\infty$ are fulfilled for each $\mu \in F(K)$, so we have the inclusions

$$K \subseteq U(K) \subseteq N(K) \subseteq F_C(K) \subseteq Q(K) \subseteq F(K) \subseteq E(K).$$

In Chapter 12 we shall consider the problem how to obtain fuzzy sets in practice with examples.

4 Operations on Fuzzy sets

4.1 SET THEORETICAL OPERATIONS

Fuzzy sets are introduced as a generalization of ordinary sets. In this chapter we are going to define an algebraic structure of the class of fuzzy sets which will be useful in the combination of vague data. If an expert is sure, for example, that the "true" value of the price of a house is in $[200,000; 300,000]$ and another expert has information that the price of the same house is in $[250,000; 350,000]$, we can calculate the intersection of these sets in order to aggregate the two expert opinions. In other situations different operations may be useful; thus, we start to discuss the basic operation performed on fuzzy sets which were originally proposed by L.A. Zadeh [212].

Definition 4.1 *In $E(\mathbb{R})$ the following operations are defined:*

$$(\mu \vee \mu')(x) \stackrel{\mathrm{d}}{=} \max\{\mu(x), \mu'(x)\} \quad \textit{union},$$

$$(\mu \wedge \mu')(x) \stackrel{\mathrm{d}}{=} \min\{\mu(x), \mu'(x)\} \quad \textit{intersection}$$

$$\overline{\mu}(x) \stackrel{\mathrm{d}}{=} 1 - \mu(x) \quad \textit{complement}\ .$$

$(E(\mathbb{R}), \vee, \wedge, \text{-})$ is a completely distributive lattice. So we can define in $E(\mathbb{R})$ unions and intersections of arbitrary families

$$\left(\bigvee_{i \in I} \mu_i\right)(x) \stackrel{\mathrm{d}}{=} \sup_{i \in I} \mu_i(x)\ , \text{ and}$$

$$\left(\bigwedge_{i \in I} \mu_i\right)(x) \stackrel{\mathrm{d}}{=} \inf_{i \in I} \mu_i(x),$$

where $\mu_i \in E(\mathbb{R})$ for all $i \in I$, and $x \in \mathbb{R}$. $E(\mathbb{R})$ is not a Boolean algebra because $\mu \wedge \overline{\mu} \neq 0 = \mathrm{I}_\emptyset$ and $\mu \vee \overline{\mu} \neq 1 = \mathrm{I}_{\mathbb{R}}$.

The justification of the choice of min and max was given by R.E. Bellman and M. Giertz [11].

Consider arbitrary functions $u : [0,1]^2 \to [0,1]$ and $v : [0,1]^2 \to [0,1]$ such that the following axioms hold:

(i) u and v are commutative, associative, and mutually distributive operators.

(ii) u and v are continuous and nondecreasing with respect to each of their arguments.

(iii) $u(x,x)$ and $v(x,x)$ are strictly increasing in x.

(iv) $\forall (x,y) \in [0,1]^2 : u(x,y) \leq \min\{x,y\}$,
$\forall (x,y) \in [0,1]^2 : v(x,y) \geq \max\{x,y\}$,

(v) $u(1,1) = 1, \; v(0,0) = 0$.

Bellman and Giertz showed that min and max are the only operators u and v that meet these axioms.

If we assume that the membership value of x in a compound fuzzy set depends on the membership value of x in the elementary fuzzy sets that form it, but not on anything else, then it is clear that the above assumptions ensure the uniqueness of the choice of the union and intersection operator.

The justification of the complement is more difficult than that of minimum and maximum. Consider an arbitrary function $h : [0,1] \to [0,1]$ such that the following axioms hold:

(vi) h is continuous and strictly monotonically decreasing,

(vii) $h(0) = 1, \; h(1) = 0$,

(viii) $\forall (x,y) \in [0,1]^2 : x - y = h(y) - h(x)$.

These axioms determine h uniquely, we have $h(x) = 1 - x$. If we assume again that the membership value of x in a complementary fuzzy set depends only on the membership value of x in the original fuzzy set, the axioms ensure the uniqueness of the complement operator.

Axiom (viii) is very strong. It means that a certain change in the membership value of μ should have the same effect on the membership function of the complement of μ.

L.W. Fung and K.S. Fu [57] and B.R. Gaines [60] use a slightly different set of assumptions for the justification of the operators.

It should be stated that the operations only make sense if the fuzzy sets μ and μ' act on the same measurement scale for the same semantic concept. For example, if A is a fuzzy concept "short" and B is the concept "fat", then it would not be meaningful to write for a particular individual $\max(A,B)$. In general the assumptions used by Bellman and Giertz are of course somewhat simple. Since fuzzy sets are only used to model imprecise reasoning, they must represent both, semantics and syntax (S. French [55]). In our restricted context, however, they are meaningful.

From a mathematical point of view the connectives and/or for fuzzy sets can be introduced as t-norms/t-conorms.

A triangular norm is a two-place function $t : [0,1]^2 \to [0,1]$ which is non-decreasing in each argument, commutative, associative, and which satisfies the boundary conditions $t(x,0) = t(0,x) = 0$, $t(x,1) = t(1,x) = x$ for each $x \in [0,1]$. The t-conorm (sometimes called s-norm) can be introduced via the formula $s(x,y) = 1 - t(1-x, 1-y), (x,y) \in [0,1]^2$. The general theory of triangular norms and their construction is based on the theory of functional equations. We will mention here only some important features of the construction of t-norms and t-conorms (compare B. Schweizer and A. Sklar [189] and S. Weber [206]).

For each continuous triangular t-norm there exists a continuous and strictly decreasing function $f : [0,1] \to [0,\infty)$ with $f(1) = 0$ such that

$$t(x,y) = \begin{cases} f^{-1}[f(x) + f(y)] & \text{, if } (f(x) + f(y)) \in [0, f(0)] \\ 0 & \text{, otherwise} \end{cases}$$

for $(x,y) \in [0,1]^2$.

For each continuous triangular t-conorm there exists a continuous and strictly increasing function $g : [0,1] \to [0,\infty)$ with $g(0) = 0$ such that

$$s(x,y) = \begin{cases} g^{-1}[g(x) + g(y)] & \text{, if } (g(x) + g(y)) \in [0, f(0)] \\ 0 & \text{, otherwise} \end{cases}$$

for $(x,y) \in [0,1]^2$.

The functions f and g are called additive generators of triangular norms and they are unique up to a positive muliplicative constant. In practical application, f.e. if one has to aggregate expert opinions, other operations of fuzzy sets may be useful.

As fuzzy sets can be seen as functions $\mu : I\!R \to [0,1]$, a t-norm $\sqcap$ and t-conorm $\sqcup$ induce connectives $\mu \sqcap \mu'$ and $\mu \sqcup \mu'$ defined pointwise.

Consider, for example, the generators $f(x) \stackrel{\mathrm{d}}{=} 1 - x^p$, $g(x) \stackrel{\mathrm{d}}{=} x^p$ for $p > 0$, by means of which we can get the Yager operations (R. Yager [209]).

$$x \sqcap_p y \stackrel{\mathrm{d}}{=} 1 - \min\left(1, ((1-x)^p + (1-y)^p)^{\frac{1}{p}}\right),$$

$$x \sqcup_p y \stackrel{\mathrm{d}}{=} \min\left(1, (x^p + y^p)^{\frac{1}{p}}\right), \quad \text{for } p > 0.$$

We have

$$x \sqcap_\infty y \stackrel{\mathrm{d}}{=} \lim_{p\to\infty} (x \sqcap_p y) = \min(x, y),$$

$$x \sqcup_\infty y \stackrel{\mathrm{d}}{=} \lim_{p\to\infty} (x \sqcup_p y) = \max(x, y),$$

$$x \sqcap_0 y \stackrel{\mathrm{d}}{=} \lim_{p\to 0} (x \sqcap_p y) = \begin{cases} x & , \text{if } y = 1 \\ y & , \text{if } x = 1 \\ 0 & , \text{otherwise} \end{cases} \quad , \text{and}$$

$$x \sqcup_0 y \stackrel{\mathrm{d}}{=} \lim_{p\to 0} (x \sqcup_p y) = \begin{cases} x & , \text{if } y = 0 \\ y & , \text{if } x = 0 \\ 1 & , \text{otherwise} \end{cases} \quad .$$

$\sqcup_0$ and $\sqcap_0$ are often called drastic operations.

It should also be noted that for any t-norm $\sqcap$ and t-conorm $\sqcup$ the following inequalities hold:

$$x \sqcap_0 y \leq x \sqcap y \leq x \sqcap_\infty y \quad , \text{and}$$

$$x \sqcup_\infty y \leq x \sqcup y \leq x \sqcup_0 y \quad .$$

In the following we will study the relations between the subclasses and the operations $\wedge, \vee, \bar{}$. The next theorem shows that our subclasses of fuzzy sets are closed under $\vee$.

Theorem 4.2 *If $\mu, \mu' \in U(\mathbb{R})$ ($N(\mathbb{R})$, $S(\mathbb{R})$, $Q(\mathbb{R})$, $F(\mathbb{R})$, $E(\mathbb{R})$ respectively), then $\mu \vee \mu' \in U(\mathbb{R})$ ($N(\mathbb{R})$, $S(\mathbb{R})$, $Q(\mathbb{R})$, $F(\mathbb{R})$, $E(\mathbb{R})$ respectively).*

Proof We restrict ourselves to prove the statement for $Q(\mathbb{R})$. Let $\{A_\alpha \mid \alpha \in (0,1)\}$ and $\{A'_\alpha \mid \alpha \in (0,1)\}$ be normal set representations of μ and μ', respectively. For all $\alpha \in (0,1)$ the inclusions,

$$(\mu \vee \mu')_\alpha = \mu_\alpha \cup \mu'_\alpha \subseteq A_\alpha \cup A'_\alpha \subseteq \mu_{\overline{\alpha}} \cup \mu'_{\overline{\alpha}} = (\mu \vee \mu')_{\overline{\alpha}}$$

can be shown by Theorem 2.3. It follows that $\{A_\alpha \cup A'_\alpha \mid \alpha \in (0,1)\}$ is a normal set representation for $\mu \vee \mu'$.

Since we have $I_{[0,1]} \wedge I_{[2,3]} = I_\emptyset$, an analogue theorem for $\wedge$ holds only for $E(\mathbb{R})$. We have to require that the intersection of the fuzzy sets is normal, the following theorem therefore holds.

Theorem 4.3 *If $\mu, \mu' \in U(\mathbb{R})$ ($N(\mathbb{R})$, $S(\mathbb{R})$, $Q(\mathbb{R})$, $F(\mathbb{R})$, $E(\mathbb{R})$ respectively), and if there is an $x \in \mathbb{R}$ such that $(\mu \wedge \mu')(x) = 1$, then $\mu \wedge \mu' \in U(\mathbb{R})$ ($N(\mathbb{R})$, $S(\mathbb{R})$, $Q(\mathbb{R})$, $F(\mathbb{R})$, $E(\mathbb{R})$ respectively).*

$E(\mathbb{R})$ is of course closed under $\bar{\ }$ but all the other classes are not. We have for example

$$\overline{g^{(0,1)}(x)} = 1 - \exp\left(-x^2\right) < 1$$

for all $x \in \mathbb{R}$ where $g^{(0,1)} \in U(\mathbb{R})$ is defined as on page 12. Therefore $\overline{g^{(0,1)}} \notin F(\mathbb{R})$ holds.

4.2 ON ZADEH'S EXTENSION PRINCIPLE

The extension principle introduced by L.A. Zadeh [214] is one of the most basic ideas of fuzzy set theory. It provides a general method for extending non-fuzzy mathematical concepts in order to deal with fuzzy quantities.

From the theorem of R.E. Bellman and M. Giertz [11], we know that it is reasonabie to use the lattice ([0,1], min,max) to describe hierarchical degrees of acceptability. We will use this lattice to motivate the extension principle. Let $T \neq \emptyset$ be an arbitrary set.
Let A be a subset of T, and let $\Phi : T \to \mathbb{R}$ be a function. Then the image of A and Φ is a subset of $\mathbb{R}$ defined by

$$\Phi(A) \stackrel{\mathrm{d}}{=} \{y \in \mathbb{R} \mid \exists x \in T : (x \in A \text{ and } \Phi(x) = y)\}.$$

Suppose $\mu : T \to [0,1]$ is a fuzzy set. What is a reasonable definition for the image of μ under Φ ? Let us evaluate the acceptability acc (a_y) of the statement

$$a_y \stackrel{\mathrm{d}}{=} \exists\, x \in T : (x \in \mu \text{ and } \Phi(x) = y),\ y \in \mathbb{R}.$$

The value of acceptability of the statement "$x \in \mu$" is, by definition, acc $(x \in \mu) = \mu(x)$ and we have $\operatorname{acc}(\Phi(x) = y) = \begin{cases} 1 & \text{, if } \Phi(x) = y, \\ 0 & \text{, otherwise.} \end{cases}$

For the existence quantor it is reasonable to choose the sup-operator, and we can evaluate the acceptability value

$$\begin{aligned} \operatorname{acc}(a_y) &= \sup\{\operatorname{acc}(x \in \mu \text{ and } \Phi(x) = y) \mid x \in T\} \\ &= \sup\{\min(\mu(x), \operatorname{acc}(\Phi(x) = y)) \mid x \in T\} \\ &= \sup\{\mu(x) \mid x \in \mathbb{R} \text{ such that } \Phi(x) = y\}. \end{aligned}$$

This heuristic consideration motivates the following definition for the image of μ under Φ:

$$\Phi(\mu)(y) \stackrel{\mathrm{d}}{=} \sup\{\mu(x) \mid x \in T \text{ such that } \Phi(x) = y\}, \quad y \in \mathbb{R}.$$

Example Let A be the vague datum "approximately 2" which is described by the fuzzy set

$$\mu_A(x) = \begin{cases} x-1 & \text{, if } 1 \le x \le 2, \\ 3-x & \text{, if } 2 \le x \le 3, \\ 0 & \text{, otherwise .} \end{cases}$$

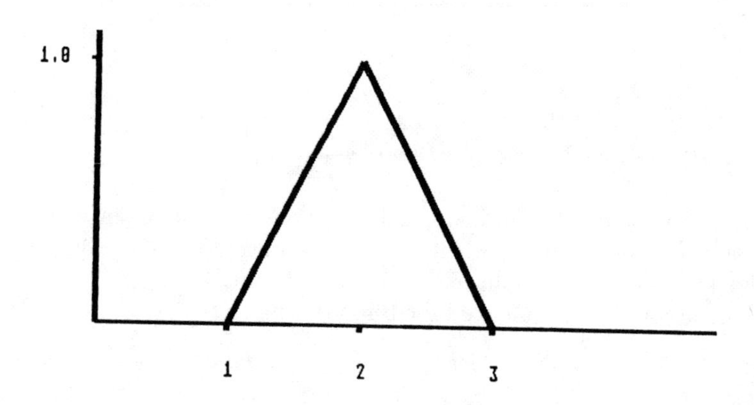

Fig. 4.1. The membership function of μ_A.

The image of μ_A under the function $\Phi(x) = x^2$ is

$$\begin{aligned} \Phi(\mu_A)(y) &= \sup\{\mu_A(x) \mid x \in \mathbb{R}, x^2 = y\} \\ &= \begin{cases} \max\{\mu_A(\sqrt{y}), \mu_A(-\sqrt{y})\} & \text{, if } y \ge 0 \\ 0 & \text{, otherwise} \end{cases} \\ &= \begin{cases} \sqrt{y}-1 & \text{, if } 1 \le y \le 4, \\ 3-\sqrt{y} & \text{, if } 4 \le y \le 9, \\ 0 & \text{, otherwise.} \end{cases} \end{aligned}$$

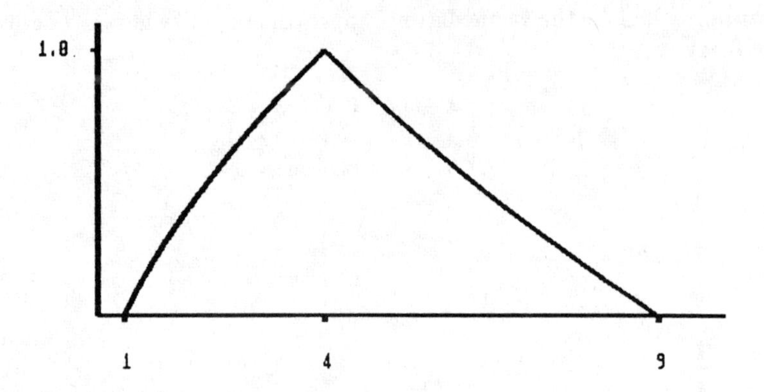

Fig. 4.2. The membership function of μ_A^2.

After these considerations, we shall continue with the definition of the so called extension principle which provides a means for extending operations on numbers to be applicable to fuzzy sets. R.R. Yager [211] gave a characterization of this principle based upon using relations.

Definition 4.4 *Let $n \in \mathbb{N}$, let $(\mu_1, \ldots, \mu_n) \in [F(\mathbb{R})]^n$, and let Φ be a mapping from $\mathbb{R}^n$ to $\mathbb{R}$. Then the fuzzy set $\Phi[\mu_1, \ldots, \mu_n]$ with*

$$\Phi[\mu_1, \ldots, \mu_n](t) \stackrel{d}{=}$$

$$\sup \{\min \{\mu_1(t_1), \ldots, \mu_n(t_n)\} \mid (t_1, \ldots, t_n) \in \mathbb{R}^n, \Phi(t_1, \ldots, t_n) = t\}$$

for all $t \in \mathbb{R}$ is called the image of $(\mu_1, \ldots, \mu_n)$ under Φ.

The image of $(\mu_1, \ldots, \mu_n)$ under Φ can be calculated with the aid of set representations.

Lemma 4.5 *Let $n \in \mathbb{N}$, let $(\mu_1, \ldots, \mu_n) \in [F(\mathbb{R})]^n$, let $\Phi : \mathbb{R}^n \to \mathbb{R}$ be a mapping. For n subsets $A_1, \ldots, A_n \subseteq \mathbb{R}$ define*

$$\Phi[A_1, \ldots, A_n] \stackrel{d}{=}$$

$$\{t \in \mathbb{R} \mid \exists (t_1, \ldots, t_n) \in A_1 \times A_2 \times \ldots \times A_n \text{ with } \Phi(t_1, \ldots, t_n) = t\}.$$

(i) If for all $i \in \{1,\ldots,n\}$ $\{(A_i)_\alpha \,|\, \alpha \in (0,1)\}$ *is a set representation of* μ_i*, then* $\{\Phi[(A_1)_\alpha,\ldots,(A_n)_\alpha] \,|\, \alpha \in (0,1)\}$ *is a set representation of* $\Phi[\mu_1,\ldots,\mu_n]$.

(ii) For all $\alpha \in [0,1)$ $(\Phi[\mu_1,\ldots,\mu_n])_\alpha = \Phi[(\mu_1)_\alpha,\ldots,(\mu_n)_\alpha]$ *is valid.*

Proof

(i) We want to show that

$$\begin{aligned}&(\Phi[\mu_1,\ldots,\mu_n])_\alpha\\ \subseteq&\Phi[(\mu_1)_\alpha,\ldots,(\mu_n)_\alpha]\\ \subseteq&\Phi[(A_1)_\alpha,\ldots,(A_n)_\alpha]\\ \subseteq&\Phi[(\mu_1)_{\overline{\alpha}},\ldots,(\mu_k)_{\overline{\alpha}}]\\ \subseteq&(\Phi[\mu_1,\ldots,\mu_n])_{\overline{\alpha}}\end{aligned}$$

is valid for $\alpha \in (0,1)$.
Let $\alpha \in (0,1)$. If $t \in (\Phi[\mu_1,\ldots,\mu_n])_\alpha$, then there exists a $(t_1,\ldots,t_n) \in I\!R^n$ with

$$\Phi(t_1,\ldots,t_n) = t \text{ and } \min_{i \in \{1,\ldots,n\}} \mu_i(t_i) > \alpha.$$

This implies $t_i \in (\mu_i)_\alpha$ for all $i \in \{1,\ldots,n\}$ and $t \in \Phi[(\mu_1)_\alpha,\ldots,(\mu_n)_\alpha]$. By Theorem 3.5 we know that $(\mu_i)_\alpha \subseteq (A_i)_\alpha \subseteq (\mu_i)_{\overline{\alpha}}$ is valid for all $i \in \{1,\ldots,n\}$. The second, third, and fourth inclusion follow immediately from this.
If $t \in \Phi[(\mu_1)_{\overline{\alpha}},\ldots,(\mu_n)_{\overline{\alpha}}]$, then there exists a

$$(t_1,\ldots,t_n) \in (\mu_1)_{\overline{\alpha}} \times \ldots \times (\mu_n)_{\overline{\alpha}}$$

with $\Phi(t_1,\ldots,t_n) = t$.
This implies

$$\min_{i \in \{1,\ldots,n\}} \mu_i(t_i) \geq \alpha,$$

i.e. $t \in (\Phi[\mu_1,\ldots,\mu_n])_{\overline{\alpha}}$.

(ii) The inclusion "$\subseteq$" has already been shown in part (i).
If $\alpha \in [0,1)$ and $t \in \Phi[(\mu_1)_\alpha,\ldots,(\mu_n)_\alpha]$, there exists a $(t_1,\ldots,t_n) \in (\mu_1)_\alpha \times \ldots \times (\mu_n)_\alpha$ with $\Phi(t_1,\ldots,t_n) = t$.
Then $\mu_i(t_i) > \alpha$ is valid for all $i \in \{1,\ldots,n\}$, i.e. $t \in (\Phi[\mu_1,\ldots,\mu_n])_\alpha$.

Using this theorem we can easily calculate fuzzy sets defined by the extension principle. Some of the functions $\Phi[A_1, A_2]$ are known in diffential geometry (see D.G. Kendall [105] and G. Matheron [135], e.g.), such as

$$A_1 + A_2 \stackrel{\mathrm{d}}{=} \{x + y \mid x \in A_1 \text{ and } y \in A_2\} \text{ and}$$
$$\lambda A \stackrel{\mathrm{d}}{=} \{\lambda x \mid x \in A\},\ \lambda \in \mathbb{R},$$

and are called the Minkowski operations.

If $n \in \mathbb{N}$, $(\mu_1, \ldots, \mu_n) \in [F(\mathbb{R})]^n$, $\Phi : \mathbb{R}^n \to \mathbb{R}$ and $\alpha \in (0,1]$, the equality

$$(\Phi[\mu_1, \ldots, \mu_n])_{\overline{\alpha}} = \Phi[(\mu_1)_{\overline{\alpha}}, \ldots, (\mu_n)_{\overline{\alpha}}]$$

in general does not hold.

For an example let $n = 2$, $\Phi =$ " $+$ " , $\alpha = 1$. Define

$$\mu(t) \stackrel{\mathrm{d}}{=} \begin{cases} 1+t & \text{, if } t \in (-1,0) \\ 1 & \text{, if } t \in [1,2] \\ 0 & \text{, otherwise} \end{cases}$$

and

$$\nu(t) \stackrel{\mathrm{d}}{=} \begin{cases} 1 & \text{, if } t \in [0,1] \\ 0 & \text{, otherwise .} \end{cases}$$

Then we have for all $u \in (0,1)$ $\mu(-u) = 1 - u$ and $\nu(u) = 1$. This implies $\min\{\mu(-u), \nu(u)\} = 1 - u$. It follows that

$$(\mu + \nu)(0) \geq \sup\{\min[\mu(-u), \nu(u)] \mid u \in (0,1)\} = 1,$$

and therefore $0 \in (\mu + \nu)_{\overline{1}}$ holds.

But $\mu_{\overline{1}} = [1,2]$ and $\nu_{\overline{1}} = [0,1]$ holds, and therefore $0 \notin [1,3] = \mu_{\overline{1}} + \nu_{\overline{1}}$.

This simple example shows that we are not able to restrict ourselves to the special set representation $\{\mu_{\overline{\alpha}} \mid \alpha \in (0,1)\}$ of μ.

4.3 ARITHMETIC OPERATIONS

In the following we shall apply the extension principle to arithmetic operations. Let us start with the k-th power of a fuzzy set. We have for $\mu \in F(\mathbb{R})$ and $t \in \mathbb{R}$:

$$\mu^k(t) = \begin{cases} \mu\left(\sqrt[k]{t}\right) & \text{, if } k \in I\!N \text{ odd} \\ \max\left\{\mu\left(\sqrt[k]{t}\right), \mu\left(-\sqrt[k]{t}\right)\right\} & \text{, if } k \in I\!N \text{ even and } t \geq 0 \\ 0 & \text{, if } k \in I\!N \text{ even and } t < 0. \end{cases}$$

If $\mu = I_{\{1,2\}}$, then we have

$$\mu^k = I_{\{1,2^k\}} \text{ , but } \prod_{i=1}^{k} \mu = I_{\{1,2,4,8,\ldots,2^k\}}$$

for $k \in I\!N$. In general, we have $\mu^k \neq \prod_{i=1}^{k} \mu$.

If $k \in I\!N$ is even, then $\mu \in Q\,(I\!R)$ does not imply: $\mu^k \in Q\,(I\!R)$. Consider, for example, the fuzzy set $\mu = I_{[-1,-0.5)\cup(+0.5,+1]} \in Q\,(I\!R)$ for which $\mu^2 = I_{(0.25,1]} \notin Q\,(I\!R)$ holds.

Another important unary operation is the multiplication with a scalar. We have

$$(\lambda\mu)\,(t) = \sup\left\{I_{\{\lambda(x)\wedge\mu(y)\}} \,\middle|\, (x,y) \in I\!R^2, x \cdot y = t\right\}$$

$$= \begin{cases} \mu(\frac{t}{\lambda}) & \text{, if } \lambda \neq 0 \\ I_{\{0\}}(t) & \text{, if } \lambda = 0. \end{cases}$$

Let us now consider the unary operation $|\ |$, the absolute value. Let $\mu \in U\,(I\!R)$ and $\{A_\alpha \mid \alpha \in (0,1)\}$ be a set representation of μ such that A_α is convex and closed for all $\alpha \in (0,1)$. Define for $\alpha \in (0,1)$

$$\underline{B}_\alpha \stackrel{\text{d}}{=} \begin{cases} 0 & \text{, if } \inf A_\alpha \leq 0 \leq \sup A_\alpha \\ |\inf A_\alpha| & \text{, if } \inf A_\alpha > 0 \\ |\sup A_\alpha| & \text{, if } \sup A_\alpha < 0 \text{ ,} \end{cases}$$

$$\overline{B}_\alpha \stackrel{\text{d}}{=} \begin{cases} \max\{|\inf A_\alpha|, |\sup A_\alpha|\} & \text{, if } \inf A_\alpha > -\infty \text{ and } \sup A_\alpha < +\infty \\ +\infty & \text{, otherwise,} \end{cases}$$

and

$$B_\alpha \stackrel{\text{d}}{=} \begin{cases} [\underline{B}_\alpha, \overline{B}_\alpha] & \text{, if } \overline{B}_\alpha < +\infty \\ [\underline{B}_\alpha, +\infty) & \text{, if } B_\alpha = +\infty. \end{cases}$$

Then the absolute value of μ, $|\mu|$ is in $U(\mathbb{R})$, since $\{B_\alpha \mid \alpha \in (0,1)\}$ is a set representation of $|\mu|$. We have the following theorem which is easily proved:

Theorem 4.6 *Let $\mu \in F(\mathbb{R})$, and let $k \in \mathbb{N}$.*
(i) If k is odd, then $\mu \in Q(\mathbb{R})$ implies $\mu^k \in Q(\mathbb{R})$.
(ii) If k is even, then $|\mu| \in Q(\mathbb{R})$ implies $\mu^k \in Q(\mathbb{R})$.

The extended binary operations $+, -, \cdot, /$ are also defined by Definition 4.4. $(F(\mathbb{R}), +)$ is not a group. We have the zero element $I_{\{0\}}$ but in general no inverse element. If $\mu \in F(\mathbb{R})$ such that $\inf \mu_0 < \sup \mu_0$, then $\nu \in F(\mathbb{R})$ with $\mu + \nu = I_{\{0\}}$ does not exist. Otherwise, from Lemma 4.5(ii) it would follow:

$$0 = \inf(\mu + \nu)_0 = \inf \mu_0 + \inf \nu_0$$

and

$$0 = \sup(\nu + \mu)_0 = \sup \mu_0 + \sup \nu_0$$

and we could conclude $\inf \nu_0 > \sup \nu_0$ which is a contradiction.

For a detailed analysis of the algebraic properties of the arithmetic operations we refer to M. Mizumato and K. Tanaka [144]. Some of the results are, at first glance, contraintuitive as the following example.

Let h be the fuzzy set $t^{(-1,+1,+1,+2)}$, the graph of which is plotted in the following figure.

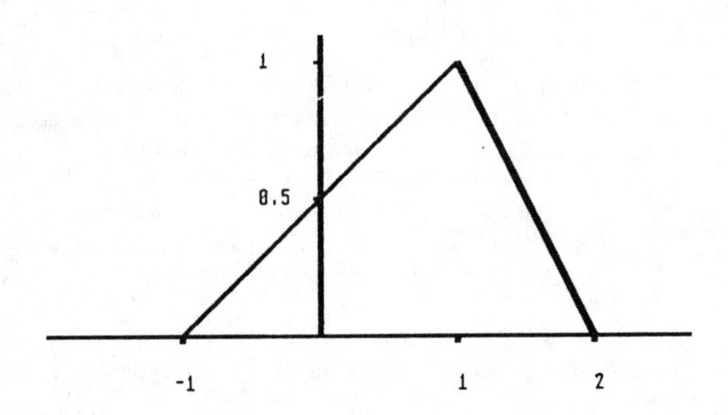

Fig. 4.3. The membership function of h.

h is in $U(\mathbb{R})$, but $1 \backslash h$ is not.

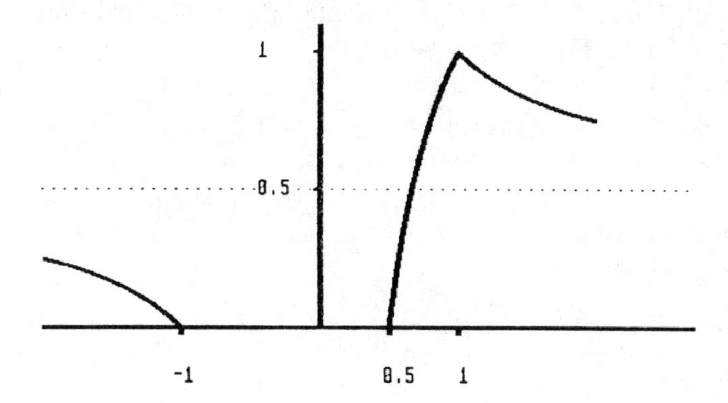

Fig. 4.4. The membership function of $1/h$.

One should be aware that we are dividing by zero with a degree of acceptability 0.5. In essence the arithmetic operations are generalized interval arithmetic operations (R.E. Moore [145]). We are interested especially in the addition and the multiplication with a scalar. The next theorem shows that our subclasses of fuzzy sets are closed against these operations.

Theorem 4.7 *Let $n \in \mathbb{N}$ and $(x_1, \ldots, x_n) \in \mathbb{R}^n$*

(i) If $(\mu_1, \ldots, \mu_n) \in [U(\mathbb{R})]^n$, then $\sum_{i=1}^{n} x_i\mu_i \in U(\mathbb{R})$

(ii) If $(\mu_1, \ldots, \mu_n) \in [U_C(\mathbb{R})]^n$, then $\sum_{i=1}^{n} x_i\mu_i \in U_C(\mathbb{R})$

(iii) If $(\mu_1, \ldots, \mu_n) \in [N(\mathbb{R})]^n$, then $\sum_{i=1}^{n} x_i\mu_i \in N(\mathbb{R})$

(iv) If $(\mu_1, \ldots, \mu_n) \in [F_C(\mathbb{R})]^n$, then $\sum_{i=1}^{n} x_i\mu_i \in F_C(\mathbb{R})$

(v) If $(\mu_1, \ldots, \mu_n) \in [Q(\mathbb{R})]^n$, then $\sum_{i=1}^{n} x_i\mu_i \in Q(\mathbb{R})$

(vi) If $(\mu_1, \ldots, \mu_n) \in [F_C(\mathbb{R})]^n$ and $\beta \in (0,1]$, then
$$\left(\sum_{i=1}^{n} x_i\mu_i\right)_{\overline{\beta}} = \sum_{i=1}^{n} x_i(\mu_i)_{\overline{\beta}}.$$

Proof For $\mu \in F(\mathbb{R})$ $0 \cdot \mu = \mathrm{I}_{\{0\}}$ and $\mu + \mathrm{I}_{\{0\}} = \mu$ is valid. Because of this we assume without loss of generality that $x_i \neq 0$ holds for all $i \in \{1, \ldots, n\}$.

(a) Define $M_1 \overset{d}{=} \{i \in \{1,\ldots,n\} \mid x_i > 0\}$. Then $M_2 \overset{d}{=} \{1,\ldots,n\} \backslash M_1 = \{i \in \{1,\ldots,n\} \mid x_i < 0\}$ follows.
Let $(\mu_1,\ldots,\mu_n) \in [F(\mathbb{R})]^n$ such that $\inf \mu_i > -\infty$ and $\sup \mu_i < +\infty$ is valid for $i \in \{1,\ldots,n\}$. Then

$$\inf \left(\sum_{i=1}^{n} x_i \mu_i \right)_0 = \inf \left(\sum_{i=1}^{n} x_i (\mu_i)_0 \right) =$$

$$= \sum_{i \in M_1} x_i \inf (\mu_i)_0 + \sum_{i \in M_2} x_i \sup (\mu_i)_0 > -\infty$$

and

$$\sup \left(\sum_{i=1}^{n} x_i \mu_i \right)_0 = \sup \left(\sum_{i=1}^{n} x_i (\mu_i)_0 \right) =$$

$$= \sum_{i \in M_1} x_i \sup (\mu_i)_0 + \sum_{i \in M_2} x_i \inf (\mu_i)_0 < +\infty$$

is valid because of Lemma 4.5(ii)

(b) Let A, B be two compact subsets of $\mathbb{R}$. We claim that $A+B$ is compact. Obviously $A+B$ is bounded.
Let $t \in \text{cl}\,(A+B)$. For all $N \in \mathbb{N}$ there exist $a_N \in A$ and $b_N \in B$ such that $\{a_N + b_N\}_{N \in \mathbb{N}} \to t$.
As A is compact, there exists a convergent subsequence $\{a_{N_k}\}_{k \in \mathbb{N}}$. Then $a \overset{d}{=} \lim_{k \to \infty} a_{N_k} \in \text{cl } A = A$ holds. For all $k \in \mathbb{N}$ $b_{N_k} = (a_{N_k} + b_{N_k}) - t + t - a_{N_k}$ holds, which implies that $\{b_{N_k}\}_{k \in \mathbb{N}}$ is convergent against $t-a$. $t - a \in \text{cl } B = B$ holds, i.e. $t \in A+B$. In the same way we can show that λA is compact if A is compact and $\lambda \in \mathbb{R}$.
By mathematical induction, we can follow: $\sum_{i=1}^{n} x_i A_i$ is compact if $A_1,\ldots,A_n$ are n compact subsets of $\mathbb{R}$.

(c) Let $(\mu,\nu) \in [F_C(\mathbb{R})]^2$. We claim that $(\mu+\nu)_{\overline{\beta}} = \mu_{\overline{\beta}} + \nu_{\overline{\beta}}$ for $\beta \in (0,1]$. If $t \in \mu_{\overline{\beta}}$ and $v \in \nu_{\overline{\beta}}$, then

$$\begin{aligned}(\mu+\nu)(t+v) &= \sup \{\min [\mu(x_1), \nu(x_2)] \mid x_1 + x_2 = t + v\} \\ &\geq \min [\mu(t), \nu(v)] \geq \beta\end{aligned}$$

follows.
This shows the inclusion "$\supseteq$".
If $t \in (\mu+\nu)_{\overline{\beta}}$, then there exists a sequence $\{u_r\}_{r \in \mathbb{N}}$ such that

$$u_r \in \mu_{\overline{\beta \cdot (1 - 1/2r)}} \text{ and } t - u_r \in \nu_{\overline{\beta \cdot (1 - 1/2r)}}$$

is valid for $r \in \mathbb{N}$. As cl μ_0 is compact and $u_r \in$ cl μ_0 holds for $r \in \mathbb{N}$, we can choose a subsequence $\{v_r\}_{r\in\mathbb{N}}$ of $\{u_r\}_{r\in\mathbb{N}}$ such that $\{v_r\}_{r\in\mathbb{N}}$ is convergent. We can choose a subsequence $\{w_r\}_{r\in\mathbb{N}}$ of $\{v_r\}_{r\in\mathbb{N}}$ such that $\{w_r\}_{r\in\mathbb{N}}$ and $\{\mu(w_r)\}_{r\in\mathbb{N}}$ are convergent. By Lemma 3.15 we can follow

$$\mu\left(\lim_{r\to\infty} v_r\right) = \mu\left(\lim_{r\to\infty} w_r\right) \geq \lim_{r\to\infty} \mu(w_r) \geq \beta.$$

We can choose another subsequence $\{z_r\}_{r\in\mathbb{N}}$ of $\{v_r\}_{r\in\mathbb{N}}$ such that $\{z_r\}_{r\in\mathbb{N}}$ and $\{\nu(t-z_r)\}_{r\in\mathbb{N}}$ are convergent; it follows:

$$\nu\left(t - \lim_{r\to\infty} v_r\right) = \nu\left(t - \lim_{r\to\infty} z_r\right) \geq \lim_{r\to\infty} \nu\left(t - z_r\right) \geq \beta.$$

This implies: $t \in \mu_{\overline{\beta}} + \nu_{\overline{\beta}}$.
The equation $\lambda \cdot \mu_{\overline{\beta}} = (\lambda \cdot \mu)_{\overline{\beta}}$ for all $\lambda \in \mathbb{R}$, $\mu \in F_C(\mathbb{R})$, $\beta \in (0,1]$ is obvious. By mathematical induction we can follow that

$$\sum_{i=1}^{n} x_i(\mu_i)_{\overline{\beta}} = \left(\sum_{i=1}^{n} x_i\mu_i\right)_{\overline{\beta}} \quad \text{for } \beta \in [0,1],$$

if $(\mu_1,\ldots,\mu_n) \in [F_C(\mathbb{R})]^n$, i.e. assertion (vi).

(d) Combining part (b) and (c), we obtain assertion (iv).

(e) Let $(\mu_1,\ldots,\mu_n) \in [U(\mathbb{R})]^n$ and $\beta \in (0,1]$.
Part (c) shows that

$$\left(\sum_{i=1}^{n} x_i\mu_i\right)_{\overline{\beta}} = \sum_{i=1}^{n} x_i(\mu_i)_{\overline{\beta}}$$

is valid.
For all $i \in \{1,\ldots,n\}$ $(\mu_i)_{\overline{\beta}}$ is a closed interval, so

$$\begin{aligned}
&\sum_{i=1}^{n} x_i(\mu_i)_{\overline{\beta}} \\
&= \left[\sum_{i\in M_1} x_i \inf(\mu_i)_{\overline{\beta}} + \sum_{i\in M_2} x_i \sup(\mu_i)_{\overline{\beta}},\right. \\
&\qquad \left.\sum_{i\in M_1} x_i \sup(\mu_i)_{\overline{\beta}} + \sum_{i\in M_2} x_i \inf(\mu_i)_{\overline{\beta}}\right]
\end{aligned}$$

is closed where

$$M_1 \stackrel{\mathrm{d}}{=} \{i \in \{1,\ldots,n\} \mid x_i > 0\}$$

and

$$M_2 \stackrel{\mathrm{d}}{=} \{1,\ldots,n\} \backslash M_1 = \{i \in \{1,\ldots,n\} \mid x_i < 0\}.$$

Assertion (i) follows by this and by Lemma 3.8.

(f) Combining part (a) and (e), we obtain (ii).

(g) Let $(\mu,\nu) \in [N(\mathbb{R})]^2$. Let $\left\{ \bigcup_{j=1}^{N_\alpha} [(a_j)_\alpha,(b_j)_\alpha] \mid \alpha \in (0,1) \right\}$ be an interval set representation of μ and $\left\{ \bigcup_{k=1}^{M_\alpha} [(c_k)_\alpha,(d_k)_\alpha] \mid \alpha \in (0,1) \right\}$ be one of ν. Define

$$A_\alpha \stackrel{\mathrm{d}}{=} \left\{ \bigcup_{j=1}^{N_\alpha} [(a_j)_\alpha,(b_j)_\alpha] \right\} + \left\{ \bigcup_{k=1}^{M_\alpha} [(c_k)_\alpha,(d_k)_\alpha] \right\}$$

for $\alpha \in (0,1)$. Lemma 4.5(i) shows that $\{A_\alpha \mid \alpha \in (0,1)\}$ is a set representation of $\mu+\nu$. For all $\alpha \in (0,1)$

$$A_\alpha = \bigcup_{j=1}^{N_\alpha} \bigcup_{k=1}^{M_\alpha} [(a_j)_\alpha + (c_k)_\alpha, (b_j)_\alpha + (d_k)_\alpha]$$

is valid, i.e. A_α can be shown as the union of disjoint compact intervals. So $\mu+\nu \in N(\mathbb{R})$.

We can follow: $\sum_{i=1}^{n} x_i\mu_i \in N(\mathbb{R})$.

(h) Let $(\mu_1,\ldots,\mu_n) \in [Q(\mathbb{R})]^n$. Let M_1, M_2 be defined as in part (a).

(h.1) Let $\{(A_i)_\alpha \mid \alpha \in (0,1)\}$ be a normal set representation of μ_i for $i \in \{1,\ldots,n\}$. Lemma 4.5(i) shows that $\left\{ \sum_{i=1}^{n} x_i(A_i)_\alpha \mid \alpha \in (0,1) \right\}$ is a set representation for $\sum_{i=1}^{n} x_i\mu_i$. We want to prove that it is a normal set representation.

Let $\alpha \in (0,1)$ arbitrary. It holds:

$$\inf\left(\sum_{i=1}^{n} x_i(A_i)_\alpha\right) = \begin{cases} \sum_{i\in M_1} x_i \inf(A_i)_\alpha + \sum_{i\in M_2} x_i \sup(A_i)_\alpha \\ \quad \text{, if } \inf(A_i)_\alpha > -\infty \text{ for } i \in M_1 \\ \quad \text{and } \sup(A_i)_\alpha < +\infty \text{ for } i \in M_2 \\ -\infty \text{ , otherwise} \end{cases}$$

and

$$\sup\left(\sum_{i=1}^{n} x_i(A_i)_\alpha\right) = \begin{cases} \sum\limits_{i\in M_1} xi\sup(A_i)_\alpha + \sum\limits_{i\in M_2} x_i \sup(A_i)_\alpha \\ \qquad , \text{if } \sup(A_i)_\alpha < +\infty \text{ for } i \in M_1 \\ \qquad \text{and } \inf(A_i)_\alpha > -\infty \text{ for } i \in M_2 \\ \\ +\infty \text{ , otherwise} \end{cases}$$

If $\inf\left(\sum\limits_{i=1}^{n} x_i(A_i)_\alpha\right) > -\infty$, then $\inf(A_i)_\alpha > -\infty$ holds for $i \in M_1$ and $\sup(A_i)_\alpha < +\infty$ holds for $i \in M_2$. By Definition 3.18(ii) it follows $\inf(A_i)_\alpha \in (A_i)_\alpha$ for $i \in M_1$ and $\sup(A_i)_\alpha \in (A_i)_\alpha$ for $i \in M_2$. So

$$\inf\left(\sum_{i=1}^{n} x_i(A_i)_\alpha\right)$$

$$= \sum_{i\in M_1} x_i \inf(A_i)_\alpha + \sum_{i\in M_2} x_i \sup(A_i)_\alpha \in \sum_{i=1}^{n} x_i(A_i)_\alpha$$

holds.

If $\sup\left(\sum\limits_{i=1}^{n} x_i(A_i)_\alpha\right) < +\infty$, we can thus conclude in a similar way that

$$\sup\left(\sum_{i=1}^{n} x_i(A_i)_\alpha\right) \in \sum_{i=1}^{n} x_i(A_i)_\alpha.$$

Let $\inf\left(\sum\limits_{i=1}^{n} x_i(A_i)_\alpha\right) = -\infty$ be valid. Then there is an $i \in M_1$ with $\inf(A_i)_\alpha = -\infty$ or an $i \in M_2$ with $\sup(\mu_i)_\alpha = +\infty$. We want to show that $t \in \sum\limits_{i=1}^{n} x_i(A_i)_\alpha$ implies: $(-\infty, t] \subseteq \sum\limits_{i=1}^{n} x_i(A_i)_\alpha$.
We know that there is an $(u_1, \ldots, u_n) \in (A_1)_\alpha \times \ldots \times (A_n)_\alpha$ with $t = \sum\limits_{i=1}^{n} x_i u_i$. If there is an $i \in M_1$ such that $\inf(A_i)_\alpha = -\infty$ holds, then $x_i > 0$ and $(-\infty, u_i] \subseteq (A_i)_\alpha$ is valid. Define for $\lambda \geq 0$

$$t_j(\lambda) \overset{\mathrm{d}}{=} \begin{cases} u_j & , \text{if } j \in \{1, \ldots, n\} \backslash \{i\} \\ u_j - \lambda & , \text{if } j = i \end{cases}$$

Then we have

$$(t_1(\lambda), \ldots, t_n(\lambda)) \in (A_1)_\alpha \times \ldots \times (A_n)_\alpha$$

for all $\lambda \geq 0$. Define $f(\lambda) \stackrel{\mathrm{d}}{=} \sum_{j=1}^{n} x_j t_j(\lambda)$; f is continuous on $[0, +\infty)$.
Obviously $t - \lambda x_i = f(\lambda) \in \sum_{j=1}^{n} x_j (A_j)_\alpha$ holds for all $\lambda \geq 0$ as well as $f(0) = t$ and $\lim_{\lambda \to +\infty} f(\lambda) = -\infty$.
Applying the first mean value theorem, we obtain:

$$(-\infty, t] \subseteq \sum_{j=1}^{n} x_j (A_j)_\alpha.$$

If there is an $i \in M_2$ with $\sup (A_i)_\alpha = +\infty$ or if $\sup \left(\sum_{i=1}^{n} x_i (A_i)_\alpha \right) = +\infty$, we can conclude in an analogous way.

(h.2) Let $\{(A_i)_\alpha \mid \alpha \in (0,1)\}$ be a set representation for μ_i $(i = 1, \ldots, n)$ such that $(A_i)_\alpha$ is convex and closed for all $\alpha \in (0,1)$ and $i \in \{1, \ldots, n\}$. We know that $\left\{ \sum_{i=1}^{n} x_i (A_i)_\alpha \mid \alpha \in (0,1) \right\}$ is a set representation of $\sum_{i=1}^{n} x_i \mu_i$. We want to show that $\sum_{i=1}^{n} x_i (A_i)_\alpha$ is closed and convex for all $\alpha \in (0,1)$.
Let $\alpha \in (0,1)$. Let $t \in \sum_{i=1}^{n} x_i (A_i)_\alpha$ and $u \in \sum_{i=1}^{n} x_i (A_i)_\alpha$. We have to show that $[t, u] \subseteq \sum_{i=1}^{n} x_i (A_i)_\alpha$. We assume without loss of generality: $t \leq u$.
There exist $(t_1, \ldots, t_n) \in (A_1)_\alpha \times \ldots \times (A_n)_\alpha$ and $(u_1, \ldots, u_n) \in (A_1)_\alpha \times \ldots \times (A_n)_\alpha$ such that $t = \sum_{i=1}^{n} x_i t_i$ and $u = \sum_{i=1}^{n} x_i u_i$ holds. As $(A_i)_\alpha$ is convex, $\lambda t_i + (1 - \lambda) u_i \in (A_i)_\alpha$ holds for $i \in \{1, \ldots, n\}$, $\lambda \in [0,1]$. A continuous function $f : [0,1] \to I\!R$ is defined by $f(\lambda) \stackrel{\mathrm{d}}{=} \sum_{i=1}^{n} x_i [\lambda t_i + (1 - \lambda) u_i]$ as $\lambda \in [0,1]$. As $f(0) = u$ and $f(1) = t$ holds, $[t, u] \subseteq \sum_{i=1}^{n} x_i (A_i)_\alpha$ follows. Obviously $\{(A_i)_\alpha \mid \alpha \in (0,1)\}$ is a normal set representation of μ_i for $i \in \{1, \ldots, n\}$. As shown in part (i) of the proof,

$$\inf\left(\sum_{i=1}^{n} x_i(A_i)_\alpha\right) > -\infty \quad \text{implies}$$

$$\inf\left(\sum_{i=1}^{n} x_i(A_i)_\alpha\right) \in \sum_{i=1}^{n} x_i(A_i)_\alpha,$$

and

$$\sup\left(\sum_{i=1}^{n} x_i(A_i)_\alpha\right) < +\infty \quad \text{implies}$$

$$\sup\left(\sum_{i=1}^{n} x_i(A_i)_\alpha\right) \in \sum_{i=1}^{n} x_i(A_i)_\alpha \quad .$$

This proves the assertion (v).

5 Representation of vague data in a digital computer

One way of storing vague data in a digital computer is to describe it with the help of parametrized families of fuzzy sets. Consider for example the class

$$g^{(a,b)}(x) \stackrel{\mathrm{d}}{=} \exp\left[-\left(\frac{x-a}{b}\right)^2\right], \; b > 0, \; a \in I\!R.$$

This class has some interesting properties [137]. For non zero scalars $(\alpha_1, \ldots, \alpha_n)$ we have the identity $\sum_{i=1}^{n} \alpha_i g^{(a_i,b_i)} = g^{(r,s)}$, ($\sum_{i=1}^{n}$ is defined by Def 4.4 on page 30) where $r = \sum_{i=1}^{n} \alpha_i a_i$, $s = \sum_{i=1}^{n} \alpha_i b_i$. But for the intersection and other operations the class is unfortunately not closed. So the restriction to parametric classes is, at least for our purposes, not suitable.

From a practical point of view it should be clear that $N(I\!R)$ is rich enough for us. $N(I\!R)$ is closed against union, scalar multiplication, addition, subtraction, and other useful operations defined by the extension principle. We cannot, of course, store all fuzzy sets in $N(I\!R)$, since the number of the different sets in a set representation may be infinite. We therefore approximate a fuzzy set by choosing a sufficiently large number of α-levels. This proceeding is also rectified by the remarks in section 2 concerning the scales of acceptability.

Several problems have to be overcome in choosing a suitable data structure:

(i) Although the complement and the intersection of two fuzzy sets in $N(I\!R)$ are in general not in $N(I\!R)$, the data structure should be flexible enough to represent (at least) approximations of the results of these operations.

(ii) The data structure should provide efficient algorithms for operations defined by the extension principle.

The classes of fuzzy sets that we choose to represent in a computer are defined as follows:

Definition 5.1 *A fuzzy set μ is in the class $C_n(\mathbb{R})$ ($n \in \mathbb{N}$), if and only if there is a set representation $\{A_\alpha \mid \alpha \in (0,1)\}$ of μ such that*

(i) $\forall i \in \{1,\ldots,n\}$, $\forall \beta \in (\frac{i-1}{n}, \frac{i}{n}] : A_\beta = A_{i/n}$

(ii) For all $i \in \{1,\ldots,n\}$ either the set $A_{i/n}$ is a finite union of N_i disjoint compact intervals, i.e.

$$A_{i/n} = \bigcup_{j=1}^{N_i} \left[(a_j)_{i/n}, (b_j)_{i/n}\right],$$

or $A_{i/n}$ is empty, i.e. $N_i = 0$.

Thereby A_1 denotes the 1-level-set $\mu_{\bar{1}}$.

If $\mu \in C_n(\mathbb{R})$ and if there is an $x \in \mathbb{R}$ such that $\mu(x) = 1$, then obviously $\mu \in N(\mathbb{R})$ holds. A fuzzy set $\mu \in C_n(\mathbb{R})$ can be represented as a linked list of intervals. Using the notions of Definition 5.1, we obtain Figure 5.1 as a representation for μ. We have to store the ends of intervals. If $\mu \in N(\mathbb{R})$ and $n \in \mathbb{N}$, then we can approximate μ by the fuzzy set $[\mu]_n \in C_n(\mathbb{R})$, for which the sets $A_{1/n}, \ldots A_{n/n}$ in a set representation coincide.

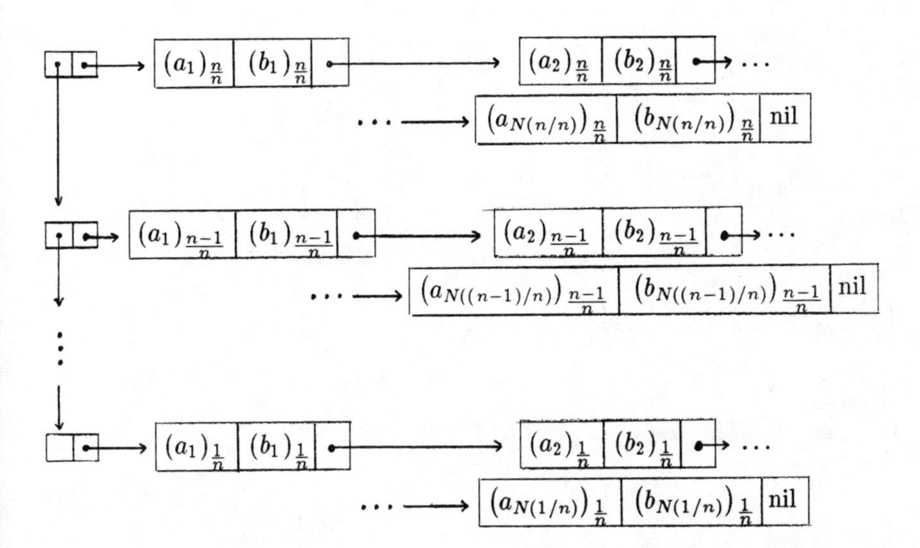

Fig. 5.1. Set representation of a fuzzy set by linked lists of intervals.

Let us consider an example which clarifies the notions.

Example 5.2 The fuzzy sets $g^{(5,1)}$ (see p. 12) and $t^{(5.5,7,7,8.5)}$ (see p. 59) have the shape of a Gaussian function and a triangle, respectively.

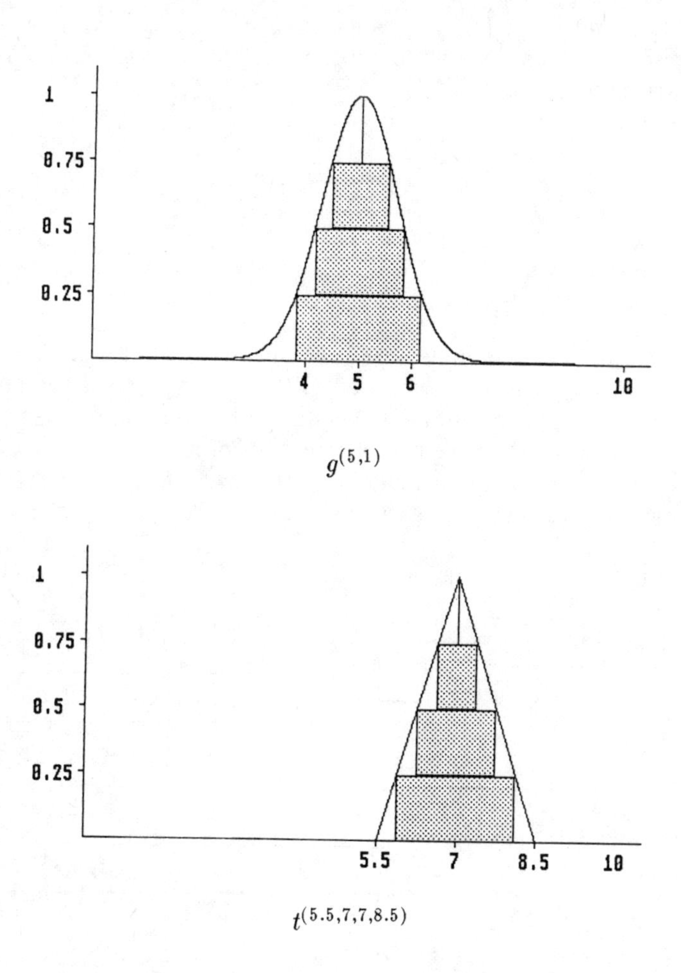

Fig. 5.2. The graphs of $g^{(5,1)}$ and $t^{(5.5,7,7,8.5)}$.

The shapes of $[g^{(5,1)}]_4$ and $[t^{(5.5,7,7,8.5)}]_4$ are hatched in Figure 5.2. We have the following representations:

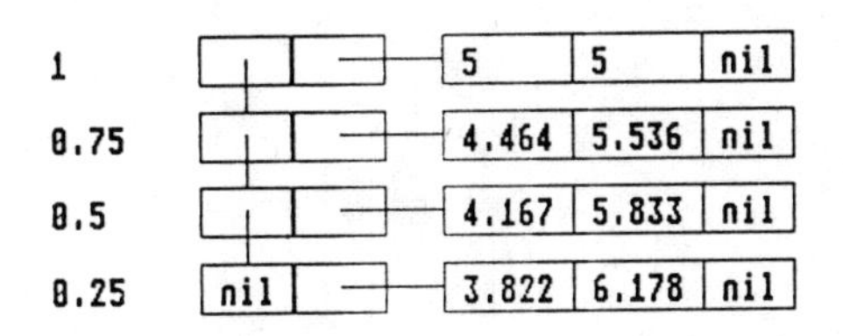

$g^{(5,1)}$

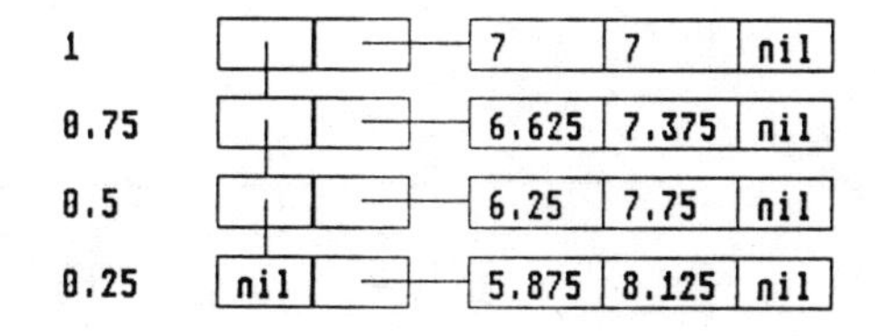

$t^{(5.5,7,7,8.5)}$

Fig. 5.3. The internal representations of $g^{(5,1)}$ and $t^{(5.5,7,7,8.5)}$.

The union $[g^{(6,1)}]_4 \vee [t^{(5.5,7,7,8.5)}]_4 = [g^{(5,1)} \vee t^{(5.5,7,7,8.5)}]_4$ has the following graph and representation:

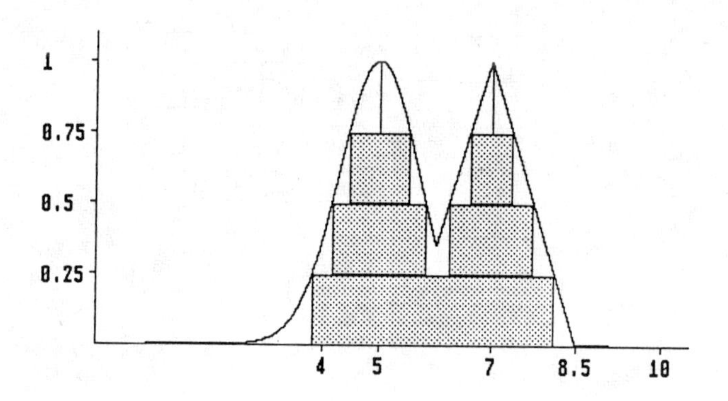

graph of $g^{(5,1)} \vee t^{(5.5,7,7,8.5)}$

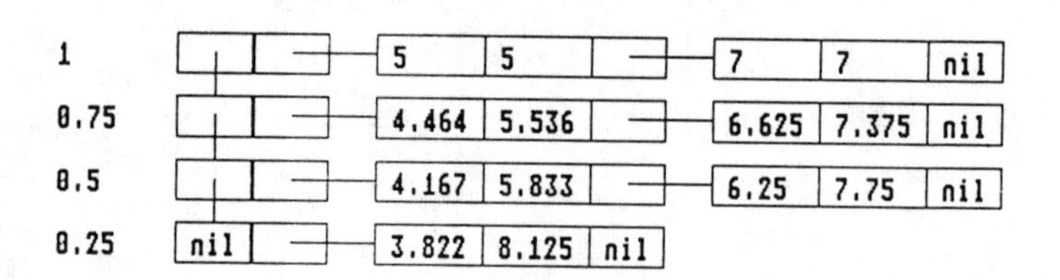

internal representation as $[g^{(5,1)} \vee t^{(5.5,7,7,8.5)}]_4$

Fig. 5.4. Graph and internal representation of $g^{(5,1)} \vee t^{(5.5,7,7,8.5)}$.

For the intersection $[g^{(5,1)}]_4 \wedge [t^{(5.5,7,7,8.5)}]_4 = [g^{(5,1)} \wedge t^{(5.5,7,7,8.5)}]_4$ we obtain:

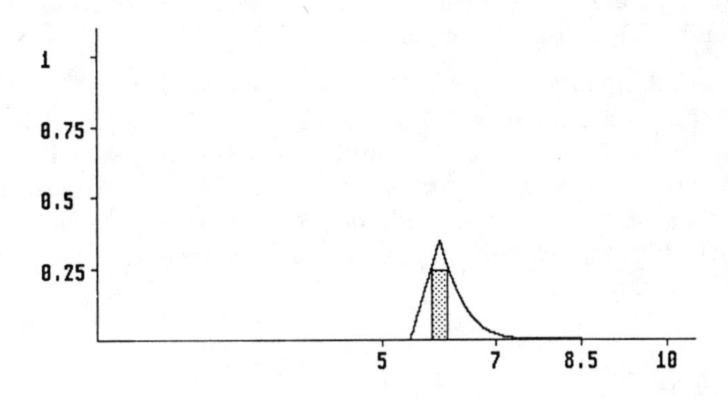

graph of $g^{(5,1)} \wedge t^{(5.5,7,7,8.5)}$

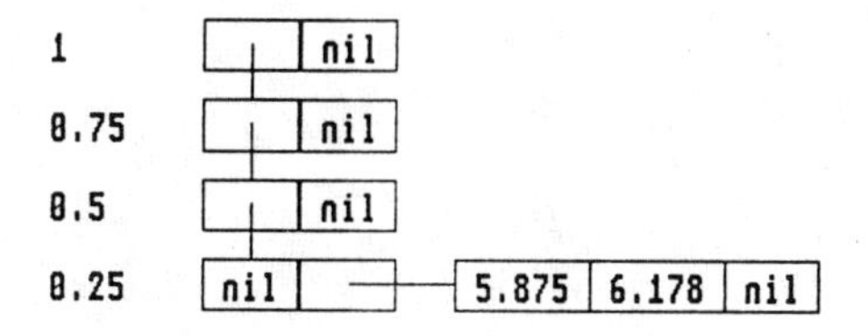

internal representation as $\left[g^{(5,1)} \wedge t^{(5.5,7,7,8.5)}\right]_4$

Fig. 5.5. Graph and internal representation of $g^{(5,1)} \wedge t^{(5.5,7,7,8.5)}$.

From the example, it should become clear that we are able to construct efficient parallel algorithms for the set theoretic operations. But from Lemma 4.5(i) we know that this structure provides efficient parallel algorithms for all operations defined by the extension principle. We do not want to give the details for all operations but we want to give some hints for implementing the arithmetic operations.

Example 5.3 In order to add the fuzzy sets $\left(g^{(5,1)} \vee t^{(5.5,7,7,8.5)}\right)$ and $g^{(5,1)}$, we have to consider the four different levels. For the level $\alpha = 0.5$ we have the sets $[4.167, 5.833] \cup [6.25, 7.75]$ and $[4.167, 5.833]$, respectively. We have to "add" these sets by the operation defined in Lemma 4.5(i) which is, in our special case, the Minkowski addition. In a first step, each interval of the first set is added to each interval of the second set, we obtain (by adding the interval borders)

$$[8.334, 11.666] \cup [10.417, 13.583].$$

Then, in a second step, these intervals are ordered and combined. We obtain

$$[8.334, 13.583]$$

as the result for the 0.5 level of $[\left(g^{(5,1)} \vee t^{(5.5,7,7,8.5)}\right) + g^{(5,1)}]_4$. This procedure is performed for each level and we obtain the following graph.

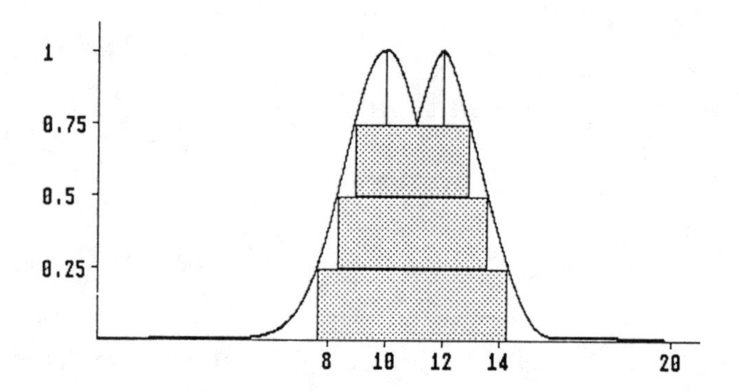

graph

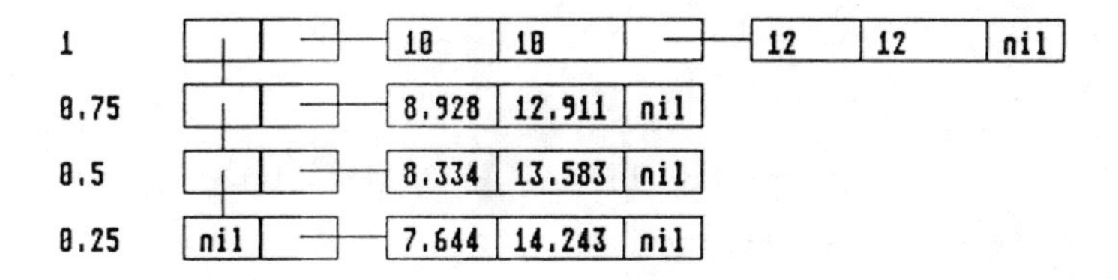

internal representation

Fig. 5.6. Graph and internal representation of $\left(g^{(5,1)} \vee t^{(5.5,7,7,8.5)}\right) + g^{(5,1)}$.

For subtraction, multiplication, and division the algorithms are similarly simple. The reason for this is that these operations are monotonous operators. Division plays a special role, since it is possible that the number zero is an element of a level set: We can, at least in principle, "divide by 0". We have to distinguish in the algorithm for the division between 9 different cases (M. Eike and J. Freckmann [41]) depending on the location of the zero.
In essence, we have a generalized interval arithmetic [145]. These considerations demonstrate that our classes $C_n(\mathbb{R})$ are flexible enough to represent all fuzzy sets which appear in practical application. The methods described here are used for an implementation in the programming language PASCAL, contact the first author for details.

6 Topological properties of fuzzy set spaces

In the preceding chapters we have studied the algebraic properties of the different classes of vague data and their representations. In the following we consider the topological properties of these classes by defining metric spaces. The metrics allow us to measure distances between fuzzy sets and to consider convergence notions of sequences of fuzzy sets.

Our starting point is the real number $x \in \mathbb{R}$, identified with its corresponding crisp fuzzy set $I_{\{x\}} \in F(\mathbb{R})$. The mapping $d : \mathbb{R}^2 \to \mathbb{R}$, $(x,y) \to |x-y|$ is the canonical distance-metric on $\mathbb{R}$. In order to generalize the so called Euclidean metric, we give the following notion.

Definition 6.1 *On the set of all nonempty subsets of* $\mathbb{R}$, *the Hausdorff pseudometric is defined by*

$$d_H[A,B] \stackrel{\mathrm{d}}{=} \max\left\{\sup_{a\in A}\inf_{b\in B}|a-b|, \sup_{b\in B}\inf_{a\in A}|a-b|\right\}.$$

for $A \subseteq \mathbb{R}, B \subseteq \mathbb{R}, A \neq \emptyset, B \neq \emptyset$.

The definition can be illustrated by the following characterization. The distance between A and B is the minimum value ϵ such that $B \subseteq A + [-\epsilon, \epsilon]$ and $A \subseteq B + [-\epsilon, \epsilon]$, where "+" is the Minkowski addition (see Lemma 4.5 and page 31). From this property it is clear that we have $d_H[\{x\},\{y\}] = |x-y|$.

If $(H, \|\ \|)$ is a normed space, we can define $d_H[A,B]$ for $A \subseteq H$ and $B \subseteq H$ being nonempty in the same way. If A and B are compact, then $d_H[A,B] = 0$ implies $A = B$.

Lemma 6.2 *If A and B are two intervals (on the real line), then*

$$d_H[A,B] = \begin{cases} \max\{|\inf A - \inf B|, |\sup A - \sup B|\} & , \text{if } A \text{ and } B \text{ are bounded} \\ |\inf A - \inf B| & , \text{if } \inf A > -\infty, \sup A = +\infty, \\ & \quad \inf B > -\infty \text{ and } \sup B = +\infty \\ |\sup A - \sup B| & , \text{if } \inf A = -\infty, \sup A < +\infty, \\ & \quad \inf B = -\infty \text{ and } \sup B < +\infty \\ 0 & , \text{if } A = B = I\!R \\ +\infty & \text{otherwise} \end{cases}$$

is valid.

Example 6.3 Let $A = [0,1] \cup [5,7]$ and $B = [b, b+5]$, $b \in I\!R$. Then

$$d_H[A,B] = \begin{cases} \max(|b-0|, |b+5-7|, 2.5) & , \text{if } b \leq 3.5 \leq b+5 \\ \max(|b-0|, |b+5-7|, |b-6|) & , \text{if } b > 3.5 \\ \max(|b-0|, |b+5-7|, |b+5-1|) & , \text{if } b < -1.5 \end{cases}$$

is valid.

For the calculations, we only have to consider the borders of the intervals (i.e. the set $\{0,1,6,7,b,b+5\}$) and the midpoints of the interval gaps (here $(1+6)/2 = 3.5$). •

The same is the case if the Hausdorff distance is calculated for sets A and B which are finite unions of compact intervals. It is also possible to calculate the Hausdorff distance with help of the interval borders and the midpoints of the interval gaps.

The collection of all nonempty compact subsets of $I\!R$ or, more generally, the collection of all nonempty compact subsets of a separable Banach space $(H, \|\ \|)$, is a complete separable metric space with respect to the metric d_H (see G. Debreu [34]). Therefore we introduce a further generalization of the Hausdorff metric by considering the space $F_C(I\!R)$ in detail.

Lemma 6.4 *Let $\mu \in F_C(I\!R)$. Let $\{A_\alpha \mid \alpha \in (0,1)\}$ be a set representation for μ such that A_α is closed for $\alpha \in (0,1)$.*

(i) Let $\beta \in (0,1]$ and $\{\beta_r\}_{r \in I\!N}$ be a sequence strictly increasing against β such that $\beta_r \in (0,\beta)$ is valid for $r \in I\!N$. Then

$$\lim_{r \to \infty} d_H\left[\mu_{\overline{\beta}}, A_{\beta_r}\right] = 0$$

follows.

(ii) *Let* $\alpha \in [0,1)$ *and* $\{\alpha_r\}_{r\in\mathbb{N}}$ *be a sequence strictly decreasing against* α *such that* $\alpha_r \in (\alpha, 1)$ *is valid for* $r \in \mathbb{N}$. *Then*

$$\lim_{r\to\infty} d_H[\mu_\alpha, A_{\alpha_r}] = 0$$

holds.

Proof

(i) By Theorem 3.4(ii) we know that $\mu_{\overline{\beta}} = \bigcap_{r=1}^{\infty} A_{\beta_r}$ is valid. So

$$d_H\left[\mu_{\overline{\beta}}, A_{\beta_r}\right] = \sup_{x\in A_{\beta_r}} \inf_{y\in\mu_{\overline{\beta}}} |x-y|$$

holds for all $r \in \mathbb{N}$.

Let $\epsilon > 0$ arbitrary. Define $t_i \overset{\mathrm{d}}{=} \frac{\epsilon}{2}\cdot i$ for $i \in \mathbb{Z}$. Define

$$B \overset{\mathrm{d}}{=} \left\{i \in \mathbb{Z} \;\middle|\; \mu_{\overline{\beta}} \cap [t_i, t_{i+1}] \neq \emptyset\right\} \quad \text{and}$$

$$B_r \overset{\mathrm{d}}{=} \{i \in \mathbb{Z} \mid A_{\beta_r} \cap [t_i, t_{i+1}] \neq \emptyset\} \quad \text{for} \quad r \in \mathbb{N}.$$

As $\inf \mu_0 > -\infty$ and $\sup \mu_0 < +\infty$ holds, B is finite, and B_r is finite for all $r \in \mathbb{N}$. Obviously $B \subseteq \bigcap_{r=1}^{\infty} B_r$ holds.

If $i \in \bigcap_{r=1}^{\infty} B_r$, then $A_{\beta_r} \cap [t_i, t_{i+1}]$ is not empty for all $r \in \mathbb{N}$. $\{A_{\beta_r} \cap [t_i, t_{i+1}]\}_{r\in\mathbb{N}}$ is a non-increasing sequence of non-empty subsets, so

$$\bigcap_{r=1}^{\infty} A_{\beta_r} \cap [t_i, t_{i+1}] = \mu_{\overline{\beta}} \cap [t_i, t_{i+1}]$$

is not empty, i.e. $i \in B$.

As we have to work with finite subsets of $\mathbb{R}$, there is an $R \in \mathbb{N}$ such that $B = B_r$ is valid for all $r \geq R$.

If $r \geq R$ and $a \in A_{\beta_r}$, we can find an $i \in B_r$ with $|t_i - a| < \frac{\epsilon}{2}$. As $i \in B$, we can find a $b \in \mu_{\overline{\beta}}$ with $|t_i - b| \leq \frac{\epsilon}{2}$. This implies $|a - b| < \epsilon$ and we can conclude the assertion.

(ii) can be shown in a similar way. •

Lemma 6.5 *Let* $\mathfrak{B}_1$ *denote the Borel* σ*-algebra of the real line and*

$$\mathfrak{B}_1 \cap [0,1] \overset{\mathrm{d}}{=} \{A \in \mathfrak{B}_1 \mid A \subseteq [0,1]\}.$$

Let $(\mu,\nu) \in [F_C(I\!R)]^2$. Then it holds:

(i) The mapping Φ with $\Phi(\beta) = d_H\left[\mu_{\overline{\beta}}, \nu_{\overline{\beta}}\right]$ for $\beta \in (0,1]$ is left-side continuous

(ii) The mapping Ψ with $\Psi(\alpha) = d_H[\mu_\alpha, \nu_\alpha]$ for $\alpha \in [0,1)$ is right-side continuous.

Proof Let $\beta \in [0,1]$. Let $\{\beta_r\}_{r \in I\!N}$ be a sequence strictly increasing against β such that $\beta_r \in (0,\beta)$ is valid for $r \in I\!N$. We can conclude with the help of Lemma 6.4:

$$\lim_{r\to\infty} d_H\left[\mu_{\overline{\beta}}, \mu_{\overline{\beta_r}}\right] = \lim_{r\to\infty} d_H\left[\nu_{\overline{\beta}}, \nu_{\overline{\beta_r}}\right] = 0.$$

Applying the triangle inequality, we obtain:

$$d_H\left[\mu_{\overline{\beta}}, \nu_{\overline{\beta}}\right] \le d_H\left[\mu_{\overline{\beta}}, \mu_{\overline{\beta_r}}\right] + d_H\left[\mu_{\overline{\beta_r}}, \nu_{\overline{\beta_r}}\right] + d_H\left[\nu_{\overline{\beta_r}}, \nu_{\overline{\beta}}\right],$$

and we conclude:

$$|d_H\left[\mu_{\overline{\beta}}, \nu_{\overline{\beta}}\right] - d_H\left[\mu_{\overline{\beta}}, \mu_{\overline{\beta_r}}\right]| \le d_H\left[\mu_{\overline{\beta_r}}, \nu_{\overline{\beta_r}}\right] + d_H\left[\nu_{\overline{\beta_r}}, \nu_{\overline{\beta}}\right],$$

which proves the first assertion.

The second one can be shown similarly. •

From the preceding theorems it is clear that the space $F_C(I\!R)$ plays an important role in topological considerations. There is no unique metric on $F_C(I\!R)$ which extends the Hausdorff distance d_H. In our book we will mainly be concerned with the two metrics d_1 and d_∞.

Definition 6.6 *We define a metric on $F_C(I\!R)$ and a pseudometric on $F(I\!R)$ by*

$$d_1(\mu,\nu) \stackrel{d}{=} \int_0^1 d_H[\mu_{\overline{\alpha}}, \nu_{\overline{\alpha}})]d\alpha$$

for $(\mu,\nu) \in [F_C(I\!R)]^2$, and

$$d_\infty(\mu,\nu) \stackrel{d}{=} \max\left\{ \sup_{\alpha\in(0,1]} d_H[\mu_{\overline{\alpha}}, \nu_{\overline{\alpha}}],\ \sup_{\alpha\in[0,1)} d_H[\mu_\alpha, \nu_\alpha] \right\}$$

for $(\mu,\nu) \in [F(I\!R)]^2$.

Lemma 6.5 shows that $d_1(\mu,\nu)$ exists for $(\mu,\nu) \in [F_C(\mathbb{R})]^2$. Obviously $d_1(\mu,\nu) \le d_\infty(\mu,\nu)$ holds for $(\mu,\nu) \in [F_C(\mathbb{R})]^2$.

d_1 is even a metric on $F_C(\mathbb{R})$ as $d_1(\mu,\nu) = 0$ implies $\mu_{\overline{\alpha}} = \nu_{\overline{\alpha}}$ for all $\alpha \in (0,1]\backslash M$ where M is a zero set. As

$$\mu(t) = \sup\left\{\alpha \mathrm{I}_{\mu_{\overline{\alpha}}}(t) \mid \alpha \in (0,1)\backslash M\right\}$$

is valid for $t \in \mathbb{R}$, we can conclude: $\mu = \nu$.

The example

$$d_\infty\left(\mathrm{I}_{(0,1)}, \mathrm{I}_{[0,1]}\right) = d_\infty\left(\mathrm{I}_{(0,1)\cap\mathbb{Q}}, \mathrm{I}_{[0,1]}\right) = 0$$

shows that d_∞ (and therefore d_1) are not metrics on $F(\mathbb{R})$. $\mathbb{Q}$ is the set of all rational numbers.

The metric d_1 was studied by M.L. Puri and D.A. Ralescu [172], d_∞ is a generalization of a metric by these authors [171,175]. From Example 6.3 and Chapter 6 we conclude that it is easy to calculate distances if the fuzzy sets are in the class $C_n(\mathbb{R})$ represented in our software tool.

Example 6.7 Let μ, ν, μ_n $(n \in \mathbb{N})$ be defined by setting

$$\mu(x) \stackrel{\mathrm{d}}{=} \begin{cases} 1, & \text{if } x = 0 \text{ or } x = 1, \\ 0, & \text{otherwise} \end{cases}$$

$$\nu(x) \stackrel{\mathrm{d}}{=} \begin{cases} 1, & \text{if } 0 \le x \le 1 \\ 0, & \text{otherwise} \end{cases},$$

and

$$\nu_n(x) = \frac{1}{n}\sum_{i=1}^{n} \mu.$$

What is the distance between ν_n and ν? We have $\nu_{\overline{\alpha}} = [0,1]$ and $(\nu_n)_{\overline{\alpha}} = \left\{0, \frac{1}{n}, \frac{2}{n}, \ldots, \frac{n-1}{n}, 1\right\}$ for $\alpha \in (0,1]$. Therefore $d_H[\nu_{\overline{\alpha}}, (\nu_n)_{\overline{\alpha}}] = \frac{1}{2n}$ holds for $\alpha \in (0,1]$. It follows

$$d_1(\nu_n,\nu) = d_\infty(\nu_n,\nu) = \frac{1}{2n}.$$

We can give an upper bound of $d_\infty(\mu,\nu)$ in terms of set representations.

Lemma 6.8 *Let $(\mu,\nu) \in [F_C(\mathbb{R})]^2$. Let $\{A_\alpha \mid \alpha \in (0,1)\}$ be a set representation of μ and $\{B_\alpha \mid \alpha \in (0,1)\}$ be one of ν. Then*

$$d_\infty(\mu,\nu) \le \sup_{\alpha\in(0,1)} d_H[A_\alpha, B_\alpha]$$

is valid.

Proof Let $\epsilon > 0$ arbitrary. Let $\alpha \in [0, 1)$.
Applying Lemma 6.4(ii), we can find a $\beta \in (\alpha, 1)$ with

$$d_H[\mu_\alpha, A_\beta] \leq \frac{\epsilon}{2} \text{ and } d_H[\nu_\alpha, B_\beta] \leq \frac{\epsilon}{2}.$$

We conclude

$$\begin{aligned} d_H[\mu_\alpha, \nu_\alpha] &\leq d_H[\mu_\alpha, A_\beta] + d_H[A_\beta, B_\beta] + d_H[B_\beta, \nu_\alpha] \\ &\leq \sup_{\alpha\in(0,1)} d_H[A_\beta, B_\beta] + \epsilon, \end{aligned}$$

which shows:

$$\sup_{\alpha\in[0,1)} d_H[\mu_\alpha, \nu_\alpha] \leq \sup_{\alpha\in(0,1)} d_H[A_\alpha, B_\alpha].$$

In a similar way, we can show with the help of Lemma 6.4(i):

$$\sup_{\alpha\in(0,1]} d_H[\mu_\alpha, \nu_\alpha] \leq \sup_{\alpha\in(0,1)} d_H[A_\alpha, B_\alpha].$$

This completes the proof.

A crucial property in connection with limit theorems is that of separability [198]. We find that the space $(F(\mathbb{R}), d_1)$ is separable. The space $(F_C(\mathbb{R}), d_\infty)$ is not separable, however, which means that the distance d_1 is in some way preferable to d_∞. These statements are proved in the following.

Theorem 6.9 *(see [109])* $(F_C(\mathbb{R}), d_1)$ *is separable.*

Proof For $R \in \mathbb{N}$, $S \in \mathbb{N}$ define

$$\Phi[R, S] \stackrel{d}{=} \left\{ \mu \in F_C(\mathbb{R}) \;\middle|\; \begin{array}{l} \mu_0 \subseteq [-S, +S] \cap \left\{ \frac{i}{R} \mid i \in \mathbb{Z} \right\} \\ \text{and } \mu(x) \in \mathbb{Q} \cap [0,1] \text{ for all } x \in \mathbb{R} \end{array} \right\}$$

and $\Phi \stackrel{d}{=} \bigcup_{R=1}^{\infty} \bigcup_{S=1}^{\infty} \Phi[R, S]$ ($\mathbb{Q}$ is the set of all rational numbers).

As μ_0 has at the most $(2RS+1)$ elements for $\mu \in \Phi[R, S]$ and as $\mathbb{Q} \cap [0, 1]$ is countable, $\Phi[R, S]$ is countable for $R \in \mathbb{N}$, $S \in \mathbb{N}$ (there is a one-to-one mapping between $\mathbb{Q}^{2RS+1}$ and $\Phi[R, S]$). So Φ is countable.
We claim that Φ is dense in $(F_C(\mathbb{R}), d_1)$.

(1) Define for $R \in \mathbb{N}$

$$\Psi[R] \stackrel{\mathrm{d}}{=} \left\{ \mu \in F_C(\mathbb{R}) \,\middle|\, \mu_0 \subseteq \left\{ \frac{i}{R} \,|\, i \in \mathbb{Z} \right\} \right\}.$$

Then $\bigcup_{R=1}^{\infty} \Psi[R]$ is dense in $(F_C(\mathbb{R}), d_\infty)$.

Let $\mu \in F_C(\mathbb{R})$ and $\epsilon > 0$ arbitrary. Define $R \stackrel{\mathrm{d}}{=} \left[\frac{1}{\epsilon}\right] + 1$ and $\nu \in \Psi[R]$ by

$$\nu(x) \stackrel{\mathrm{d}}{=} \begin{cases} \sup \left\{ \mu(t) \,|\, t \in \left[\frac{i}{R}, \frac{i+1}{R}\right) \right\}, & \text{if } x = \frac{i}{R} \text{ for an } i \in \mathbb{Z} \\ 0, & \text{otherwise.} \end{cases}$$

($\mathbb{Z}$ is the set of all integers.)

Let $\alpha \in [0,1)$. We want to show that $d_H[\mu_\alpha, \nu_\alpha] \leq \epsilon$.

If $x = \mu_\alpha$, choose $i \in \mathbb{Z}$ with $x \in \left[\frac{i}{R}, \frac{i+1}{R}\right)$. Then $|x - \frac{i}{R}| \leq \epsilon$ and $\nu\left(\frac{i}{R}\right) \geq \mu(x) > \alpha$ is valid, i.e. $\frac{i}{R} \in \nu_\alpha$. It follows:

$$\sup_{x \in \mu_\alpha} \inf_{y \in \nu_\alpha} |x - y| \leq \epsilon.$$

If $y \in \nu_\alpha$, there is an $i \in \mathbb{Z}$ with $y = \frac{i}{R}$. As $\nu\left(\frac{i}{R}\right) > \alpha$, there is an $x \in \left[\frac{i}{R}, \frac{i+1}{R}\right)$ with $\mu(x) > \alpha$, i.e. $x \in \mu_\alpha$, and $|x - \frac{i}{R}| \leq \epsilon$. We conclude

$$\sup_{y \in \nu_\alpha} \inf_{x \in \mu_\alpha} |x - y| \leq \epsilon.$$

All together we obtain by Lemma 6.8:

$$d_\infty(\mu, \nu) = \sup_{\alpha \in [0,1)} d_H[\mu_\alpha, \nu_\alpha] \leq \epsilon.$$

(2) We claim that Ψ is dense in $\left(\bigcup_{R=1}^{\infty} \Psi[R], d_1 \right)$.

Let $R \in \mathbb{N}$ and $\nu \in \Psi[R]$. Let $\epsilon > 0$ arbitrary.

Choose $S \in \mathbb{N}$ such that $\nu_0 \subseteq [-S, +S]$. Define $\{\alpha_0, \ldots, \alpha_l\} \subseteq [0,1]$ such that

(i) $\alpha_0 = 0$ and $\alpha_l = 1$,

(ii) $\{\alpha_0, \ldots, \alpha_l\}$ is the codomain of ν, and

(iii) $\alpha_{k-1} < \alpha_k$ for $k \in \{1, \ldots, l\}$

is valid.
For $k \in \{1, \ldots, l\}$ choose $\beta_k \in \mathbb{Q} \cap (\max\{\alpha_{k-1}, \alpha_k - \frac{\epsilon}{2 \cdot l \cdot S}\}, \alpha_k]$ and define $\chi \in \Phi[R, S]$ by

$$\chi(t) \stackrel{\mathrm{d}}{=} \begin{cases} \beta_k, & \text{if } t = \frac{i}{R} \text{ for an } i \in \mathbb{Z} \\ & \text{and } \nu\left(\frac{i}{R}\right) = \alpha_k \text{ for a } k \in \{1, \ldots, l\} \\ 0, & \text{otherwise.} \end{cases}$$

If $\alpha \in (\alpha_{k-1}, \beta_k)$ for a $k \in \{1, \ldots, l\}$, then

$$\chi_{\overline{\alpha}} = \nu_{\overline{\alpha}} = \left\{ t \in \mathbb{R} \;\middle|\; \exists i \in \mathbb{Z} : \; t = \frac{i}{R} \wedge \nu\left(\frac{i}{R}\right) \in \{\alpha_k, \alpha_{k+1}, \ldots, 1\} \right\}$$

is valid, i.e. $d_H[\chi_{\overline{\alpha}}, \nu_{\overline{\alpha}}] = 0$.
If $d_H[\chi_{\overline{\alpha}}, \nu_{\overline{\alpha}}] > 0$, then $\alpha \in \bigcup\limits_{k=1}^{l} [\beta_k, \alpha_k]$ follows.
As $\chi_0 \subseteq [-s, +s]$ and $\nu_0 \subseteq [-s, +s]$ holds, $d_H[\chi_{\overline{\alpha}}, \nu_{\overline{\alpha}}] \leq 2 \cdot S$ is valid for $\alpha \in (0, 1]$. We conclude:

$$d_1(\nu, \chi) = \int_0^1 d_H[\nu_{\overline{\alpha}}, \chi_{\overline{\alpha}}] d\alpha =$$

$$\sum_{k=1}^{l} \int_{\beta_k}^{\alpha_k} d_H[\nu_{\overline{\alpha}}, \chi_{\overline{\alpha}}] d\alpha \leq \sum_{k=1}^{l} 2 \cdot S \cdot (\alpha_k - \beta_k) \leq \epsilon.$$

(3) As $d_1(\mu, \nu) \leq d_\infty(\mu, \nu)$ holds for $(\mu, \nu) \in [F_C(\mathbb{R})]^2$ and by part (1), we know that $\bigcup\limits_{R=1}^{\infty} \Psi(\mathbb{R})$ is dense in $(F(\mathbb{R}), d_1)$. Combining this with part (2), we obtain the assertion.

For later applications it is necessary to embed $U_C(\mathbb{R})$ in a metric space. This can be done in the following way (compare M. Puri, D.A. Ralescu [171]):

Theorem 6.10 *There exists a normed space $(Z, \|\ \|)$ with a linear isometry*

$$\Phi : (U_C(\mathbb{R}), d_1) \longrightarrow (Z, \|\ \|), \text{ i.e.}$$

(i) $\|\Phi(\mu) - \Phi(\nu)\| = d_1(\mu, \nu)$,
(ii) $\Phi(\mu + \nu) = \Phi(\mu) + \Phi(\nu)$, and
(iii) $\Phi(\lambda\mu) = \lambda\Phi(\mu)$

is valid for $(\mu, \nu) \in [U_C(\mathbb{R})]^2$ *and* $\lambda \in \mathbb{R}$.

For proving this, a general embedding theorem of H. Radström [176] can be used. Z is the space of all equivalence classes in $[U_C(\mathbb{R})]^2$ with respect to $\sim$ where

$$(\mu_1, \nu_1) \sim (\mu_2, \nu_2) \overset{\text{d}}{\Leftrightarrow} \mu_1 + \nu_2 = \mu_2 + \nu_1.$$

$\| \ \|$ is defined by

$$\|[\mu, \nu]\| \overset{\text{d}}{=} d_1(\mu, \nu)$$

for $(\mu, \nu) \in [U_C(\mathbb{R})]^2$, and Φ by

$$\Phi(\mu) \overset{\text{d}}{=} [\mu, \mathrm{I}_{\{0\}}]$$

for $\mu \in U_C(\mathbb{R})$. It follows: $(Z, \| \ \|)$ is separable.

Theorem 6.11 *$(F_C(\mathbb{R}), d_\infty)$, $(N(\mathbb{R}), d_\infty)$, and $(U_C(\mathbb{R}), d_\infty)$ are not separable.*

Proof Let us assume that the set $\{\mu_i \mid i \in \mathbb{N}\}$ is dense in $(F_C(\mathbb{R}), d_\infty)$. For $\alpha \in (0,1)$ define $\nu[\alpha] \in U_C(\mathbb{R})$ by

$$\nu[\alpha](t) \overset{\text{d}}{=} \begin{cases} 1, & \text{if } t = 0 \\ \alpha, & \text{if } t \in [-1,+1] \backslash \{0\} \\ 0, & \text{if } t \in \mathbb{R} \backslash [-1,+1]. \end{cases}$$

If $0 < \alpha < \beta < 1$, $d_\infty(\nu[\alpha], \nu[\beta]) = d_H[[-1,+1], \{0\}] = 1$ is valid. For $\alpha \in (0,1)$ there is an $i(\alpha) \in \mathbb{N}$ with $d_\infty(\mu_{i(\alpha)}, \nu[\alpha]) \leq \frac{1}{4}$.

If $0 < \alpha < \beta < 1$, the triangle inequality yields

$$d_\infty(\mu_{i(\alpha)}, \mu_{i(\beta)}) \geq d_\infty(\nu[\alpha], \nu[\beta]) - d_\infty(\nu[\alpha], \mu_{i(\alpha)}) - d_\infty(\nu[\beta], \mu_{i(\beta)}) \geq \frac{1}{2},$$

i.e. $i(\alpha) \neq i(\beta)$. It follows: there is a one-to-one mapping between $(0,1)$ and a subset of $\mathbb{N}$. This is a contradiction. •

On $F(\mathbb{R})$ resp. $F_C(\mathbb{R})$ two convergence notions are induced in a natural way which are stated for completeness.

Definition 6.12

(i) Let $\{\mu_n\}_{n \in \mathbb{N}}$ be a sequence of elements of $F(\mathbb{R})$ and $\mu \in F(\mathbb{R})$. $\{\mu_n\}_{n \in \mathbb{N}}$ is called Hausdorff-convergent with respect to d_∞

– in signs: $\{\mu_n\}_{n \in \mathbb{N}} \overset{d_\infty}{\rightarrow} \mu$ –,

if $\lim_{n \to \infty} d_\infty(\mu_n, \mu) = 0$ is valid.

(ii) Let $\{\mu_n\}_{n \in \mathbb{N}}$ be a sequence of elements of $F_C(\mathbb{R})$ and $\mu \in F_C(\mathbb{R})$. $\{\mu_n\}_{n \in \mathbb{N}}$ is called Hausdorff-convergent with respect to d_1

– in signs: $\{\mu_n\}_{n \in \mathbb{N}} \overset{d_1}{\rightarrow} \mu$ –,

if $\lim_{n \to \infty} d_1(\mu_n, \mu) = 0$ is valid.

The sequence $\{\nu_n\}_{n\in\mathbb{N}}$ defined in Example 6.7 is obviously Hausdorff-convergent with respect to d_∞ and d_1 against ν. In the following we consider a typical example concerning convex fuzzy sets.

Example 6.13 The class $\{t^{(a,b,c,d)} \mid a,b,c,d \in \mathbb{R}, a \leq b \leq c \leq d\} \subseteq U_C(\mathbb{R})$ of fuzzy sets is defined by

$$t^{(a,b,c,d)}(x) \overset{\mathrm{d}}{=} \begin{cases} 0, & \text{if } x \leq \text{ or } x > d \\ \frac{x-a}{b-a}, & \text{if } a \leq x < b, a \neq b \\ 1, & \text{if } b \leq x \leq c \text{ or } a = b = x \text{ or } c = d = x, \\ \frac{x-d}{c-d}, & \text{if } c < x \leq d, c \neq d. \end{cases}$$

Fuzzy sets in this class are called trapezoids. We have

$$t^{(a,b,c,d)}{}_{\overline{\alpha}} = [a + \alpha(b-a), d - \alpha(d-c)]$$

for $\alpha > 0$.

Consider the sequence $(\mu_n)_{n\in\mathbb{N}}$ where

$$\mu_n \overset{\mathrm{d}}{=} t^{(0,1-1/n,1-1/n,1+1/n)}.$$

Then it follows

$$(\mu_n)_{\overline{\alpha}} = \left[\alpha\left(1 - \frac{1}{n}\right), 1 + \frac{1}{n} - \alpha\left(\frac{2}{n}\right)\right]$$

for $\alpha > 0$. Define $\nu \overset{\mathrm{d}}{=} t^{(0,1,1,1)}$.

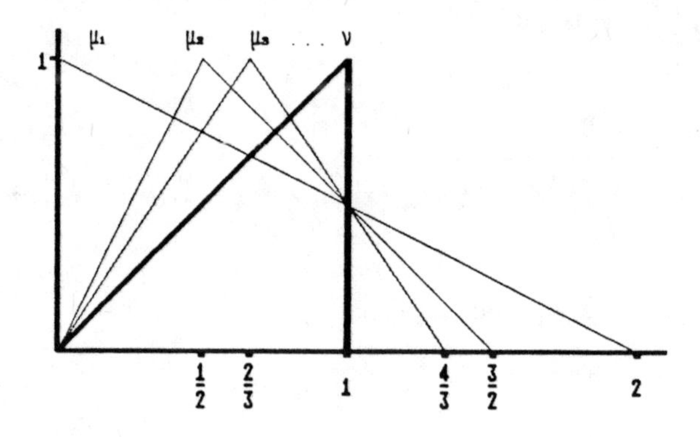

Fig. 6.1. The membership functions of μ_1, μ_2, μ_3, and ν.

We can calculate the Hausdorff distance

$$d_H[(\mu_n)_{\overline{\alpha}}, \nu_{\overline{\alpha}}] = d_H[(\mu_n)_{\alpha}, \nu_{\alpha}]$$

$$= \max\left(|\alpha\left(1-\frac{1}{n}\right)-\alpha|, |1+\frac{1}{n}-\alpha\frac{2}{n}-1|\right) = \max\left(\frac{\alpha}{n}, |\frac{1}{n}-\alpha\frac{2}{n}|\right) \leq \frac{1}{n},$$

and it follows $\{\mu_n\}_{n\in\mathbb{N}} \xrightarrow{d_1} \nu$ and $\{\mu_n\}_{n\in\mathbb{N}} \xrightarrow{d_\infty} \nu$.

It happens that these convergence notions are too strong in several cases. Therefore we use two further convergence definitions.

Definition 6.14 *Let* $\{\mu_n\}_{n\in\mathbb{N}}$ *be a sequence of elements of* $F(\mathbb{R})$ *and* $\mu \in F(\mathbb{R})$.
$\{\mu_n\}_{n\in\mathbb{N}}$ *is called Hausdorff-convergent with respect to* $\mathbb{Q}$ *(*$\mathbb{Q}$ *is the set of all rational numbers)*
– in signs: $\{\mu_n\}_{n\in\mathbb{N}} \xrightarrow{d_H,\mathbb{Q}} \mu$ *– ,*
if $\lim_{n\to\infty} d_H[\mu_n, \mu] = 0$ *holds for all* $\alpha \in [0,1) \cap \mathbb{Q}$.

Definition 6.15 *Let $D \subseteq \mathbb{R}$ be a nonempty set. Let $F : D \to F(\mathbb{R})$ be a mapping and $F_n : D \to F(\mathbb{R})$ be a mapping for all $n \in \mathbb{N}$. Then $\{F_n\}_{n\in\mathbb{N}}$ is called uniformly Hausdorff-convergent with respect to $\mathbb{Q}$*
– in signs: $\{F_n\}_{n\in\mathbb{N}} \overset{d_H,\mathbb{Q}}{\longrightarrow} F$ –,
if $\lim\limits_{n\to\infty} \sup\limits_{x\in D} d_H[(F_n(x))_\alpha, (F(x))_\alpha] = 0$ holds for all $\alpha \in [0,1) \cap \mathbb{Q}$.

If $\{\mu_n\}_{n\in\mathbb{N}} \overset{d_H,\mathbb{Q}}{\longrightarrow} \mu$ is valid, we must not conclude:

$$\lim_{n\to\infty} d_H[(\mu_n)_\alpha, \mu_\alpha] = 0$$

for all $\alpha \in [0,1)$, even if $\mu_n \in U_C(\mathbb{R})$ $(n \in \mathbb{N})$ and $\mu \in U_C(\mathbb{R})$ is valid. This shows the following example.

Example 6.16 Let $q \in (0,1)\backslash\mathbb{Q}$. Define $q_n \overset{\text{d}}{=} q + \frac{1}{2n}(1-q)$ and

$$\mu_n(t) \overset{\text{d}}{=} \begin{cases} 1, & \text{if } t = 0 \\ q_n, & \text{if } t \in [-1,+1]\backslash\{0\} \\ 0, & \text{if } t \in \mathbb{R}\backslash[-1,+1] \end{cases}$$

for $n \in \mathbb{N}$, and

$$\mu(t) \overset{\text{d}}{=} \begin{cases} 1, & \text{if } t = 0 \\ q, & \text{if } t \in [-1,+1]\backslash\{0\} \\ 0, & \text{if } t \in \mathbb{R}\backslash[-1,+1]. \end{cases}$$

It follows

$$(\mu_n)_\alpha = \begin{cases} [-1,+1], & \text{if } \alpha \in [0,q_n) \\ \{0\}, & \text{if } \alpha \in [q_n,1) \end{cases}, n \in \mathbb{N}, \text{ and}$$

$$\mu_\alpha = \begin{cases} [-1,+1], & \text{if } \alpha \in [0,q) \\ \{0\}, & \text{if } \alpha \in [q,1). \end{cases}$$

Then $\{\mu_n\}_{n\in\mathbb{N}} \overset{d_H,\mathbb{Q}}{\longrightarrow} \mu$ is valid but not : $\lim\limits_{n\to\infty} d_H\left[(\mu_n)_q, \mu_q\right] = 0$.
If $\alpha \in (q,1)$, then there is an $N \in \mathbb{N}$ such that $\alpha \in [q_n,1)$ is valid for all $n \geq N$, and therefore

$$(\mu_n)_\alpha = \mu_\alpha = \{0\}$$

holds for $n \geq N$.

If $\alpha \in [0,q)$, then $(\mu_n)_\alpha = \mu_\alpha = [-1,+1]$ is valid for $n \in \mathbb{N}$. So $\lim\limits_{n\to\infty} d_H[(\mu_n)_\alpha, \mu_\alpha] = 0$ holds for all $\alpha \in [0,1)\backslash\{q\}$, i.e.

$$\{\mu_n\}_{n \in \mathbb{N}} \xrightarrow{d_H, Q} \mu.$$

But $(\mu_n)_q = [-1, +1]$ holds for all $n \in \mathbb{N}$, whereas $\mu_q = \{0\}$. So $d_H\left[(\mu_n)_q, \mu_q\right] = 1$ is valid for all $n \in \mathbb{N}$.

7 Random sets and fuzzy random variables

In this section we will consider situations in which two different types of uncertainty — vagueness and randomness, appear simultaneously. A random experiment is described as usual by a probability space $(\Omega, \mathfrak{A}, P)$, where Ω is the set of all possible outcomes of the experiment, $\mathfrak{A}$ is a σ-algebra of subsets of Ω (the set of all possible events), and the set function P, defined on $\mathfrak{A}$, is a probability measure. In the simplest case we have a finite set $\Omega = \{\omega_1, \ldots, \omega_n\}$ of possible outcomes, where the element $\omega_i \in \Omega$ has the probability p_i.

Consider a random experiment which generates a electrical current impuls, measured by an ammeter. We can describe the random choice by a probability space $(\Omega, \mathfrak{A}, P)$. The results, which are the current impulses, are specified by a mapping, $U : \Omega \to I\!R$, which assigns to each random choice its random value. We assume that U is measurable with respect to the σ-algebra $\mathfrak{A}$ and the Borel σ-algebra $\mathfrak{B}_1$ of the real line. U is then called a random variable. If we only have a digital ammeter with the possible values $v_1, \ldots, v_n$, then we perceive the random variable U through a set of windows $V_i \subseteq I\!R$, where $v_i \in V_i$ and each window is representing an interval of the real line. We have $V_i \cap V_j = \emptyset$ for $i \neq j$, and $V_1 \cup \ldots \cup V_m = I\!R$. "Perceiving" the random variable through this windows means that for each $\omega \in \Omega$ we can only establish whether $U(\omega) \in V_i$ for some i.

We can describe this view of an experiment by another mapping $X : \Omega \to \mathfrak{P}(I\!R)$, where

$$X(\omega) = V_i \overset{\mathrm{d}}{\Leftrightarrow} V(\omega) \in V_i.$$

This means that we associate with each $\omega \in \Omega$ not a real number $U(\omega)$ as in the case of ordinary random variables, but a set. X is called a random set.

The concept of a random set, although known for a long time in connections with set-valued functions (J. Aumann [3] and G. Debreu [34], e.g.) and geometric probabilities, was recently formalized by G. Matheron [135] and

D.G. Kendall [105]. A random set L is a Borel measurable function from Ω to the class $\mathfrak{C}(\mathbb{R})$ where $\mathfrak{C}(\mathbb{R})$ denotes the set of all nonempty compact subsets of $\mathbb{R}$. $\mathfrak{C}(\mathbb{R})$ becomes a complete and separable metric space when endowed with the Hausdorff metric d_H. For random sets, incidence functions play the role of the distribution functions of random vectors. Specifically, the incidence function (also called trapping function) T_L of the random set L, a special type of the Choquèt' capacity [18], is defined as follows:

$$T_L : \mathfrak{C}(\mathbb{R}) \to [0,1], \quad T_L(K) \stackrel{\mathrm{d}}{=} P(\{\omega \in \Omega \,|\, L(\omega) \cap K \neq \emptyset\}) \text{ for } K \in \mathfrak{C}(\mathbb{R}).$$

G. Debreu [34] showed that a function $L : \Omega \to \mathfrak{C}(\mathbb{R})$ is a random set, if and only if $\{\omega \in \Omega \,|\, L(\omega) \cap K \neq \emptyset\} \in \mathfrak{A}$ for all $K \in \mathfrak{C}(\mathbb{R})$. T_L is also called lower probability [36] and plays a central role as belief function [192] in the uncertainty management of knowledge based systems (expert systems). Various properties related to the calculus of set valued functions are discussed in [3,34]. A comprehensive theory of random sets is given in G. Matheron [135].

In our example described above, L is a random set. The random variable U, of which L is a perception, is called an original of L.

If a random set L is given, then the corresponding original U is not known in many cases. We only have a possible set of originals.

If no further information is available, then each random variable U with the property $U(\omega) \in L(\omega)$ for all $\omega \in \Omega$ is a possible candidate for being the original. A random variable U satisfying this property is called a selector (J. Aumann [3]) of L. Selectors always exist as shown by K. Kuratowski and C. Ryll-Nardzewski [126].

Consider next an opinion poll, during which a number of individuals are questioned on their opinions concerning "the maximum temperature in Braunschweig on September, 12th, 1952". Randomness occurs when the individuals are selected at random. Often the answers given by the individuals are vague statements such as "the temperature was between 23°C or 24°C" or "the temperature was approximately 27°C". We can represent such answers of individuals by fuzzy sets of the real line, and we describe our opinion poll by a fuzzy "random variable"

$$X : \Omega \to E(\mathbb{R}).$$

Fuzzy random variables are introduced as generalized random variables and random sets, on which we impose some measurability criteria.

Definition 7.1 *A mapping $X : \Omega \to Q(\mathbb{R})$ is called a fuzzy random variable (f.r.v.) if there is a system $\{A_\alpha(\omega) \,|\, \omega \in \Omega, \alpha \in (0,1)\}$ of subsets of $\mathbb{R}$ with the following properties:*

(i) $\{A_\alpha(\omega) \mid \alpha \in (0,1)\}$ *is a normal set representation of* $X_\omega \stackrel{d}{=} X(\omega)$ *for all* $\omega \in \Omega$.

(ii) The mappings $\underline{A}_\alpha$ *and* $\overline{A}_\alpha$ *are* $\mathfrak{A} \longrightarrow \mathfrak{B}_1$*-measurable for all* $\alpha \in (0,1)$ *.* *(*$\mathfrak{B}_1$ *is the Borel* σ*-algebra on* $I\!R \cup \{-\infty, +\infty\}$*).*

We define hereby

$$\underline{A}_\alpha(\omega) \stackrel{d}{=} \inf A_\alpha(\omega) \quad and \quad \overline{A}_\alpha(\omega) \stackrel{d}{=} \sup A_\alpha(\omega) \ .$$

for all $\alpha \in (0,1)$ *and all* $\omega \in \Omega$.

The system $\{A_\alpha(\omega) \mid \omega \in \Omega, \alpha \in (0,1)\}$ *is called a measurable set representation of* X.

Remark In our work we assume that each random set L, described here by the associated mapping $X_L : \Omega \to U_C(I\!R)$, $(X_L)_\omega \stackrel{d}{=} \mathrm{I}_{L(\omega)}$, is a fuzzy random variable in the sense mentioned above. We have to postulate this since the measurability concepts differ. The notion of a fuzzy random variable and the related notion probabilistic set were introduced by several authors in different ways. Our approach is similar to the one of H. Kwakernaak [127]. R. Kruse in [112] and M. Miyakoshi and M. Shimbo [142] restrict themselves to F-convex fuzzy sets. M. Miyakoshi and M. Shimbo in [141] also consider fuzzy random variables in a wider sense.

A different approach to the combined treatment of fuzziness and randomness is due to E.P. Klement, M.L. Puri and D.A. Ralescu [108,109]. It is a generalized theory of random sets. A fuzzy random variable is a measurable mapping X from a probability space $(\Omega, \mathfrak{A}, P)$ to the metric space $(F_C(I\!R), d_\infty)$. P. Billingsley [12] showed that it is appropriate to talk about i.i.d. sequences of f.r.v.'s in the sense of these authors.

A combination of these works leads to our unified view of fuzzy random variables. For other approaches we refer to K. Hirota [86 to 96] and E. Czogala [22 to 30] and their concept of a probabilistic set as well as to L.A. Zadeh [214 to 219], R.R. Yager [209,210], and R. Feron [44 to 51].

As in the case of random sets, a fuzzy random variable is considered as a (fuzzy) perception of an unknown, usual random variable $U : \Omega \to I\!R$, which is called an original of X.

We denote the set of all possible originals by χ,

$$\chi \stackrel{d}{=} \{U \mid U : \Omega \to I\!R, U \ \text{ is } \ \mathfrak{A} \longrightarrow \mathfrak{B}_1 - \ \text{measurable} \ \} .$$

If only vague data are available, it is of course impossible to answer the question: "Which of the possible originals is the true original ? " But we can evaluate the acceptability acc (b_U^X) of the vague statement

$$b_U^X \stackrel{\mathrm{d}}{=} \text{" U is original of } X \text{ "}$$

in the following way:

$$\begin{aligned} \text{acc } (b_U^X) &= \text{acc } \{ \forall \omega \in \Omega : \text{ the value of the original } X \text{ at } \omega \text{ is } U(\omega) \ \} \\ &= \inf \{ \text{acc } (U(\omega) \in X(\omega)) \,|\, \omega \in \Omega \} \\ &= \inf \{ (X(\omega))(U(\omega)) \,|\, \omega \in \Omega \}. \end{aligned}$$

We have thus defined a fuzzy set of χ:

$$\begin{aligned} \mu_X : \quad \chi &\rightarrow \qquad [0,1] \\ U &\mapsto \inf \{ (X(\omega))(U(\omega)) \,|\, \omega \in \Omega \}. \end{aligned}$$

This fuzzy set consists of all possible originals of the fuzzy random variable in question.

This method can also be used when additional information is available. If e.g. we know that the originals are in a subset $\mathfrak{Y}$ of χ, then the fuzzy set

$$\mu_X^{\mathfrak{Y}} : \mathfrak{Y} \rightarrow [0,1], \quad U \mapsto \mu_X(U),$$

is appropriate for the description of the 'class of the originals'.

The picture that has just been sketched of a fuzzy random variable is complicated by the fact that the probability space $(\Omega, \mathfrak{B}, P)$ on which the fuzzy random variable manifests itself may be a reduction of a richer probability space $(\tilde{\Omega}, \tilde{\mathfrak{B}}, \tilde{P})$.

Let us first consider an example (see [127]) in which a reduction process is described.

Example 7.2 Let $(\hat{\Omega}, \hat{\mathfrak{B}}, \hat{P})$ be the probability space and let $\hat{\mathfrak{A}}$ be a subsigma algebra of $\hat{\mathfrak{B}}$. Define an equivalence relation $\sim$ on $\hat{\Omega}$ as follows:

$$\hat{\omega}_1 \sim \hat{\omega}_2 \stackrel{\mathrm{d}}{\Leftrightarrow} \left[\forall \hat{A} \in \hat{\mathfrak{A}} : \ \hat{\omega}_1 \in \hat{A} \stackrel{\mathrm{d}}{\Leftrightarrow} \hat{\omega}_2 \in \hat{A} \right].$$

Then we take Ω as the quotient set of $\hat{\Omega}$ under $\sim$. Let $c : \hat{\Omega} \rightarrow \Omega$ be the canonical projection with respect to $\sim$. Then we define $(\Omega, \mathfrak{A}, P)$ as the image probability space of $(\hat{\Omega}, \hat{\mathfrak{B}}, \hat{P})$ under c, that is

$$\mathfrak{A} \stackrel{\mathrm{d}}{=} \left\{ c(\hat{A}) \,\middle|\, \hat{A} \in \hat{\mathfrak{B}} \right\}, \quad P(A) \stackrel{\mathrm{d}}{=} \hat{P}\left(c^{-1}(A)\right) \quad \text{for} \quad A \in \mathfrak{A}.$$

$(\Omega, \mathfrak{A}, P)$ is called a reduction of $(\hat{\Omega}, \hat{\mathfrak{B}}, \hat{P})$. Clearly $(\Omega, \mathfrak{A}, P)$ is a space with less detail than $(\hat{\Omega}, \hat{\mathfrak{B}}, \hat{P})$. $(\Omega, \mathfrak{A}, P)$ may still be used to carry certain random variables, originally defined on $(\hat{\Omega}, \hat{\mathfrak{B}}, \hat{P})$.

Let $\hat{U}$ be an $\hat{\mathfrak{A}}$-measurable random variable defined on $(\hat{\Omega}, \hat{\mathfrak{B}}, \hat{P})$. Then for any $\omega \in \Omega$, $\hat{U}$ is constant on the set $c^{-1}(\omega)$. Thus we can define the random variable U on $(\Omega, \mathfrak{A}, P)$ by

$$U(\omega) \stackrel{\mathrm{d}}{=} \hat{U}(\hat{\omega}), \quad \text{where} \quad \hat{\omega} \in c^{-1}(\omega).$$

U and $\hat{U}$ have the same probability distributions. •

Thus we see that given a probability space $(\Omega, \mathfrak{A}, P)$, there may exist at least one richer probability space $(\hat{\Omega}, \hat{\mathfrak{B}}, \hat{P})$ of which $(\Omega, \mathfrak{A}, P)$ is a reduction. Which of these richer probability spaces have relevance to practical applications? We consider the case where the random mechanism depends on the probability space $(\Omega, \mathfrak{A}, P)$ and independently on a further probability space $(\Omega', \mathfrak{A}', P')$ which takes measure errors, noise, genetic differences, etc. into account. So that it is rich enough, we require

$$\forall \lambda \in [0,1] \; \exists A' \in \mathfrak{A}' : \; P'(A') = \lambda.$$

The probability space, from which $(\Omega, \mathfrak{A}, P)$ is the perception, is the product space

$$(\hat{\Omega}, \hat{\mathfrak{B}}, \hat{P}) \stackrel{\mathrm{d}}{=} (\Omega \times \Omega', \mathfrak{A} \otimes \mathfrak{A}', P \otimes P'),$$

where $\mathfrak{A} \otimes \mathfrak{A}'$ is the product σ-algebra, i.e. the smallest σ-algebra including all sets of the form $A \times A'$ with $A \in \mathfrak{A}$ and $A' \in \mathfrak{A}'$. We have in this case $\tilde{P}(A \times A') = P(A) \cdot P(A')$ (see P.R. Halmos [81,82] and E. Henze [84]). $(\Omega, \mathfrak{A}, P)$ is the reduction of $(\hat{\Omega}, \hat{\mathfrak{B}}, \hat{P})$ with respect to the σ-algebra $\hat{\mathfrak{A}} \subseteq \hat{\mathfrak{B}}$ consisting of all cylinder sets of the form $\tilde{A} = A \times \Omega'$ with $A \in \mathfrak{A}$.

In this case the possible originals of a given fuzzy random variable are random variables $U : \Omega \times \Omega' \to \mathbb{R}$. We denote the set of all possible originals in this case by

$$\tilde{\chi} \stackrel{\mathrm{d}}{=} \{U : \Omega \times \Omega' \to \mathbb{R} \,|\, U \text{ is } \mathfrak{A} \otimes \mathfrak{A}' - \mathfrak{B}_1 \text{ - measurable } \}.$$

Using the same methods described above, we calculate the acceptability of the statement " U is the original of X" $\stackrel{\mathrm{d}}{=} c_U^X$, and we obtain in this situation

$$\text{acc } (c_U^X) = \inf \{X(\omega)(U(\omega, \omega')) \,|\, \omega \in \Omega, \omega' \in \Omega'\}.$$

We have thus defined a fuzzy set of $\tilde{\chi}$

$$\begin{array}{rccc} \tilde{\mu}_X : & \tilde{\chi} & \to & [0,1], \\ & U & \mapsto & \inf \{X(\omega)(U(\omega, \omega')) \,|\, \omega \in \Omega, \omega' \in \Omega'\}, \end{array}$$

consisting of all possible originals of the fuzzy random variable in question if we know that the originals act on the probability space $(\hat{\Omega}, \hat{\mathfrak{B}}, \hat{P})$. It should be clear that, if we have further information of the probability space, we can use this information to restrict the class of originals and to modify the mappings μ_X resp. $\tilde{\mu}_X$.

Definition 7.3 *Let $n \in \mathbb{N}$ and let $X_1, \ldots, X_n$ be fuzzy random variables. Then $(X_1, \ldots, X_n) : \Omega \to [Q(\mathbb{R})]^n$ is called a fuzzy random vector (f.r.ve.) . If $U_1, \ldots, U_n$ are originals with the acceptability degrees $\mu_{X_i}(U_i)$, then*

$$\mu_{(X_1,\ldots,X_n)}(U_1, \ldots, U_n) \stackrel{\mathrm{d}}{=} \min\{\mu_{X_i}(U_i) \mid i \in \{1, \ldots, n\}\}$$

is a reasonable acceptability degree for the statement "$(U_1, \ldots, U_n)$ is the original of $(X_1, \ldots, X_n)$". If the originals are in $\tilde{\chi}$, then an analogue degree is chosen by

$$\tilde{\mu}_{(X_1,\ldots,X_n)}(U_1, \ldots, U_n) \stackrel{\mathrm{d}}{=} \min\{\tilde{\mu}_{X_i}(U_i) \mid i \in \{1, \ldots, n\}\}.$$

If $n \in \mathbb{N}$, if $(X_1, \ldots, X_n)$ is a fuzzy random vector and if

$$\Phi : [Q(\mathbb{R})]^n \to Q(\mathbb{R})$$

is a mapping, then $\Phi[X_1, \ldots, X_n]$ is defined pointwise by

$$\Phi[X_1, \ldots, X_n](\omega) \stackrel{\mathrm{d}}{=} \Phi[(X_1)_\omega, \ldots, (X_n)_\omega], \ \omega \in \Omega.$$

The proof of the following theorem follows from the theorems 4.6 and 4.7.

Theorem 7.4

(i) If $n \in \mathbb{N}$, if $(X_1, \ldots, X_n)$ is a fuzzy random vector, and if $(x_1, \ldots, x_n) \in \mathbb{R}^n$, then $\sum_{i=1}^{n} x_i X_i$ is a fuzzy random variable.

(ii) If $k \in \mathbb{N}$ is odd, and if X is a fuzzy random variable, then X^k is a fuzzy random variable.

(iii) If $k \in \mathbb{N}$ is even, and if $X : \Omega \to F(\mathbb{R})$ is a mapping such that $|X|$ is a fuzzy random variable, then X^k is a fuzzy random variable.

In the sequel, we need the following lemma.

Lemma 7.5 *Let Ω be a nonempty set. Let $A : \Omega \to \mathbb{R}$ and $B : \Omega \to \mathbb{R}$ be two mappings. Let $A_r : \Omega \to \mathbb{R}$ and $B_r : \Omega \to \mathbb{R}$ be two mappings for $r \in \mathbb{N}$. Let $\{A_r(\omega)\}_{r \in \mathbb{N}}$ be monotonously non-increasing against $A(\omega)$, and $\{B_r(\omega)\}_{r \in \mathbb{N}}$ be monotonously non-decreasing against $B(\omega)$.*

Then for all $x \in \mathbb{R}$

(i) $[A \geq x] = \bigcap_{r=1}^{\infty} [A_r \geq x]$ *and* $[B \leq x] = \bigcap_{r=1}^{\infty} [B_r \leq x]$ *and*

(ii) $[A \leq x] \supseteq \bigcup_{r=1}^{\infty} [A_r \leq x]$ *and* $[B \geq x] \supseteq \bigcup_{r=1}^{\infty} [B_r \geq x]$;

is valid. In assertion (ii) the identity in general does not hold.

Hereby, we use the abbreviations

$$[C \geq x] \stackrel{\text{d}}{=} \{\omega \in \Omega \,|\, C(\omega) \geq x\},$$

$$[C \leq x] \stackrel{\text{d}}{=} \{\omega \in \Omega \,|\, C(\omega) \leq x\}, \text{ and}$$

$$[C = x] \stackrel{\text{d}}{=} \{\omega \in \Omega \,|\, C(\omega) = x\}$$

for a mapping $C : \Omega \to \mathbb{R}$.

Let $X : \Omega \to Q(\mathbb{R})$ be a fuzzy random variable. We consider the quantities $\underline{Y}_\alpha, \overline{Y}_\alpha$ for $\alpha \in [0,1)$ and $\underline{X}_\beta, \overline{X}_\beta$ for $\beta \in (0,1]$ defined by

$$\underline{Y}_\alpha(\omega) \stackrel{\text{d}}{=} \inf{(X_\omega)_\alpha}\ , \ \overline{Y}_\alpha(\omega) \stackrel{\text{d}}{=} \sup{(X_\omega)_\alpha},$$

$$\underline{X}_\beta(\omega) \stackrel{\text{d}}{=} \inf{(X_\omega)_{\overline{\beta}}}\ , \ \overline{X}_\beta(\omega) \stackrel{\text{d}}{=} \sup{(X_\omega)_{\overline{\beta}}},$$

for $\omega \in \Omega$.

These "borderline functions" are useful for numerical calculation. Consider first the question, if they are random variables, i.e., if they are measurable.

Lemma 7.6 *Let* $X : \Omega \to Q(\mathbb{R})$ *be a f.r.v.*

(i) $\underline{Y}_\alpha$ *and* $\overline{Y}_\alpha$ *are* $\mathfrak{A} \longrightarrow \overline{\mathfrak{B}}_1$*–measurable for all* $\alpha \in [0,1)$.

(ii) *If* $X_\omega \in F_C(\mathbb{R})$ *for* $\omega \in \Omega$, *then* $\underline{X}_\beta$ *and* $\overline{X}_\beta$ *are* $\mathfrak{A} \longrightarrow \overline{\mathfrak{B}}_1$*–measurable for all* $\beta \in (0,1]$.

Proof Let $\{A_\alpha(\omega) \,|\, \omega \in \Omega,\ \alpha \in (0,1)\}$ be a measurable set representation of X.

(i) Let $\alpha \in [0,1)$. Define $\alpha_r \stackrel{\text{d}}{=} \alpha + (1-\alpha)/(2r)$ for $r \in \mathbb{N}$.
$\{\inf A_{\alpha_r}(\omega)\}_{r \in \mathbb{N}}$ is monotonously non-increasing against $\underline{Y}_\alpha(\omega)$ and $\{\sup A_{\alpha_r}(\omega)\}_{r \in \mathbb{N}}$ is monotonously non-decreasing against $\overline{Y}_\alpha(\omega)$ for $\omega \in \Omega$ because of Theorem 3.6.
Applying Lemma 7.5, we obtain for all $x \in \mathbb{R}$:

$$[\underline{Y}_\alpha \geq x] = \bigcap_{r=1}^{\infty} [\underline{A}_{\alpha_r} \geq x] \in \mathfrak{A} \text{ and}$$

$$[\overline{Y}_\alpha \leq x] = \bigcap_{r=1}^{\infty} [\overline{A}_{\alpha_r} \leq x] \in \mathfrak{A}\ .$$

(ii) Let $\beta \in (0,1]$. Define $\beta_r \stackrel{d}{=} \beta \cdot (1 - 1/2r)$ for $r \in I\!N$. In a similar way then in (i), we obtain for $x \in I\!R$ (by Lemma 3.15 and Lemma 7.5)

$$[\underline{X}_\beta \leq x] = \bigcap_{r=1}^{\infty} [\underline{A}_{\beta_r} \leq x] \in \mathfrak{A} \text{ and}$$

$$[\overline{X}_\beta \geq x] = \bigcap_{r=1}^{\infty} [\overline{A}_{\beta_r} \geq x] \in \mathfrak{A}. \bullet$$

In statistics, the discrete random variables play a central role. A fuzzy random variable is called finite if we have only a finite number of different vague observations. It is not assumed that the underlying originals U resp. $\tilde{U}$ are discrete, since it may be possible that different realizations of U or $\tilde{U}$ may lead to the same vague datum.

Definition 7.7 *Let $X : \Omega \to Q(I\!R)$ be a fuzzy random variable, let $n \in I\!N$, and let $\{\mu_1, \ldots, \mu_n\} \subseteq Q(I\!R)$. X is called finite with codomain $\{\mu_1, \ldots, \mu_n\}$, if the following conditions are fulfilled:*

(i) $\forall i \in \{1, \ldots, n\} : \{\omega \in \Omega \mid X_\omega = \mu_i\} \in \mathfrak{A}$,
(ii) $\forall i \in \{1, \ldots, n\} : P(X = \mu_i) > 0$,
(iii) $\sum_{i=1}^{n} P(X = \mu_i) = 1$.

8 Descriptive statistics with vague data

The study of probability distributions of a random variable is essentially the study of some numerical characteristics associated with them. These so-called parameters of the distribution play a key role in mathematical statistics. We generalize these methods and we define the notions of moments of fuzzy random variables. The reason for considering moments is that one is often interested in a short description of a population. This can be achieved with statistical characteristics. There are a lot of useful parameters defined in the literature on statistics. To answer the question "What is the 'centre' of the observed values $x_1, \ldots, x_n$? ", we can use location parameters, such as arithmetic mean, geometric mean, harmonic mean, modal value, median, etc. It depends on the situation which parameter is appropriate. Since questions of this kind have been answered for those cases in which only randomness is involved, we can restrict ourselves to consider the further problem that arises if randomness and vagueness appear simultaneously. It follows then that we are able to extend the known results of random variables to random sets and to fuzzy random variables. We will concentrate on moments and their properties.

Let $(\Omega, \mathfrak{B}, P)$ be a probability space and let $V : \Omega \to I\!R$ be a usual random variable. If $E\,V^k = \int_\Omega V^k dP$ exists for the positive integer k, we call $E\,V^k$ the k-th moment of V and $D\,^kV = E\,(V - E\,V)^k$ the central moment of order k of V.

In the case of vague data we cannot use these parameters directly, since no integral definitions are available. In order to derive a reasonable notion for the expected value of a fuzzy random variable X, we consider the originals $U \in \chi$ and their degrees of acceptability $\mu_X(U)$.

The acceptability degree for a given real number t to be the expected value of X is defined to be the largest value $\mu_X(U)$ such that U is a possible original of X with $E\,U = t$.

We obtain analogue definitions for the moments with respect to χ, $\tilde{\chi}$.

Definition 8.1 *Let $X : \Omega \to Q(\mathbb{R})$ be a fuzzy random variable and let k be a positive integer. We define the following fuzzy sets of $F(\mathbb{R})$.*

(i) $(E\,X^k)(t) \stackrel{\mathrm{d}}{=} \sup\{\mu_X(U) \mid U \in \chi, E\,|U^k| < \infty, E\,U^k = t\},\ t \in \mathbb{R}$, *$E\,X^k$ is called moment of order k of X with respect to χ.*

(ii) $\left(\tilde{E}\,X^k\right)(t) \stackrel{\mathrm{d}}{=} \sup\{\tilde{\mu}_X(U) \mid U \in \tilde{\chi},\ E\,|U^k| < \infty, E\,U^k = t\},\ t \in \mathbb{R}$, *$\tilde{E}\,X^k$ is called moment of order k of X with respect to $\tilde{\chi}$.*

(iii) $(D^k X)(t) \stackrel{\mathrm{d}}{=}$
$\sup\left\{\mu_X(U) \,\middle|\, U \in \chi, E\,|U - E\,U|^k < \infty, E\,(U - E\,U)^k = t\right\},\ t \in \mathbb{R}$, *$D^k X$ is called central moment of order k of X with respect to χ*

(iv) $\left(\tilde{D}^k X\right)(t) \stackrel{\mathrm{d}}{=}$
$\sup\left\{\tilde{\mu}_X(U) \,\middle|\, U \in \tilde{\chi},\ E\,|U - E\,U|^k < \infty,\ E\,(U - E\,U)^k = t\right\},\ t \in \mathbb{R}$, *$\tilde{D}^k X$ is called central moment of order k of X with respect to $\tilde{\chi}$.*

(v) $\mathrm{Var}\,X \stackrel{\mathrm{d}}{=} D^2 X$ *and* $\widetilde{\mathrm{Var}}\,X \stackrel{\mathrm{d}}{=} \tilde{D}^2 X$ *are called the variances of X with respect to χ (to $\tilde{\chi}$).*

In essence the moment $E\,X^k$ of a random variable X is defined as the image of the fuzzy subset $\mu_X : \chi \to [0,1]$ under the mapping $E^k : \chi \to \mathbb{R}$, $U \mapsto E\,U^k$, by using the extension principle. An analogue statement holds for the other notions.

In the following, we restrict ourselves to the consideration of the expected values $E\,X \stackrel{\mathrm{d}}{=} E\,X^1$, $\tilde{E}\,X \stackrel{\mathrm{d}}{=} \tilde{E}^1 X$, and the variances of X. Theorems for the other moments are obtained similarly.

8.1 EXPECTED VALUE

Let us start with the expected value, which is the most commonly used location parameter for the 'centre' of a fuzzy random variable. It depends on the situation whether the originals are in χ or in $\tilde{\chi}$.

Example 8.2

a) If $U : \Omega \to \mathbb{R}$ is a random variable with $E\,|U| < \infty$, then $X_U : \Omega \to Q(\mathbb{R})$, $\omega \mapsto \mathrm{I}_{U(\omega)}$, is its associated fuzzy random variable. We have $\tilde{E}\,X_U = E\,X_U = \mathrm{I}_{\{E\,U\}}$, that is, we have an extension of the notion expected value.

b) Let $\Omega = \{\omega_1, \omega_2\}$, $\mathfrak{B} = P(\Omega)$, $P(\{\omega_1\}) = P(\{\omega_2\}) = \frac{1}{2}$ be a finite probability space and let X be a fuzzy random variable defined by $X(\omega_1) = X(\omega_2) = \mathrm{I}_{\{0,1\}}$. Then we have $E\,X = \mathrm{I}_{\{0,\frac{1}{2},1\}} \neq \mathrm{I}_{[0,1]} = \tilde{E}\,X$.

c) Let $(R, S) : \Omega \to \mathbb{R} \times (0, \infty)$ be a random vector such that $E\,R$ and $E\,S$ exist, and let X be the fuzzy random variable defined by (compare p. 12)

$$X_\omega(t) \stackrel{\mathrm{d}}{=} g^{R(\omega),S(\omega)}(t) = \exp\left(-\left(\frac{t - R(\omega)}{S(\omega)}\right)^2\right), \; t \in \mathbb{R},\; \omega \in \Omega.$$

It follows that $E\,X = \tilde{E}\,X = g^{ER,ES}$.

The next theorem shows how to calculate $\tilde{E}\,X$ in the general case.

Theorem 8.3 *Let $X : \Omega \to Q(\mathbb{R})$ be a f.r.v. such that $E\,|\underline{Y}_o|$ and $E\,|\overline{Y}_o|$ exist. Then*

(i) $E\,|\underline{Y}_\alpha| < \infty$ and $E\,|\overline{Y}_\alpha| < \infty$ holds for all $\alpha \in [0, 1)$

(ii) $\tilde{E}\,X \in U_C(\mathbb{R})$

(iii) $\left(\tilde{E}\,X\right)_\alpha$ is convex, and $\inf\left(\tilde{E}\,X\right)_\alpha = E\,\underline{Y}_\alpha$ and $\sup\left(\tilde{E}\,X\right)_\alpha = E\,\underline{Y}_\alpha$ is valid for all $\alpha \in [0, 1)$.

(iv) $\left\{\left[E\,\underline{Y}_\alpha, E\,\overline{Y}_\alpha\right] \mid \alpha \in (0, 1)\right\}$ is a set representation of $\tilde{E}\,X$.

(v) If $X_\omega \in F_C(\mathbb{R})$ for all $\omega \in \Omega$, then $\left(\tilde{E}\,X\right)_{\overline{\beta}} = \left[E\,\underline{X}_{\overline{\beta}}, E\,\overline{X}_{\overline{\beta}}\right]$ is valid for $\beta \in (0, 1]$.

Proof

(a) For $\alpha \in [0, 1)$ and $\omega \in \Omega$ we have

$$0 \le |\underline{Y}_\alpha(\omega)| \le \max\left\{|\underline{Y}_0(\omega)|, |\overline{Y}_0(\omega)|\right\} \le |\underline{Y}_0(\omega)| + |\overline{Y}_0(\omega)| \;\text{ and}$$
$$0 \le |\overline{Y}_\alpha(\omega)| \le \max\left\{|\underline{Y}_0(\omega)|, |\overline{Y}_0(\omega)|\right\} \le |\underline{Y}_0(\omega)| + |\overline{Y}_0(\omega)|,$$

as

$$\underline{Y}_0(\omega) \le \underline{Y}_\alpha(\omega) \le \overline{Y}_\alpha(\omega) \le \overline{Y}_0(\omega) \;\text{ holds.}$$

We conclude:

$$E\,|\underline{Y}_\alpha| \le E\,|\underline{Y}_0| + E\,\left|\overline{Y}_0\right| < \infty \;\text{ and}$$
$$E\,\left|\overline{Y}_\alpha\right| \le E\,|\underline{Y}_0| + E\,\left|\overline{Y}_0\right| < \infty.$$

(b) Let $\{A_\alpha(\omega) \mid \omega \in \Omega, \alpha \in (0, 1)\}$ be a measurable set representation of X. Then $E\underline{A}_\alpha$ and $E\overline{A}_\alpha$ are existing for $\alpha \in (0, 1)$, if we define for $\omega \in \Omega$ und $\alpha \in (0, 1)$

$$\underline{A}_\alpha(\omega) \stackrel{\mathrm{d}}{=} \inf A_\alpha(\omega) \;\text{ and }\; \overline{A}_\alpha(\omega) \stackrel{\mathrm{d}}{=} \sup A_\alpha(\omega).$$

Let $\alpha \in (0,1)$ arbitrary. We want to show that

$$\left(\tilde{E}\,X\right)_{\alpha} \subseteq [E\,\underline{A}_{\alpha}, E\,\overline{A}_{\alpha}] \subseteq \left(\tilde{E}\,X\right)_{\overline{\alpha}}$$

is valid.
If $t \in \left(\tilde{E}\,X\right)_{\alpha}$, there is an $U \in \tilde{\chi}$ with $E\,U = t$ such that $U(\omega,\omega') \in (X_{\omega})_{\alpha} \subseteq A_{\alpha}(\omega)$ holds for $\omega \in \Omega$ and $\omega' \in \Omega'$. It follows for $\omega \in \Omega, \omega' \in \Omega'$:

$$\underline{A}_{\alpha}(\omega) \leq U(\omega,\omega') \leq \overline{A}_{\alpha}(\omega)$$

and therefore

$$E\,\underline{A}_{\alpha} \leq E\,U = t \leq E\,\overline{A}_{\alpha}$$

is valid.
Let $t \in \left[E\,\underline{A}_{\alpha}, E\,\overline{A}_{\alpha}\right]$.
Choose $A_{\lambda} \in \mathfrak{A}'$ with $P'(A_{\lambda}) = \lambda$. Define $U_{\lambda} \in \tilde{\chi}$ for $\lambda \in [0,1]$ by

$$U_{\lambda}(\omega,\omega') \overset{\mathrm{d}}{=} \begin{cases} \overline{A}_{\alpha}(\omega) & \text{, if } \omega \in \Omega \text{ and } \omega' \in \Omega' \backslash A_{\lambda} \\ \underline{A}_{\alpha}(\omega) & \text{, if } \omega \in \Omega \text{ and } \omega' \in A_{\lambda}. \end{cases}$$

Then $U_{\lambda}(\omega,\omega') \in A_{\alpha}(\omega) \subseteq (X_{\omega})_{\overline{\alpha}}$ holds for $\omega \in \Omega, \omega' \in \Omega', \lambda \in [0,1]$, and therefore $\tilde{\mu}_X(U_{\lambda}) \geq \alpha$ is valid for $\lambda \in [0,1]$.
Define $f(\lambda) \overset{\mathrm{d}}{=} E\,U_{\lambda} = \lambda E\,\underline{A}_{\alpha} + (1-\lambda)E\,\overline{A}_{\alpha}$ for $\lambda \in [0,1]$. f is continuous on $[0,1]$, $f(0) \geq t$ and $f(1) \leq t$ holds. We apply the first mean value theorem and obtain:

$$\exists \lambda^* \in [0,1] \text{ with } E\,U_{\lambda^*} = t.$$

It follows:

$$t \in \left(\tilde{E}\,X\right)_{\overline{\alpha}}.$$

(c) Let $\alpha \in [0,1)$. Define $\alpha_r \overset{\mathrm{d}}{=} \alpha + 1/2r(1-\alpha)$ for $r \in \mathbb{N}$. Theorem 3.6 shows that

$$\underline{Y}_{\alpha}(\omega) = \lim_{r\to\infty} \underline{A}_{\alpha_r}(\omega) \text{ and } \overline{Y}_{\alpha}(\omega) = \lim_{r\to\infty} \overline{A}_{\alpha_r}(\omega)$$

is valid for $\omega \in \Omega$. We have for $\omega \in \Omega$ and $r \in \mathbb{N}$:

$$\left|\underline{A}_{\alpha_r}(\omega)\right| \leq |\underline{Y}_o(\omega)| + \left|\overline{Y}_o(\omega)\right| \text{ and}$$
$$\left|\overline{A}_{\alpha_r}(\omega)\right| \leq |\underline{Y}_o(\omega)| + \left|\overline{Y}_o(\omega)\right|.$$

We can apply Lebesgue's theorem of the dominated convergence (compare E. Henze [84], e.g.) and obtain (with the help of Theorem 3.6 and part (b)):

$$\inf\left(\tilde{E}\,X\right)_\alpha = \lim_{r\to\infty} E\,\underline{A}_{\alpha_r} = E\,\lim_{r\to\infty}\underline{A}_{\alpha_r} = E\,\underline{Y}_\alpha \quad \text{and}$$

$$\sup\left(\tilde{E}\,X\right)_\alpha = \lim_{r\to\infty} E\,\overline{A}_{\alpha_r} = E\,\lim_{r\to\infty}\overline{A}_{\alpha_r} = E\,\overline{Y}_\alpha .$$

This shows assertion (iii).
Especially we have:

$$\inf\left(\tilde{E}\,X\right)_0 = E\,\underline{Y}_o > -\infty \quad \text{and}$$

$$\sup\left(\tilde{E}\,X\right)_0 = E\,\overline{Y}_o < +\infty .$$

Together with part (b) this shows the assertion (ii).

(d) Let $\beta \in (0,1]$. Define $\beta_r \stackrel{\mathrm{d}}{=} \beta \cdot (1 - 1/2r)$ for $r \in I\!N$. By Theorem 3.6 we know that

$$\underline{X}_\beta(\omega) = \lim_{r\to\infty}\underline{A}_{\beta_r}(\omega) \quad \text{and}$$

$$\overline{X}_\beta(\omega) = \lim_{r\to\infty}\overline{A}_{\beta_r}(\omega)$$

is valid for $\omega \in \Omega$ if $X_\omega \in F_C\,(I\!R)$. It follows:

$$\inf\left(\tilde{E}\,X\right)_{\overline{\beta}} = \lim_{r\to\infty} E\,\underline{A}_{\beta_r} = E\,\underline{X}_\beta \quad \text{and}$$

$$\sup\left(\tilde{E}\,X\right)_{\overline{\beta}} = \lim_{r\to\infty} E\,\overline{A}_{\beta_r} = E\,\overline{X}_\beta .$$

We know that

$$\tilde{E}\,X \in U_C\,(I\!R) \subseteq F_C\,(I\!R).$$

So $\left(\tilde{E}\,X\right)_{\overline{\beta}}$ is convex and closed. All together this proves assertion (v).

(e) As

$$\tilde{E}\,X \in U_C\,(I\!R) \subseteq F_C\,(I\!R)$$

holds and $\left\{\left(\tilde{E}\,X\right)_\alpha \mid \alpha \in (0,1)\right\}$ is a set representation of $\tilde{E}\,X$, Lemma 3.17(i) shows that $\left\{\mathrm{cl}\,\left(\tilde{E}\,X\right)_\alpha \mid \alpha \in (0,1)\right\}$ is a set representation for $\tilde{E}\,X$. Applying assertion (iii), we obtain:

$$\mathrm{cl}\,(E\,X)_\alpha = \left[E\,\underline{Y}_\alpha, E\,\overline{Y}_\alpha\right]$$

for $\alpha \in (0,1)$. •

The expected value of a fuzzy random variable X with respect to $\tilde{\chi}$ is always a convex fuzzy set, which is therefore easy to calculate. Since the space χ is not so rich, the expected value of X with respect to χ is not always convex (compare Example 8.2(b)).The relation between these notions is clarified by the following theorem. If $X : \Omega \to Q(\mathbb{R})$ is a fuzzy random variable, then we can make this f.r.v. convex by setting co $X : \Omega \to U(\mathbb{R})$

$$(\text{co } X)_\omega \stackrel{\text{d}}{=} \text{co } (X_\omega)$$

for $\omega \in \Omega$.

Theorem 8.4 *Let $X : \Omega \to Q(\mathbb{R})$ be a f.r.v. such that $E\,|\underline{Y}_o|$ and $E\,|\overline{Y}_o|$ exist. Then $E\,(\text{co } X) = \tilde{E}\,X$ is valid.*

Proof

(a) Theorem 8.3(i) shows that $E\,|\underline{Y}_\alpha| < \infty$ and $E\,|\overline{Y}_\alpha| < \infty$ holds for all $\alpha \in [0,1)$. Let $\{A_\alpha(\omega)\,|\,\omega \in \Omega, \alpha \in (0,1)\}$ be a measurable set representation of X. Then $E\,\underline{A}_\alpha$ and $E\,\overline{A}_\alpha$ are existing for $\alpha \in (0,1)$. Let $\alpha \in (0,1)$ arbitrary. We want to show that

$$(E\,(\text{co } X))_\alpha \subseteq [E\,\underline{A}_\alpha, E\,\overline{A}_\alpha] \subseteq (E\,(\text{co } X))_{\overline{\alpha}}$$

is valid.

If $t \in (E\,(\text{co } X))_\alpha$, there is an $U \in \chi$ with $E\,U = t$ such that

$$U(\omega) \in ((\text{co } X)_\omega)_\alpha \subseteq \text{co } (A_\alpha(\omega))$$

holds for $\omega \in \Omega$. It follows for $\omega \in \Omega$:

$$\underline{A}_\alpha(\omega) = \inf A_\alpha(\omega) =$$
$$\inf \text{co } (A_\alpha(\omega)) \le U(\omega) \le \sup \text{co } (A_\alpha(\omega)) = \overline{A}_\alpha(\omega),$$

and therefore

$$E\,\underline{A}_\alpha \le E\,U = t \le E\,\overline{A}_\alpha$$

is valid.

Let $t \in [E\,\underline{A}_\alpha, E\,\overline{A}_\alpha]$. For $\lambda \in [0,1]$ define $U_\lambda \stackrel{\text{d}}{=} \lambda \underline{A}_\alpha + (1-\lambda)\overline{A}_\alpha$. Then $U_\lambda \in \chi$ holds as well as

$$U_\lambda(\omega) \in \text{co } (A_\alpha(\omega)) \subseteq ((\text{co } X)_\omega)_{\overline{\alpha}}.$$

Therefore $\mu_X(U_\lambda) \ge \alpha$ is valid for $\lambda \in [0,1]$.

Define $f(\lambda) \stackrel{\text{d}}{=} E\,U_\lambda = \lambda E\,\underline{A}_\alpha + (1-\lambda)E\,\overline{A}_\alpha$ for $\lambda \in [0,1]$. As f is continuous on $[0,1]$ and $f(0) \ge t$ and $f(1) \le t$ holds, there is a $\lambda^* \in [0,1]$ with $E\,U_{\lambda^*} = t$. It follows:

$$t \in (E \text{ co } X)_{\overline{\alpha}}.$$

(b) We conclude: $\{[E\,\underline{A}_\alpha, E\,\overline{A}_\alpha]\,|\,\alpha \in (0,1)\}$ is set representation for E co X. Together with Theorem 8.3, this shows the assertion. •

How are these results related to the theory of random sets? A random set L is used to describe experiments where the results are described by subsets of the real line. J. Aumann [3] defines the notion of the expected value $E\,L$ as the set of all values $E\,U$, where $U : \Omega \to I\!R$ are random variables with existing expected value $E\,U$ and the property $U(\omega) \in L(\omega)$ almost everywhere. A mapping U with this property is called a selector of L. This terminology corresponds to our notion of an original with respect to χ. If all the sets $L(\omega)$ are convex and compact, and if $E\,\|L\| < \infty$, where $\|L\|(\omega) \overset{d}{=} d_H[L(\omega), \{0\}]$ for $\omega \in \Omega$, then $E\,L$ is also convex and compact (compare [3]). For the corresponding fuzzy random variable X_L, defined by $X_L(\omega) \overset{d}{=} \mathrm{I}_{L(\omega)}$, the equality $E\,X_L = \tilde{E}\,X_L = \mathrm{I}_{E\,L}$ holds; that is, we have generalized the notion of the expectation of random sets.

The functional $\tilde{E}$ has the following property:

Lemma 8.5 *The expectation value $\tilde{E}$ is a linear functional on the space*

$$\Sigma \overset{d}{=} \left\{ X : \Omega \to Q\,(I\!R) \,\middle|\, X \text{ is f.r.v.},\; E\,|\underline{Y}_o| < \infty, E\,\left|\overline{Y}_o\right| < +\infty \right\}.$$

Proof

(i) We claim that $X_1 \in \Sigma$ and $X_2 \in \Sigma$ implies $X_1 + X_2 \in \Sigma$.
From Lemma 7.4(i) we know that $X_1 + X_2$ is a f.r.v. Lemma 4.5(ii) shows that for all $\omega \in \Omega$

$$((X_1)_\omega)_0 + ((X_2)_\omega)_0 = ((X_1)_\omega + (X_2)_\omega)_0$$

holds. Define

$$(\underline{Y}_i)_\alpha(\omega) \overset{d}{=} \inf\,((X_i)_\omega)_\alpha \quad \text{and}$$
$$\left(\overline{Y}_i\right)_\alpha(\omega) \overset{d}{=} \sup\,((X_i)_\omega)_\alpha \quad \text{for } \omega \in \Omega, \alpha \in [0,1), i \in \{1,2\}.$$

Then

$$\inf\,((X_1)_\omega + (X_2)_\omega)_\alpha = (\underline{Y}_1)_\alpha(\omega) + (\underline{Y}_2)_\alpha(\omega) \quad \text{and}$$
$$\sup\,((X_1)_\omega + (X_2)_\omega)_\alpha = \left(\overline{Y}_1\right)_\alpha(\omega) + \left(\overline{Y}_2\right)_\alpha(\omega)$$

is valid for $\omega \in \Omega, \alpha \in [0,1)$.
It follows :

$$X_1 + X_2 \in \Sigma.$$

(ii) Combining Theorem 8.3 and part (i), we obtain:

$$\left\{ \left[E\,(\underline{Y}_1)_\alpha + E\,(\underline{Y}_2)_\alpha, E\,\left(\overline{Y}_1\right)_\alpha + E\,\left(\overline{Y}_2\right)_\alpha \right] \,|\, \alpha \in (0,1) \right\}$$

is a set representation of $\tilde{E}\ (X_1 + X_2)$, and Lemma 4.5(i) shows that

$$\left\{\left[E\ (\underline{Y}_1)_\alpha, E\ (\overline{Y}_1)_\alpha\right] + \left[E\ (\underline{Y}_1)_\alpha, E\ (\overline{Y}_2)_\alpha\right] \mid \alpha \in (0,1)\right\}$$

is a set representation of $\tilde{E}\ (X_1) + \tilde{E}\ (X_2)$. As these two set representations are identical,

$$\tilde{E}\ (X_1 + X_2) = \tilde{E}\ X_1 + \tilde{E}\ X_2$$

follows.
In the same way $\tilde{E}\ (\lambda X) = \lambda \tilde{E}\ X$ can be shown for $X \in \Sigma$ and $\lambda \in \mathbb{R}$.
•

An analogous statement for E only holds for convex fuzzy random variables.
To conclude we demonstrate how to calculate the expected value for finite f.r.v.'s.

Theorem 8.6 *Let X be a finite f.r.v. with the codomain $\{\mu_1, \ldots, \mu_n\} \subseteq Q\,(\mathbb{R})$. Let $\inf\,(\mu_i)_0 > -\infty$ and $\sup\,(\mu_i)_0 < +\infty$ be valid for $i \in \{1, \ldots, n\}$. Define*

$$p_i \stackrel{\mathrm{d}}{=} P(\{\omega \in \Omega \mid X_\omega = \mu_i\}) \quad \text{for } i \in \{1, \ldots, n\}.$$

(i) If $\{(A_i)_\alpha \mid \alpha \in (0,1)\}$ is a normal set representation of μ_i for $i \in \{1, \ldots, n\}$, then

$$\left\{\left[\sum_{i=1}^{n} p_i \inf\,(A_i)_\alpha, \sum_{i=1}^{n} p_i \sup\,(A_i)_\alpha\right] \mid \alpha \in (0,1)\right\}$$

is a set representation of E (co X).

(ii)

$$\left\{\left[\sum_{i=1}^{n} p_i \inf\,(\mu_i)_\alpha, \sum_{i=1}^{n} p_i \sup\,(\mu_i)_\alpha\right] \mid \alpha \in (0,1)\right\}$$

is a set representation of E (co X).

Proof From the assertion follows that

$$E\ |\underline{Y}_o| < +\infty \ \text{ and } \ E\ |\overline{Y}_o| < +\infty$$

is valid where

$$\underline{Y}_o(\omega) \stackrel{\mathrm{d}}{=} \inf\,(X_\omega)_0 \ \text{ and } \ \overline{Y}_o(\omega) \stackrel{\mathrm{d}}{=} \sup\,(X_\omega)_0$$

for $\omega \in \Omega$.

(i) There exists a normal set representation $\{A_\alpha(\omega) \,|\, \omega \in \Omega, \alpha \in (0,1)\}$ of X such that $A_\alpha(\omega) = (A_i)_\alpha$ holds for all $\alpha \in (0,1)$ and $\omega \in \Omega$ if $X_\omega = \mu_i$ for an $i \in \{1,\ldots,n\}$. Theorem 8.4 shows that $\{[E\,\underline{A}_\alpha, E\,\overline{A}_\alpha] \,|\, \alpha \in (0,1)\}$ is a set representation of E (co X) where

$$\underline{A}_\alpha(\omega) \stackrel{\mathrm{d}}{=} \inf A_\alpha(\omega) \text{ and } \overline{A}_\alpha(\omega) \stackrel{\mathrm{d}}{=} \sup A_\alpha(\omega)$$

for $\omega \in \Omega$ and $\alpha \in (0,1)$. As X is finite,

$$E\,\underline{A}_\alpha = \sum_{i=1}^{n} p_i \cdot \inf (A_i)_\alpha \text{ and } E\,\overline{A}_\alpha = \sum_{i=1}^{n} p_i \cdot \sup (A_i)_\alpha$$

holds for $\alpha \in (0,1)$.

(ii) Let $\alpha \in [0,1)$. Define $\alpha_r \stackrel{\mathrm{d}}{=} \alpha + 1/2r(1-\alpha)$ for $r \in I\!N$. By Theorem 3.4(i) and Theorem 8.3 we know that $(E\ (\text{co } X))_\alpha$ is convex. Applying part (i) we obtain:

$$\inf (E\ (\text{co } X))_\alpha = \lim_{r\to\infty} \sum_{i=1}^{n} p_i \cdot \inf (A_i)_{\alpha_r} = \sum_{i=1}^{n} p_i \cdot \inf (\mu_i)_\alpha \text{ and}$$

$$\sup (E\ (\text{co } X))_\alpha = \lim_{r\to\infty} \sum_{i=1}^{n} p_i \cdot \sup (A_i)_{\alpha_r} = \sum_{i=1}^{n} p_i \cdot \sup (\mu_i)_\alpha .$$

By Lemma 3.17(i) we know that

$$\{\text{co } (E\ (\text{co } X))_\alpha \,|\, \alpha \in (0,1)\} =$$

$$\left\{ \left[\sum_{i=1}^{n} p_i \cdot \inf (\mu_i)_\alpha, \sum_{i=1}^{n} p_i \cdot \sup (\mu_i)_\alpha \right] \,|\, \alpha \in (0,1) \right\}$$

is a set representation of E (co X).

How can the expected value be calculated in a digital computer? Let, for example, the probability space $\Omega = \{\omega_1, \ldots, \omega_m\}$, $m \in I\!N$, be finite such that the element ω_i has the probability p_i ($i \in \{1,\ldots,m\}$). If the results of the experiments are in $C_n(I\!R)$, that is, if we have a f.r.v. $X : \Omega \to C_n(I\!R)$, then

$$E\,X = \sum_{i=1}^{m} p_i X(\omega_i) \text{ , and}$$

$$\tilde{E}\,X = \text{co}\;E\,X = E\;(\text{co}\;X) = \sum_{i=1}^{m} p_i\,\text{co}\;(X(\omega_i)) \in U_C\,(\mathbb{R})$$

holds. As demonstrated in Chapter 5, we can efficiently implement arithmetic operations on fuzzy sets in $C_n\,(\mathbb{R})$. The operation needed to assign to each fuzzy set its convex hull is easy to implement.

8.2 VARIANCE

Location parameters alone give very little information about a random experiment. They describe the centre of a distribution but they do not give hints about the possible deviation of a concrete measure from such a centre. Parameters that describe the deviation of a centre are called dispersion measures. We restrict ourselves to consider the variance of finite f.r.v. If U is a finite (usual) random variable, which assumes the values x_i with the probabilities p_i, then the quantity

$$\text{Var}\;U \stackrel{\text{d}}{=} \sum_{i=1}^{n} p_i \left(x_i - \sum_{j=1}^{n} p_j x_j \right)^2$$

is called the variance of U.
In this section we assume that $X : \Omega \to Q\,(\mathbb{R})$ is a finite f.r.v. $\{\mu_1, \ldots, \mu_n\} \subseteq N\,(\mathbb{R})$ contains the codomain of X. We allow that $P\,(X = \mu_i) = 0$ is valid for an $i \in \{1, \ldots, n\}$.

We define

$$\Gamma_i \stackrel{\text{d}}{=} \{\omega \in \Omega \,|\, X_\omega = \mu_i\}$$

for $i \in \{1, \ldots, n\}$, we know that $\Gamma_i \in \mathfrak{A}$ is valid for $i \in \{1, \ldots, n\}$ and $\sum_{i=1}^{n} P(\Gamma_i) = 1$.

We assume that $\{(A_i)_\alpha \,|\, \alpha \in (0,1)\}$ is an interval set representation of μ_i for $i \in \{1, \ldots, n\}$. This means that for all $\alpha \in (0,1)$ and $i \in \{1, \ldots, n\}$ there are integers $(N_i)_\alpha \geq 1$ and systems

$$\left\{ (a_j)_\alpha^{(i)} \,|\, j \in \{1, \ldots, (N_i)_\alpha\} \right\} \quad \text{and} \quad \left\{ (b_j)_\alpha^{(i)} \,|\, j \in \{1, \ldots, (N_i)_\alpha\} \right\}$$

of real numbers such that

$$\begin{aligned} \inf\,(A_i)_\alpha = &(a_1)_\alpha^{(i)} \leq (b_1)_\alpha^{(i)} < (a_2)_\alpha^{(i)} \leq (b_2)_\alpha^{(i)} < \ldots \\ &\ldots < (a_{(N_i)_\alpha})_\alpha^{(i)} \leq \left(b_{(N_i)_\alpha}\right)_\alpha^{(i)} = \sup\,(A_i)_\alpha \end{aligned}$$

and

$$(A_i)_\alpha = \bigcup_{j=1}^{(N_i)_\alpha} \left[(a_j)_\alpha^{(i)}, (b_j)_\alpha^{(i)}\right]$$

is valid for all $\alpha \in (0,1)$ and $i \in \{1,\ldots,n\}$. Define

$$D \stackrel{\mathrm{d}}{=} \max_{i \in \{1,\ldots,n\}} \max\{|\inf(\mu_i)_0|, |\sup(\mu_i)_o|\} < +\infty.$$

We have for $\alpha \in (0,1)$, $i \in \{1,\ldots,n\}$: $(A_i)_\alpha \subseteq [-D,+D]$.

Let $\{A_\alpha(\omega) \mid \alpha \in (0,1)\}$ be a normal set representation of X_ω for $\omega \in \Omega \setminus \bigcup_{i=1}^{n} \Gamma_i$. Define for $\omega \in \Omega$ and $\alpha \in (0,1)$

$$Z_\alpha(\omega) \stackrel{\mathrm{d}}{=} \begin{cases} \inf A_\alpha(\omega) & \text{, if } \omega \in \Omega \setminus \bigcup_{i=1}^{n} \Gamma_i \text{ and } \inf A_\alpha(\omega) > -\infty \\ \sup A_\alpha(\omega) & \text{, if } \omega \in \Omega \setminus \bigcup_{i=1}^{n} \Gamma_i \text{ and} \\ & \quad \inf A_\alpha(\omega) = -\infty \text{ and } \sup A_\alpha(\omega) < +\infty \\ 0 & \text{, otherwise .} \end{cases}$$

Then Z_α is $\mathfrak{A} \otimes \mathfrak{A}' \to \mathfrak{B}_1$-measurable for $\alpha \in (0,1)$. For all $\omega \in \Omega \setminus \bigcup_{i=1}^{n} \Gamma_i$ and $\alpha \in (0,1)$ $Z_\alpha(\omega) \in A_\alpha(\omega)$ is valid. Obviously we have $P(Z_\alpha = 0) = 1$ for $\alpha \in (0,1)$. •

In the next theorem we show how to get a set representation of $\tilde{\mathrm{Var}}\, X$.

Lemma 8.7 *Define for* $\alpha \in (0,1)$

$$\tilde{Y}_\alpha \stackrel{\mathrm{d}}{=} \left\{ U \in \tilde{\mathcal{X}} \;\middle|\; \begin{array}{l} \forall i \in \{1,\ldots,n\},\ \forall \omega \in \Gamma_i,\ \forall \omega' \in \Omega' :\ U(\omega,\omega') \in (A_i)_\alpha, \\ \forall \omega \in \Omega \setminus \bigcup_{i=1}^{n} \Gamma_i,\ \forall \omega' \in \Omega' :\ U(\omega,\omega') = Z_\alpha(\omega) \end{array} \right\}.$$

and

$$\tilde{B}_\alpha \stackrel{\mathrm{d}}{=} \left\{ t \in \mathbb{R} \;\middle|\; \exists\, U \in \tilde{Y}_\alpha \text{ with } t = \mathrm{Var}\, U \right\}$$

Then

(i) $\left\{ \tilde{B}_\alpha \mid \alpha \in (0,1) \right\}$ *is a set representation of* $\widetilde{\mathrm{Var}}\, X$.

(ii) $\inf \left(\widetilde{\mathrm{Var}}\, X \right)_0 \geq 0$ *and* $\sup \left(\widetilde{\mathrm{Var}}\, X \right)_0 \leq +D^2$ *holds.*

Proof

(a) If $\alpha \in (0,1)$ and $U \in \tilde{Y}_\alpha$, we have for $\omega \in \bigcup_{i=1}^{n} \Gamma_i$ and $\omega' \in \Omega'$:

$$U(\omega,\omega') \in [-D,+D].$$

As $\sum_{i=1}^{n} P(\Gamma_i) = 1$ holds, Var U exists.

(b) Let $\alpha \in (0,1)$.

(b.i) Let $t \in (\text{ Var } X)_\alpha$. Then there exists an $U \in \tilde{\chi}$ with Var $U = t$ such that $U(\omega,\omega') \in (X_\omega)_\alpha$ is valid for all $\omega \in \Omega, \omega' \in \Omega'$. Define $\tilde{U} \in \tilde{\chi}$ by

$$\tilde{U}(\omega,\omega') \stackrel{\mathrm{d}}{=} \begin{cases} U(\omega,\omega') & \text{, if } \omega \in \bigcup_{i=1}^{n} \Gamma_i \text{ and } \omega' \in \Omega' \\ Z_\alpha(\omega) & \text{, if } \omega \in \Omega \setminus \bigcup_{i=1}^{n} \Gamma_i \text{ and } \omega' \in \Omega'. \end{cases}$$

Then $P(U = \tilde{U}) = 1$, and therefore Var $\tilde{U} = t$ is valid. For all $i \in \{1,\ldots,n\}$ and $\omega \in \Gamma_i$, $\omega' \in \Omega'$

$$\tilde{U}(\omega,\omega') \in (X_\omega)_\alpha = (\mu_i)_\alpha \subseteq (A_i)_\alpha$$

is valid. So $\tilde{U} \in \tilde{Y}_\alpha$ and $t \in \tilde{B}_\alpha$ follows.

(b.ii) Let $t \in \tilde{B}_\alpha$. Then there is an $U \in \tilde{Y}_\alpha$ with $t =$ Var U such that $U(\omega,\omega') \in (A_i)_\alpha$ is valid for $i \in \{1,\ldots,n\}$ and $\omega \in \Gamma_i$, $\omega' \in \Omega'$, and $U(\omega,\omega') = Z_\alpha(\omega)$ for $\omega \in \Omega \setminus \bigcup_{i=1}^{n} \Gamma_i$, $\omega' \in \Omega'$. For $i \in \{1,\ldots,n\}$ and $\omega \in \Gamma_i$ we have

$$(A_i)_\alpha \subseteq (\mu_i)_{\overline{\alpha}} = (X_\omega)_{\overline{\alpha}} \quad .$$

For $\omega \in \Omega \setminus \bigcup_{i=1}^{n} \Gamma_i$ and $\omega' \in \Omega'$ the relation $Z_\alpha(\omega) \in A_\alpha(\omega) \subseteq (X_\omega)_{\overline{\alpha}}$ is valid. So $\tilde{\mu}_X(U) \geq \alpha$ follows. We conclude:

$$t \in \left(\widetilde{\text{Var}}\ X\right)_{\overline{\alpha}}.$$

(b.iii) We have shown that

$$\left(\widetilde{\text{Var}}\ X\right)_\alpha \subseteq \tilde{B}_\alpha \subseteq \left(\widetilde{\text{Var}}\ X\right)_{\overline{\alpha}}$$

is valid for $\alpha \in (0,1)$. Theorem 3.5 shows that $\left\{\tilde{B}_\alpha \mid \alpha \in (0,1)\right\}$ is a set representation of $\widetilde{\text{Var}}\, X$, i.e. assertion (i) is shown.

(c) If $\alpha \in (0,1)$ and $t \in \tilde{B}_\alpha$, then $t \geq 0$ is obvious. It follows:

$$\inf\left(\widetilde{\text{Var}}\, X\right)_0 = \lim_{r\to\infty} \inf \tilde{B}_{1/2^r} \geq 0.$$

If $\alpha \in (0,1)$ and $t \in \tilde{B}_\alpha$, then there exists a $U \in \tilde{Y}_\alpha$ with $\text{Var}\, U = t$. For all $\omega \in \bigcup_{i=1}^{n} \Gamma_i$ and $\omega' \in \Omega'$ we have

$$U(\omega,\omega') \in \bigcup_{i=1}^{n} (A_i)_\alpha \subseteq [-D,+D],$$

i.e. $P(U \in [-D,+D]) = 1$. It follows:

$$t = \text{Var}\, U \leq E\, U^2 \leq D^2,$$

and therefore

$$\sup\left(\widetilde{\text{Var}}\, X\right)_0 = \lim_{r\to\infty} \sup \tilde{B}_{1/2^r} \leq D^2$$

is valid.

An analogous lemma for Var X is valid. We make that statement without proving it.

Lemma 8.8 *Define for* $\alpha \in (0,1)$

$$Y_\alpha \stackrel{\text{d}}{=} \left\{ U \in \tilde{\chi} \;\middle|\; \begin{array}{l} \forall i \in \{1,\ldots,n\},\ \forall \omega \in \Gamma_i : \ U(\omega) \in (A_i)_\alpha, \\ \forall \omega \in \Omega \backslash \bigcup_{i=1}^{n} \Gamma_i : \ U(\omega) = Z_\alpha(\omega) \end{array} \right\}$$

and

$$B_\alpha \stackrel{\text{d}}{=} \{t \in \mathbb{R} \mid \exists\, U \in Y_\alpha \ \text{with} \ t = \text{Var}\, U\}$$

Then

(i) $\{B_\alpha \mid \alpha \in (0,1)\}$ *is set representation of* $\widetilde{\text{Var}}\, X$.

(ii) $\inf(\text{Var}\, X)_0 \geq 0$ *and* $\sup(\text{Var}\, X)_0 \leq +D^2$ *holds.*

In the Theorem 8.30 we will show that $\tilde{B}_\alpha$ is an interval. We determine $\sup \tilde{B}_\alpha$ first and give some efficient algorithms for the practical calculation.

Lemma 8.9

For $A \in \mathfrak{A} \otimes \mathfrak{A}'$ define $U_A \in \tilde{Y}_\alpha$ by setting

$$U_A(\omega,\omega') \stackrel{\mathrm{d}}{=} \begin{cases} \sup(A_i)_\alpha & \text{, if there is an } i \in \{1,\dots,n\} \text{ with } \\ & \quad \omega \in \Gamma_i \text{ and } (\omega,\omega') \in A \\ \inf(A_i)_\alpha & \text{, if there is an } i \in \{1,\dots,n\} \text{ with } \\ & \quad \omega \in \Gamma_i \text{ and } (\omega,\omega') \in \Omega \times \Omega' \setminus A \\ Z_\alpha(\omega) & \text{, if } \omega \in \Omega \setminus \bigcup_{i=1}^{n} \Gamma_i \text{ and } \omega' \in \Omega' \end{cases}$$

Define

$$\tilde{S}_\alpha \stackrel{\mathrm{d}}{=} \sup\{t \in \mathbb{R} \mid \exists\, A \in \mathfrak{A} \otimes \mathfrak{A}' \text{ with } t = \operatorname{Var} U_A\}.$$

Then

(i) $\forall\, U \in \tilde{Y}_\alpha : \quad \operatorname{Var} U \le \operatorname{Var} U_{[U \ge E\, U]}$ *and*

(ii) $\sup \tilde{B}_\alpha = \tilde{S}_\alpha$ *is valid.*

Proof (a) Let $U \in \tilde{Y}_\alpha$. For abbreviation define

$$B \stackrel{\mathrm{d}}{=} \{(\omega,\omega') \in \Omega \times \Omega' \mid U(\omega,\omega') \ge E\, U\}$$

and

$$U_1(\omega,\omega') \stackrel{\mathrm{d}}{=} \begin{cases} \sup(A_i)_\alpha & \text{, if there is an } i \in \{1,\dots,n\} \text{ with } \\ & \quad \omega \in \Gamma_i \text{ and } (\omega,\omega') \in A \\ U(\omega,\omega') & \text{, if there is an } i \in \{1,\dots,n\} \text{ with } \\ & \quad \omega \in \Gamma_i \text{ and } (\omega,\omega') \in \Omega \times \Omega' \setminus A \\ Z_\alpha(\omega) & \text{, if } \omega \in \Omega \setminus \bigcup_{i=1}^{n} \Gamma_i \text{ and } \omega' \in \Omega'. \end{cases}$$

We claim that $\operatorname{Var} U \le \operatorname{Var} U_1$ is valid. We calculate:

$$\operatorname{Var} U_1 - \operatorname{Var} U = E\, U_1^2 - [E\, U_1]^2 - E\, U^2 + [E\, U]^2 =$$

$$= \int_B U_1^2 dP \otimes P' + \int_{\Omega \times \Omega' \setminus B} U^2 dP \otimes P' - \left(\int_B U_1 dP \otimes P' \right)^2 -$$

$$- 2 \int_B U_1 dP \otimes P' \int_{\Omega \times \Omega' \setminus B} U dP \otimes P' - \left(\int_{\Omega \times \Omega' \setminus B} U dP \otimes P' \right)^2 -$$

$$- \int_B U^2 dP \otimes P' - \int_{\Omega \times \Omega' \setminus B} U^2 dP \otimes P' + \left(\int_B U dP \otimes P' \right)^2 +$$

$$+ 2 \int_B U dP \otimes P' \int_{\Omega \times \Omega' \setminus B} U dP \otimes P' + \left(\int_{\Omega \times \Omega' \setminus B} U dP \otimes P' \right)^2$$

$$= \int_B U_1^2 dP \otimes P' - \int_B U^2 dP \otimes P' - 2E\, U \int_B (U_1 - U)\, dP \otimes P' +$$

$$+ 2 \int_B U dP \otimes P' \int_B (U_1 - U)\, dP \otimes P' - \left(\int_B U_1 dP \otimes P' \right)^2 +$$

$$+ \left(\int_B U dP \otimes P' \right)^2$$

$$= \int_B U_1^2 dP \otimes P' - \int_B U^2 dP \otimes P' - 2E\, U \int_B (U_1 - U)\, dP \otimes P'$$

$$- \left\{ \int_B (U_1 - U)\, dP \otimes P' \right\}^2$$

$$\geq \int_B U_1^2 dP \otimes P' - \int_B U^2 dP \otimes P' - 2E\, U \int_B (U_1 - U)\, dP \otimes P'$$

$$- \int_B (U_1 - U)^2\, dP \otimes P'.$$

For all $(\omega, \omega') \in B$ we have $U(\omega, \omega') \geq E\, U$. It follows for $(\omega, \omega') \in B$: If there is an $i \in \{1, \ldots, n\}$ with $\omega \in \Gamma_i$, then

$$U_1(\omega,\omega') + U(\omega,\omega') - 2E\,U \geq U_1(\omega,\omega') + U(\omega,\omega') - 2U(\omega,\omega')$$
$$= U_1(\omega,\omega') - U(\omega,\omega') = \sup(A_i)_\alpha - U(\omega,\omega') \geq 0.$$

If $\omega \in \Omega \setminus \bigcup_{i=1}^{n} \Gamma_i$, then

$$U_1(\omega,\omega') + U(\omega,\omega') - 2EU \geq U_1(\omega,\omega') + U(\omega,\omega') - 2U(\omega,\omega') = 0.$$

It follows for $(\omega,\omega') \in B$, as $U_1(\omega,\omega') \geq U(\omega,\omega')$:

$$[U_1(\omega,\omega') + U(\omega,\omega') - 2E\,U][U_1(\omega,\omega') - U(\omega,\omega')] \geq$$
$$\geq [U_1(\omega,\omega') - U(\omega,\omega')]^2.$$

By integrating on B, we obtain:

$$\mathrm{Var}\, U_1 \geq \mathrm{Var}\, U.$$

In a similar way we can show: $\mathrm{Var}\, U_B \geq \mathrm{Var}\, U_1$. This proves (i).

(b) On the one hand, $U_A \in \tilde{Y}_\alpha$ holds for $A \in \mathfrak{A} \otimes \mathfrak{A}'$, and therefore $\tilde{S}_\alpha \leq \sup \tilde{B}_\alpha$ is valid.
On the other hand, let $U \in \tilde{Y}_\alpha$. Define

$$B \stackrel{\mathrm{d}}{=} \{(\omega,\omega') \in \Omega \times \Omega' \mid U(\omega,\omega') \geq E\,U\} \in \mathfrak{A} \otimes \mathfrak{A}'.$$

We obtain with part (a):

$$\mathrm{Var}\, U \leq \mathrm{Var}\, U_B \leq \tilde{S}_\alpha$$

and conclude:

$$\sup \tilde{B}_\alpha \leq \tilde{S}_\alpha. \bullet$$

Define $P_n \stackrel{\mathrm{d}}{=} \left\{ (p_1, \ldots, p_n) \in [0,1]^n \,\middle|\, \sum_{i=1}^{n} p_i = 1 \right\}$ for $n \in \mathbb{N}$.

Theorem 8.10 *Define for* $\alpha \in (0,1)$ *and* $(p_1, \ldots, p_n) \in P_n$*:*

$$E_{\alpha, p_1, \ldots, p_n} : [0,1]^n \to I\!R$$
$$(c_1, \ldots, c_n) \mapsto \sum_{i=1}^n c_i p_i \sup(A_i)_\alpha + \sum_{i=1}^n (1 - c_i) p_i \inf(A_i)_\alpha$$

$$E^2_{\alpha, p_1, \ldots, p_n} : [0,1]^n \to I\!R$$
$$(c_1, \ldots, c_n) \mapsto \sum_{i=1}^n c_i p_i [\sup(A_i)_\alpha]^2 + \sum_{i=1}^n (1 - c_i) p_i [\inf(A_i)_\alpha]^2$$

$$D^2_{\alpha, p_1, \ldots, p_n} : [0,1]^n \to I\!R$$
$$(c_1, \ldots, c_n) \mapsto E^2_{\alpha, p_1, \ldots, p_n}(c_1, \ldots, c_n) - [E_{\alpha, p_1, \ldots, p_n}(c_1, \ldots, c_n)]^2$$

$$\Psi_\alpha[p_1, \ldots, p_n] \stackrel{d}{=} \sup\left\{ D^2_{\alpha, p_1, \ldots, p_n}[c_1, \ldots, c_n] \mid (c_1, \ldots, c_n) \in [0,1]^n \right\}.$$

Then we have:

(i)

$$D^2_{\alpha, p_1, \ldots, p_n}(c_1, \ldots, c_n) =$$
$$= \sum_{i=1}^n p_i c_i \left[\sup(A_i)_\alpha - E_{\alpha, p_1, \ldots, p_n}(c_1, \ldots, c_n)\right]^2 + \sum_{i=1}^n p_i (1 - c_i) \left[\inf(A_i)_\alpha - E_{\alpha, p_1, \ldots, p_n}(c_1, \ldots, c_n)\right]^2$$

and

(ii)

$$\sup \tilde{B}_\alpha = \Psi_\alpha[P(\Gamma_1), \ldots, P(\Gamma_n)].$$

Proof

(i) This equality can easily be checked by using the equation $\sum_{i=1}^n p_i = 1$

(ii) For $A \in \mathfrak{A} \otimes \mathfrak{A}'$ and $\omega \in \Omega$ define $A_\omega \stackrel{\mathrm{d}}{=} \{\omega' \in \Omega' \mid (\omega, \omega') \in A\} \in \mathfrak{A}'$. A_ω is called the ω-cut-set of A (compare E. Henze [84]). Applying Fubini's theorem ([76], e.g.), we obtain for $A \in \mathfrak{A} \otimes \mathfrak{A}'$:

$$
\begin{aligned}
&\operatorname{Var} U_A \\
&= \sum_{i=1}^{n} E\left\{(U_A \mathrm{I}_{\Gamma_i})^2\right\} - \left\{\sum_{i=1}^{n} E\,(U_A \mathrm{I}_{\Gamma_i})\right\}^2 \\
&= \sum_{i=1}^{n} \left\{\int_{\Gamma_i}\int_{\Omega'} U_A^2 \, dP \otimes P'\right\} - \left\{\sum_{i=1}^{n} \int_{\Gamma_i}\int_{\Omega'} U_A \, dP \otimes P'\right\}^2 \\
&= \sum_{i=1}^{n} \Bigg\{ [\sup(A_i)_\alpha]^2 \int_{\Gamma_i} P'(A_\omega)\, dP \\
&\quad + [\inf(A_i)_\alpha]^2 \left[P(\Gamma_i) - \int_{\Gamma_i} P'(A_\omega)\, dP\right]\Bigg\} \\
&\quad - \Bigg\{\sum_{i=1}^{n} \Bigg[\sup(A_i)_\alpha \int_{\Gamma_i} P'(A_\omega)\, dP \\
&\quad + \inf(A_i)_\alpha \left(P(\Gamma_i) - \int_{\Gamma_i} P'(A_\omega)\, dP\right)\Bigg]\Bigg\}^2.
\end{aligned}
$$

On the one hand, let $A \in \mathfrak{A} \otimes \mathfrak{A}'$ arbitrary. Then we have for $i \in \{1, \ldots, n\}$: $0 \leq \int_{\Gamma_i} P'(A_\omega)\, dP \leq P(\Gamma_i)$.

Therefore we can find $(c_1, \ldots, c_n) \in [0,1]^n$ with $\int_{\Gamma_i} P'(A_\omega)\, dP = c_i P(\Gamma_i)$ for all $i \in \{1, \ldots, n\}$.
It follows:

$$
\operatorname{Var} U_A = D^2_{\alpha, P(\Gamma_1), \ldots, P(\Gamma_n)}(c_1, \ldots, c_n) \leq \Psi_\alpha[P(\Gamma_1), \ldots, P(\Gamma_n)].
$$

Applying Lemma 8.9, we obtain:

$$
\sup \tilde{B}_\alpha = \sup\{\operatorname{Var} U_A \mid A \in \mathfrak{A} \otimes \mathfrak{A}'\} \leq \Psi_\alpha[P(\Gamma_1), \ldots, P(\Gamma_n)].
$$

On the other hand, let $(c_1, \ldots, c_n) \in [0,1]^n$ arbitrary. We can find for $i \in \{1, \ldots, n\}$ sets $B_i \in \mathfrak{A}'$ with $P'(B_i) = c_i$. Define

$$B \stackrel{\mathrm{d}}{=} \bigcup_{i=1}^{n} \Gamma_i \times B_i \in \mathfrak{A} \otimes \mathfrak{A}'.$$

For all $\omega \in \Omega$ $B_\omega = B_i$ is valid if and only if $\omega \in \Gamma_i$ holds. It follows for $i \in \{1, \ldots, n\}$

$$\int_{\Gamma_i} P'(B_\omega) dP = P'(B_i) \int_{\Gamma_i} dP = c_i P(\Gamma_i),$$

and therefore

$$D^2_{\alpha, P(\Gamma_1), \ldots, P(\Gamma_n)}(c_1, \ldots, c_n) = \text{Var}\, U_B \leq \sup \tilde{B}_\alpha$$

is valid. This completes the proof. •

Let $\alpha \in (0,1)$. As $[0,1]^n$ is compact and $D^2_{\alpha, P(\Gamma_1), \ldots, P(\Gamma_n)}$ is continuous on $\mathbb{R}^n$, it takes its maximum on $[0,1]^n$, say in $(c_1^*, \ldots, c_n^*)$. Choose $A_i \in \mathfrak{A}'$ with $P'(A_i) = c_i^*$ for $i \in \{1, \ldots, n\}$. Define

$$B \stackrel{\mathrm{d}}{=} \bigcup_{i=1}^{n} \Gamma_i \times A_i \in \mathfrak{A} \otimes \mathfrak{A}'.$$

Then $U_B \in \tilde{Y}_\alpha$ and Var $U_B = \Psi_\alpha[P(\Gamma_1), \ldots, P(\Gamma_n)]$ holds. So we can conclude: $\sup \tilde{B}_\alpha \in \tilde{B}_\alpha$.

Lemma 8.11 *Let $\alpha \in (0,1)$ and $(p_1, \ldots, p_n) \in P_n$ be given. Let $T \subseteq \{1, \ldots, n\}$ be a nonempty set. Let $c_i^* \in [0,1]$ be given for all $i \in \{1, \ldots, n\} \backslash T$.*

If we can find $c_i^* \in [0,1]$ for all $i \in T$ such that

$$2E_{\alpha, p_1, \ldots, p_n}(c_1^*, \ldots, c_n^*) = \inf(A_i)_\alpha + \sup(A_i)_\alpha$$

holds for $i \in T$, then we can follow:

$$D^2_{\alpha, p_1, \ldots, p_n}(c_1^*, \ldots, c_n^*) =$$

$$\sup \left\{ D^2_{\alpha, p_1, \ldots, p_n}(c_1, \ldots, c_n) \;\middle|\; \begin{array}{l} (c_1, \ldots, c_n) \in [0,1]^n \text{ and} \\ \forall i \in \{1, \ldots, n\} \backslash T: \; c_i = c_i^* \end{array} \right\}.$$

Proof Let $(c_1, \ldots, c_n) \in [0,1]^n$ be arbitrary such that $c_i = c_i^*$ holds for $i \in \{1, \ldots, n\} \backslash T$. Then it follows:

$$D^2_{\alpha, p_1, \ldots, p_n}(c_1^*, \ldots, c_n^*) - D^2_{\alpha, p_1, \ldots, p_n}(c_1, \ldots, c_n)$$

$$= \sum_{i \in T} (c_i^* - c_i) p_i \left\{ [\sup(A_i)_\alpha]^2 - [\inf(A_i)_\alpha]^2 \right\}$$
$$- \{E_{\alpha, p_1, \ldots, p_n}(c_1, \ldots, c_n) + E_{\alpha, p_1, \ldots, p_n}(c_1^*, \ldots, c_n^*)\} \cdot$$
$$\{E_{\alpha, p_1, \ldots, p_n}(c_1^*, \ldots, c_n^*) - E_{\alpha, p_1, \ldots, p_n}(c_1, \ldots, c_n)\}$$

$$= \sum_{i \in T} (c_i^* - c_i) p_i [\sup(A_i)_\alpha - \inf(A_i)_\alpha][\sup(A_i)_\alpha + \inf(A_i)_\alpha]$$
$$- \left\{ 2E_{\alpha, p_1, \ldots, p_n}(c_1^*, \ldots, c_n^*) - \sum_{i \in T} (c_i^* - c_i) p_i [\sup(A_i)_\alpha - \inf(A_i)_\alpha] \right\} \cdot$$
$$\left\{ \sum_{i \in T} (c_i^* - c_i) p_i [\sup(A_i)_\alpha - \inf(A_i)_\alpha] \right\}$$

$$= 2E_{\alpha, p_1, \ldots, p_n}(c_1^*, \ldots, c_n^*) \sum_{i \in T} (c_i^* - c_i) p_i [\sup(A_i)_\alpha - \inf(A_i)_\alpha]$$
$$- 2E_{\alpha, p_1, \ldots, p_n}(c_1^*, \ldots, c_n^*) \sum_{i \in T} (c_i^* - c_i) p_i [\sup(A_i)_\alpha - \inf(A_i)_\alpha]$$
$$+ \left\{ \sum_{i \in T} (c_i^* - c_i) p_i [\sup(A_i)_\alpha - \inf(A_i)_\alpha] \right\}^2 \geq 0. \quad \bullet$$

Lemma 8.12 *Let $\alpha \in (0,1)$ and $(p_1, \ldots, p_n) \in P_n$ be given. Define*

$$I_\alpha[p_1, \ldots, p_n] \stackrel{\mathrm{d}}{=} \{i \in \{1, \ldots, n\} \mid p_i = 0 \ \text{ or } \ \inf(A_i)_\alpha = \sup(A_i)_\alpha\}$$

and

$$C_\alpha[p_1, \ldots, p_n] \stackrel{\mathrm{d}}{=}$$
$$\left\{ (\tilde{c}_1, \ldots, \tilde{c}_n) \in [0,1]^n \;\middle|\; D^2_{\alpha, p_1, \ldots, p_n}(\tilde{c}_1, \ldots, \tilde{c}_n) = \Psi_\alpha[p_1, \ldots, p_n] \right\}.$$

Then there exists a $(\tilde{c}_1, \ldots, \tilde{c}_n) \in C_\alpha[p_1, \ldots, p_n]$ and an $i_0 \in \{1, \ldots, n\} \backslash I_\alpha[p_1, \ldots, p_n]$ such that $c_i \in \{0,1\}$ is valid for all $i \in \{1, \ldots, n\} \backslash \{i_0\}$.

Proof

(a) As $[0,1]^n$ is compact and $D^2_{\alpha,p_1,\dots,p_n}$ is continuous on $I\!R^n$, $C_\alpha[p_1,\dots,p_n]$ is not empty. If $(c_1,\dots,c_n) \in C_\alpha[p_1,\dots,p_n]$ arbitrary, define $(\tilde{c}_1,\dots,\tilde{c}_n) \in [0,1]^n$ by

$$\tilde{c}_i \stackrel{\mathrm{d}}{=} \begin{cases} c_i & \text{, if } i \in \{1,\dots,n\} \backslash I_\alpha[p_1,\dots,p_n] \\ 0 & \text{, if } i \in I_\alpha[p_1,\dots,p_n] . \end{cases}$$

Then

$$D^2_{\alpha,p_1,\dots,p_n}(c_1,\dots,c_n) = D^2_{\alpha,p_1,\dots,p_n}(\tilde{c}_1,\dots,\tilde{c}_n)$$

follows. So there exists a $(\tilde{c}_1,\dots,\tilde{c}_n) \in C_\alpha[p_1,\dots,p_n]$ with $c_i = 0$ for all $i \in I_\alpha[p_1,\dots,p_n]$.

(b) Let $(\tilde{c}_1,\dots,\tilde{c}_n) \in C_\alpha[p_1,\dots,p_n]$.
If there is an $i \in \{1,\dots,n\} \backslash I_\alpha[p_1,\dots,p_n]$ with $\tilde{c}_i \in (0,1)$, we can conclude:

$$0 = \frac{\partial D^2_{\alpha,p_1,\dots,p_n}}{\partial c_i}(\tilde{c}_1,\dots,\tilde{c}_n),$$

as the function f with

$$f(c) \stackrel{\mathrm{d}}{=} D^2_{\alpha,p_1,\dots,p_n}(\tilde{c}_1,\dots,\tilde{c}_{i-1},c,\tilde{c}_{i+1},\dots,\tilde{c}_n) \quad \text{for} \quad c \in [0,1]$$

has a local maximum in $\tilde{c}_i$. It follows:

$$\begin{aligned} &2E_{\alpha,p_1,\dots,p_n}(\tilde{c}_1,\dots,\tilde{c}_n)\; p_i\; [\sup(A_i)_\alpha - \inf(A_i)_\alpha] \\ &= p_i \Big\{ [\sup(A_i)_\alpha]^2 - [\inf(A_i)_\alpha]^2 \Big\}. \end{aligned}$$

As $i \in \{1,\dots,n\} \backslash I_\alpha[p_1,\dots,p_n]$ holds, $p_i \neq 0$ and $\inf(A_i)_\alpha < \sup(A_i)_\alpha$ is valid. Therefore:

$$2E_{\alpha,p_1,\dots,p_n}(\tilde{c}_1,\dots,\tilde{c}_n) = \inf(A_i)_\alpha + \sup(A_i)_\alpha \quad \text{holds.}$$

Note that the right side of this equation depends on i, whereas the left does not.

(c) We know that there exists a

$$(c_1^*,\dots,c_n^*) \in C_\alpha[p_1,\dots,p_n]$$

such that $c_i^* = 0$ holds for $i \in I_\alpha[p_1,\dots,p_n]$. Define

$$T \stackrel{\mathrm{d}}{=} \{ i \in \{1,\dots,n\} \mid c_i^* \in (0,1) \} .$$

Then $T \cap I_\alpha[p_1, \dots, p_n]$ is empty.
If T is empty or has exactly one element i_0, the assertion is fulfilled. Let us consider the case that $T = \{j_1, \dots, j_k\}$ has at least two elements. Step (b) shows that

$$2E\alpha, p_1, \dots, p_n(c_1^*, \dots, c_n^*) = \inf(A_i)_\alpha + \sup(A_i)_\alpha$$

holds for all $i \in T$. This equation is equivalent to

$$\sum_{j \in T} c_j^* p_j [\sup(A_j)_\alpha - \inf(A_j)_\alpha] = J \stackrel{\mathrm{d}}{=}$$

$$\frac{1}{2}[\inf(A_i)_\alpha + \sup(A_i)_\alpha] - \sum_{j=1}^{n} p_j \inf(A_j)_\alpha$$

$$- \sum_{j \in \{1, \dots, n\} \setminus T} c_j^* p_j [\sup(A_j) - \inf(A_j)_\alpha]$$

for $i \in T$.
Define $d_0 \stackrel{\mathrm{d}}{=} 0$ and

$$d_l \stackrel{\mathrm{d}}{=} \sum_{r=1}^{l} p_{j_r} \left[\sup(A_{j_r})_\alpha - \inf(A_{j_r})_\alpha\right]$$

for $l \in \{1, \dots, k\}$. Obviously

$$0 = d_0 < d_1 < d_2 < \dots < d_k$$

is valid. As $c_i^* \in (0,1)$ for $i \in T$ and $T \cap I_\alpha[p_1, \dots, p_n] = \emptyset$ holds,

$$J = \sum_{j \in T} c_j^* p_j \left[\sup(A_j)_\alpha - \inf(A_j)_\alpha\right] \in (0, d_k)$$

is valid. So we can find an $l_0 \in \{1, \dots, k\}$ with $T \in [d_{l_0 - 1}, d_{l_0})$. Define $i_0 \stackrel{\mathrm{d}}{=} j_{l_0}$ and

$$\tilde{c}_{j_l} \stackrel{\mathrm{d}}{=} \begin{cases} 1 & , \text{if } l \in \{1, \dots, l_0 - 1\} \\ \dfrac{J - d_{l_0 - 1}}{d_{l_0} - d_{l_0 - 1}} & , \text{if } l = l_0 \\ 0 & , \text{if } l \in \{l_0 + 1, \dots, k\}. \end{cases}$$

Obviously $\tilde{c}_{i_{l_0}} = \tilde{c}_{i_0} \in [0,1]$ holds as well as

$$\begin{aligned}
&\sum_{j \in T} \tilde{c}_j p_j [\sup(A_j)_\alpha - \inf(A_j)_\alpha] = \\
&= \sum_{l=1}^{l_0-1} p_{j_l} [\sup(A_{j_l})_\alpha - \inf(A_{j_l})_\alpha] \\
&\quad + \frac{J - d_{l_0-1}}{d_{l_0} - d_{l_0-1}} p_{j_{l_0}} \left[\sup(A_{j_{l_0}})_\alpha - \inf(A_{j_{l_0}})_\alpha\right] \\
&= d_{l_0-1} + \frac{J - d_{l_0-1}}{d_{l_0} - d_{l_0-1}} [d_{l_0} - d_{l_0-1}] = J.
\end{aligned}$$

Define for $j \in \{1, \ldots, n\} \backslash T \quad \tilde{c}_j \stackrel{\mathrm{d}}{=} c_j^*$. It follows:

$$2E_{\alpha, p_1, \ldots, p_n}(\tilde{c}_1, \ldots, \tilde{c}_n) = \inf(A_i)_\alpha + \sup(A_i)_\alpha$$

for $i \in T$.
As $c_i^* = \tilde{c}_i$ holds for $i \in \{1, \ldots, n\} \backslash T$, step (c) together with Lemma 8.11 shows

$$\begin{aligned}
&\sup \left\{ D^2_{\alpha, p_1, \ldots, p_n}(c_1, \ldots, c_n) \mid (c_1, \ldots, c_n) \in [0,1]^n \right\} \\
&\geq D^2_{\alpha, p_1, \ldots, p_n}(\tilde{c}_1, \ldots, \tilde{c}_n) \\
&= \sup \Big\{ D^2_{\alpha, p_1, \ldots, p_n}(c_1, \ldots, c_n) \Big| \\
&\qquad \forall\, i \in T: \ c_i \in [0,1] \ \wedge \ \forall\, i \in \{1, \ldots, n\} \backslash T: \ c_i = \tilde{c}_i \} \\
&\geq D^2_{\alpha, p_1, \ldots, p_n}(c_1^*, \ldots, c_n^*) \\
&= \sup \left\{ D^2_{\alpha, p_1, \ldots, p_n}(c_1, \ldots, c_n) \mid (c_1, \ldots, c_n) \in [0,1]^n \right\}
\end{aligned}$$

Therefore $(c_1^*, \ldots, c_n^*) \in C_\alpha[p_1, \ldots, p_n]$ holds.
For all $i \in \{1, \ldots, n\} \backslash \{i_0\}$ $\tilde{c}_i \in \{0,1\}$ is valid.

Lemma 8.13 *Let $\alpha \in (0,1)$ and $(p_1, \ldots, p_n) \in P_n$ be given. Define*

$$A_\alpha[p_1, \ldots, p_n] \stackrel{\mathrm{d}}{=}$$

$$\left\{ i \in \{1, \ldots, n\} \backslash I_\alpha[p_1, \ldots, p_n] \,\middle|\, \inf(A_i)_\alpha + \sup(A_i)_\alpha \leq 2 \sum_{j=1}^{n} p_j \inf(A_j)_\alpha \right\}$$

and

$$B_\alpha[p_1,\ldots,p_n] \stackrel{\mathrm{d}}{=}$$

$$\left\{ i \in \{1,\ldots,n\}\backslash I_\alpha[p_1,\ldots,p_n] \;\middle|\; \inf(A_i)_\alpha + \sup(A_i)_\alpha \geq 2\sum_{j=1}^{n} p_j \sup(A_j)_\alpha \right\}.$$

Let $i_0 \in \{1,\ldots,n\}$.

Let $(c_1,\ldots,c_n) \in [0,1]^n$ *and* $(\tilde{c}_1,\ldots,\tilde{c}_n) \in [0,1]^n$ *such that* $c_i = \tilde{c}_i$ *holds for all* $i \in \{1,\ldots,n\}\backslash\{i_0\}$.

(i) $i_0 \in A_\alpha[p_1,\ldots,p_n]$ *and* $\tilde{c}_{i_0} = 0$ *implies*

$$D^2_{\alpha,p_1,\ldots,p_n}(c_1,\ldots,c_n) \leq D^2_{\alpha,p_1,\ldots,p_n}(\tilde{c}_1,\ldots,\tilde{c}_n).$$

(ii) $i_0 \in B_\alpha[p_1,\ldots,p_n]$ *and* $\tilde{c}_{i_0} = 1$ *implies*

$$D^2_{\alpha,p_1,\ldots,p_n}(c_1,\ldots,c_n) \leq D^2_{\alpha,p_1,\ldots,p_n}(\tilde{c}_1,\ldots,\tilde{c}_n).$$

Proof It holds (compare the proof of Lemma 8.11):

$$\begin{aligned}
&D^2_{\alpha,p_1,\ldots,p_n}(\tilde{c}_1,\ldots,\tilde{c}_n) - D^2_{\alpha,p_1,\ldots,p_n}(c_1,\ldots,c_n)\\
&= (\tilde{c}_{i_0} - c_{i_0})\, p_{i_0} \left\{ [\sup(A_{i_0})_\alpha]^2 - [\inf(A_{i_0})_\alpha]^2 \right\}\\
&\quad - \left\{ 2E_{\alpha,p_1,\ldots,p_n}(\tilde{c}_1,\ldots,\tilde{c}_n) - (\tilde{c}_{i_0} - c_{i_0})\, p_{i_0}[\sup(A_{i_0})_\alpha - \inf(A_{i_0})_\alpha] \right\}\cdot\\
&\quad \cdot (\tilde{c}_{i_0} - c_{i_0})\, p_{i_0}[\sup(A_{i_0})_\alpha - \inf(A_{i_0})_\alpha]\\
&= (\tilde{c}_{i_0} - c_{i_0})\, p_{i_0}[\sup(A_{i_0})_\alpha - \inf(A_{i_0})_\alpha]\cdot\\
&\quad \cdot \left\{ \begin{array}{l} [\sup(A_{i_0})_\alpha + \inf(A_{i_0})_\alpha] - 2E_{\alpha,p_1,\ldots,p_n}(\tilde{c}_1,\ldots,\tilde{c}_n) \\ \qquad\qquad + (\tilde{c}_{i_0} - c_{i_0})\, p_{i_0}[\sup(A_{i_0})_\alpha + \inf(A_{i_0})_\alpha] \end{array} \right\}.
\end{aligned}$$

(i) If $\tilde{c}_{i_0} = 0$ and $i_0 \in A_\alpha[p_1,\ldots,p_n]$, then $\tilde{c}_{i_0} - c_{i_0} \leq 0$ holds as well as

$$\begin{aligned}
&[\sup(A_{i_0})_\alpha + \inf(A_{i_0})_\alpha] - 2E_{\alpha,p_1,\ldots,p_n}(\tilde{c}_1,\ldots,\tilde{c}_n)\\
&\leq [\sup(A_{i_0})_\alpha + \inf(A_{i_0})_\alpha] - 2\sum_{j=1}^{n} p_j \inf(A_j)_\alpha \leq 0.
\end{aligned}$$

The assertion (i) follows.

(ii) Assertion (ii) can be shown in a similar way.

Theorem 8.14 *Let* $\alpha \in (0,1)$ *and* $(p_1,\ldots,p_n) \in P_n$. *Define for* $i_0 \in \{1,\ldots,n\}\backslash I_\alpha[p_1,\ldots,p_n]$ *and* $A \subseteq \{1,\ldots,n\}\backslash\{i_0\}$:

$$g_{i_0,A}(p_1,\ldots,p_n) \stackrel{\mathrm{d}}{=} \frac{\inf(A_{i_0})_\alpha + \sup(A_{i_0})_\alpha - 2\sum_{j\in A} p_j \sup(A_j)_\alpha - 2\sum_{j\in\{1,\ldots,n\}\setminus A} p_j \inf(A_j)_\alpha}{2p_{i_0}[\sup(A_{i_0})_\alpha - \inf(A_{i_0})_\alpha]}$$

and for $i \in \{1,\ldots,n\}$, $A \subseteq \{1,\ldots,n\}\setminus\{i_0\}$:

$$c_{i,A}^{(i_0)}(p_1,\ldots,p_n) \stackrel{\mathrm{d}}{=} \begin{cases} g_{i_0,A}(p_1,\ldots,p_n) & \text{, if } i = i_0 \text{ and } g_{i_0,A}(p_1,\ldots,p_n) \in [0,1] \\ 0 & \text{, if } (i = i_0 \text{ and } g_{i_0,A}(p_1,\ldots,p_n) < 0) \text{ or } i \in \{1,\ldots,n\}\setminus(\{i_0\}\cup A) \\ 1 & \text{, if } (i = i_0 \text{ and } g_{i_0,A}(p_1,\ldots,p_n) > 1) \text{ or } i \in A \end{cases}$$

and

$$f_{i_0,A}(p_1,\ldots,p_n) \stackrel{\mathrm{d}}{=} D^2_{\alpha,p_1,\ldots,p_n}\left(c_{1,A}^{(i_0)}(p_1,\ldots,p_n),\ldots,c_{n,A}^{(i_0)}(p_1,\ldots,p_n)\right).$$

Then

(i) $$E_{\alpha,p_1,\ldots,p_n}\left(c_{1,A}^{(i_0)}(p_1,\ldots,p_n),\ldots,c_{n,A}^{(i_0)}(p_1,\ldots,p_n)\right) = \frac{1}{2}[\inf(A_{i_0})_\alpha + \sup(A_{i_0})_\alpha],$$
$$\text{if } g_{i_0,A}(p_1,\ldots,p_n) \in [0,1]$$

and

(ii) $$\Psi_\alpha[p_1,\ldots,p_n] = \beta \stackrel{\mathrm{d}}{=}$$

$$\begin{cases} \max\left\{ f_{i_0,A}(p_1,\ldots,p_n) \;\middle|\; \begin{array}{l} i_0 \in \{1,\ldots,n\}\setminus I_\alpha[p_1,\ldots,p_n] \\ \quad \cup A_\alpha[p_1,\ldots,p_n] \cup B_\alpha[p_1,\ldots,p_n] \\ \text{and } B_\alpha[p_1,\ldots,p_n] \subseteq A \subseteq \\ \{1,\ldots,n\}\setminus(\{i_0\}\cup I_\alpha[p_1,\ldots,p_n] \cup A_\alpha[p_1,\ldots,p_n]) \end{array} \right\} \\ \qquad \text{, if } I_\alpha[p_1,\ldots,p_n] \cup A_\alpha[p_1,\ldots,p_n] \cup B_\alpha[p_1,\ldots,p_n] \neq \{1,\ldots,\ \} \\ D^2_{\alpha,p_1,\ldots,p_n}(c_1,\ldots,c_n) \quad \text{, where } c_i = \begin{cases} 0 & \text{, if } i \in I_\alpha[p_1,\ldots,p_n] \cup A_\alpha[p_1,\ldots,p_n] \\ 1 & \text{, if } i \in B_\alpha[p_1,\ldots,p_n] \end{cases} \\ \qquad \text{, if } I_\alpha[p_1,\ldots,p_n] \cup A_\alpha[p_1,\ldots,p_n] \cup B_\alpha[p_1,\ldots,p_n] = \{1,\ldots,\ \} \end{cases}$$

Proof

(i) can be proved by calculating in a straight forward way

(ii) If $I_\alpha[p_1,\dots,p_n] \cup A_\alpha[p_1,\dots,p_n] \cup B_\alpha[p_1,\dots,p_n] = \{1,\dots,n\}$ is valid, then step (a) of the proof of Lemma 8.12 and Lemma 8.13 show that there exists a $(c_1^*,\dots,c_n^*) \in C_\alpha[p_1,\dots,p_n]$ such that $c_i^* = 0$ holds for $i \in I_\alpha[p_1,\dots,p_n] \cup A_\alpha[p_1,\dots,p_n]$ and $c_i^* = 1$ for $i \in B_\alpha[p_1,\dots,p_n]$. Otherwise, Lemma 8.12 shows that we can find an

$$(c_1^*,\dots,c_n^*) \in C_\alpha[p_1,\dots,p_n] \text{ and an } i_0 \in \{1,\dots,n\}\backslash I_\alpha[p_1,\dots,p_n]$$

such that $c_i \in \{0,1\}$ holds for $i \in \{1,\dots,n\}\backslash\{i_0\}$ and $c_i^* = 0$ for $i \in I_\alpha[p_1,\dots,p_n]$.

If $i_0 \in A_\alpha[p_1,\dots,p_n] \cup B_\alpha[p_1,\dots,p_n]$, we may choose an other $i_0 \in \{1,\dots,n\}$ with the property $c_i^* \in \{0,1\}$ for $i \in \{1,\dots,n\}\backslash\{i_0\}$ because of Lemma 8.13.

So $i_0 \in \{1,\dots,n\}\backslash(I_\alpha[p_1,\dots,p_n] \cup A_\alpha[p_1,\dots,p_n] \cup B_\alpha[p_1,\dots,p_n])$ holds for a suitable $(c_1^*,\dots,c_n^*) \in C_\alpha[p_1,\dots,p_n]$. Define

$$A \stackrel{\mathrm{d}}{=} \{i \in \{1,\dots,n\}\backslash\{i_0\} \mid c_i^* = 1\} \quad \text{and}$$

$$B \stackrel{\mathrm{d}}{=} \{i \in \{1,\dots,n\}\backslash\{i_0\} \mid c_i^* = 0\} \; .$$

Because of Lemma 8.13, we know that we may assume the inclusions $B_\alpha[p_1,\dots,p_n] \subseteq A$ and $A_\alpha[p_1,\dots,p_n] \subseteq B \subseteq \{1,\dots,n\}\backslash A$. It follows

$$A \subseteq \{1,\dots,n\}\backslash A_\alpha[p_1,\dots,p_n].$$

Moreover, $A \subseteq \{1,\dots,n\}\backslash(\{i_0\} \cup I_\alpha[p_1,\dots,p_n])$ holds.

Let $g_{i_0,A}(p_1,\dots,p_n) \in [0,1]$ is valid. By part (i) we can follow:

$$2E\alpha, p_1,\dots,p_n \left(c_{1,A}^{(i_0)}(p_1,\dots,p_n),\dots,c_{n,A}^{(i_0)}(p_1,\dots,p_n)\right) =$$

$$\frac{1}{2}[\inf(A_{i_0})_\alpha + \sup(A_{i_0})_\alpha].$$

Because of Lemma 8.11 (with $T \stackrel{\mathrm{d}}{=} \{i_0\}$), we know that

$$f_{i_0,A}(p_1,\dots,p_n) \geq D^2_{\alpha,p_1,\dots,p_n}(c_1^*,\dots,c_n^*) = \Psi_\alpha[p_1,\dots,p_n] \geq f_{i_0,A}(p_1,\dots,p_n).$$

If $g_{i_0,A}(p_1,\dots,p_n) < 0$ then

$$2E\alpha, p_1,\dots,p_n \left(c_{1,A}^{(i_0)}(p_1,\dots,p_n),\dots,c_{n,A}^{(i_0)}(p_1,\dots,p_n)\right) >$$

$$\frac{1}{2}[\inf(A_{i_0})_\alpha + \sup(A_{i_0})_\alpha]$$

follows. We can conclude:

$$\begin{aligned}
& f_{i_0,A}(p_1,\ldots,p_n) - D^2_{\alpha,p_1,\ldots,p_n}(c_1^*,\ldots,c_n^*) \\
= & \left(-c_{i_0}^*\right) p_{i_0} \left\{[\sup(A_{i_0})_\alpha]^2 - [\inf(A_{i_0})_\alpha]^2\right\} - \\
& - \left\{2E_{\alpha,p_1,\ldots,p_n}\left(c_{1,A}^{(i_0)}(p_1,\ldots,p_n),\ldots,c_{n,A}^{(i_0)}(p_1,\ldots,p_n)\right) - \right. \\
& \left. \left(-c_{i_0}^*\right) p_{i_0}[\sup(A_{i_0})_\alpha - \inf(A_{i_0})_\alpha]\right\} \cdot \\
& \cdot \left(-c_{i_0}^*\right) p_{i_0}[\sup(A_{i_0})_\alpha - \inf(A_{i_0})_\alpha] \\
= & \left(-c_{i_0}^*\right) p_{i_0}[\sup(A_{i_0})_\alpha - \inf(A_{i_0})_\alpha] \cdot \\
& \cdot \{[\sup(A_{i_0})_\alpha + \inf(A_{i_0})_\alpha] \\
& - 2E_{\alpha,p_1,\ldots,p_n}\left(c_{1,A}^{(i_0)}(p_1,\ldots,p_n),\ldots,c_{n,A}^{(i_0)}(p_1,\ldots,p_n)\right) \\
& - c_{i_0}^* p_{i_0}[\sup(A_{i_0})_\alpha - \inf(A_{i_0})_\alpha]\} \geq 0,
\end{aligned}$$

so $f_{i_0,A}[p_1,\ldots,p_n] = \Psi_\alpha[p_1,\ldots,p_n]$ follows.
If $g_{i_0,A}(p_1,\ldots,p_n) > 1$, we can demonstrate that

$$\sup(A_{i_0})_\alpha \; + \; \inf(A_{i_0})_\alpha - p_{i_0}[\sup(A_{i_0})_\alpha - \inf(A_{i_0})_\alpha]$$

$$-2E_{\alpha,p_1,\ldots,p_n}\left(c_{1,A}^{(i_0)}(p_1,\ldots,p_n),\ldots,c_{n,A}^{(i_0)}(p_1,\ldots,p_n)\right) > 0$$

Similar to the last calculation, we show

$$f_{i_0,A}(p_1,\ldots,p_n) = \Psi_\alpha[p_1,\ldots,p_n].$$

Obviously $f_{i_0,A}(p_1,\ldots,p_n) \leq \beta$ and $f_{i_0,A}(p_1,\ldots,p_n) \geq \beta$ are valid, therefore the proof is complete. ∎

We summarize the results concerning $\sup \tilde{B}_\alpha$ in the following theorem:

Theorem 8.15 *For all* $\alpha \in (0,1)$

$$\sup \tilde{B}_\alpha =$$

$$\begin{cases} \max\left\{ f_{i_0,A}[P(\Gamma_1),\ldots,P(\Gamma_n)] \;\middle|\; \begin{array}{l} i_0 \in \{1,\ldots,n\} \setminus I_\alpha[P(\Gamma_1),\ldots,P(\Gamma_n)] \\ \quad \cup A_\alpha[P(\Gamma_1),\ldots,P(\Gamma_n)] \\ \quad \cup B_\alpha[P(\Gamma_1),\ldots,P(\Gamma_n)] \\ \text{and } B_\alpha[P(\Gamma_1),\ldots,P(\Gamma_n)] \subseteq A \\ \quad \subseteq \{1,\ldots,n\} \setminus (\{i_0\} \\ \qquad \cup I_\alpha[P(\Gamma_1),\ldots,P(\Gamma_n)]) \\ \qquad \cup A_\alpha[P(\Gamma_1),\ldots,P(\Gamma_n)]) \end{array} \right\} \\ \quad , \text{if } I_\alpha[P(\Gamma_1),\ldots,P(\Gamma_n)] \cup A_\alpha[P(\Gamma_1),\ldots,P(\Gamma_n)] \\ \qquad \cup B_\alpha[P(\Gamma_1),\ldots,P(\Gamma_n)] \neq \{1,\ldots,n\} \\[2ex] D^2_{\alpha, P(\Gamma_1),\ldots,P(\Gamma_n)}(c_1,\ldots,c_n) \\ \quad , \text{whereas } c_i = \begin{cases} 0 & , \text{if } i \in I_\alpha[P(\Gamma_1),\ldots,P(\Gamma_n)] \\ & \quad \cup A_\alpha[P(\Gamma_1),\ldots,P(\Gamma_n)] \\ 1 & , \text{if } i \in B_\alpha[P(\Gamma_1),\ldots,P(\Gamma_n)] \end{cases} , \\ \quad , \text{if } I_\alpha[P(\Gamma_1),\ldots,P(\Gamma_n)] \cup A_\alpha[P(\Gamma_1),\ldots,P(\Gamma_n)] \\ \qquad \cup B_\alpha[P(\Gamma_1),\ldots,P(\Gamma_n)] = \{1,\ldots,n\} \end{cases}$$

is valid.

In the following we calculate $\sup B_\alpha$.

Lemma 8.16 *For* $A \in \mathfrak{A}$ *define* $U_A \in Y_\alpha$ *by*

$$U_A(\omega) \stackrel{\text{d}}{=} \begin{cases} \sup(A_i)_\alpha & , \text{if there is an } i \in \{1,\ldots,n\} \text{ with} \\ & , \omega \in \Gamma_i \cap A \\ \inf(A_i)_\alpha & , \text{if there is an } i \in \{1,\ldots,n\} \text{ with} \\ & , \omega \in \Gamma_i \cap (\Omega \setminus A) \\ Z_\alpha(\omega) & , \text{if } \omega \in \Omega \setminus \bigcup_{i=1}^{n} \Gamma_i \,. \end{cases}$$

Define

$$S_\alpha \stackrel{\text{d}}{=} \sup\{t \in I\!R \mid \exists\, A \in A \text{ with } t = \operatorname{Var} U_A\}.$$

Then
(i) $\forall\, U \in \tilde{Y}_\alpha : \quad \operatorname{Var} U \leq \operatorname{Var} U_{[U \geq E\, U]}$ *and*
(ii) $\sup B_\alpha = S_\alpha$ *is valid.*

This lemma can be proved in a similar way to Lemma 8.9.

Lemma 8.17 *Define for $\alpha \in (0,1)$ and $(c_1,\dots,c_n) \in [0,1]^n$*

$$(i) \quad E_\alpha(c_1,\dots,c_n) \stackrel{\mathrm{d}}{=} \sum_{i=1}^{n} c_i \sup(A_i)_\alpha + \sum_{i=1}^{n} [P(\Gamma_i) - c_i] \inf(A_i)_\alpha$$

$$(ii) \quad E_\alpha^2(c_1,\dots,c_n) \stackrel{\mathrm{d}}{=} \sum_{i=1}^{n} [\sup(A_i)_\alpha]^2 \cdot c_i + \sum_{i=1}^{n} [P(\Gamma_i) - c_i][\inf(A_i)_\alpha]^2$$

$$(iii) \quad D_\alpha^2(c_1,\dots,c_n) \stackrel{\mathrm{d}}{=} E_\alpha^2(c_1,\dots,c_n) - \{E_\alpha(c_1,\dots,c_n)\}^2$$

$$(iv) \quad \Lambda_\alpha(c_1,\dots,c_n) \stackrel{\mathrm{d}}{=}$$

$$\left\{ (c_1,\dots,c_n) \in [0,1]^n \,\middle|\, \forall i \in \{1,\dots,n\} : \begin{array}{l} \exists A_i \in \mathfrak{A} \ \text{such that} \\ A_i \subseteq \Gamma_i \ \text{and} \ P(A_i) = c_i \end{array} \right\}$$

Then we have:

$$\sup B_\alpha = \sup \left\{ D_\alpha^2(c_1,\dots,c_n) \,|\, (c_1,\dots,c_n) \in \Lambda(\Gamma_1,\dots,\Gamma_n) \right\}.$$

Proof Let $A \in \mathfrak{A}$. Then

$$\begin{aligned}
\operatorname{Var} U_A &= \sum_{i=1}^{n} E\left\{ (U_A I_{\Gamma_i})^2 \right\} - \left\{ \sum_{i=1}^{n} E\, U_A I_{\Gamma_i} \right\}^2 \\
&= \sum_{i=1}^{n} \left\{ \int_{\Gamma_i} U_A^2 dP \right\} - \left\{ \sum_{i=1}^{n} \int_{\Gamma_i} U_A dP \right\}^2 \\
&= \sum_{i=1}^{n} [\sup(A_i)_\alpha]^2 P(A \cap \Gamma_i) + \sum_{i=1}^{n} [\inf(A_i)_\alpha]^2 [P(\Gamma_i) - P(A \cap \Gamma_i)] \\
&\quad - \left\{ \sum_{i=1}^{n} \sup(A_i)_\alpha P(A \cap \Gamma_i) + \inf(A_i)_\alpha [P(\Gamma_i) - P(A \cap \Gamma_i)] \right\}^2 \\
&= D_\alpha^2 (P(A \cap \Gamma_1), \cdots, P(A \cap \Gamma_n)).
\end{aligned}$$

On the one hand, let $A \in \mathfrak{A}$ be given. Then $A \cap \Gamma_i \subseteq \Gamma_i$ holds for all $i \in \{1,\dots,n\}$ as well as $A \cap \Gamma_i \in \mathfrak{A}$. It follows:

$$(P(A \cap \Gamma_1),\dots,P(A \cap \Gamma_n)) \in \Lambda(\Gamma_1,\dots,\Gamma_n)$$

Therefore

$$\text{Var } U_A = D^2_\alpha \left(P(A \cap \Gamma_1, \ldots, P(A \cap \Gamma_n)\right) \leq$$

$$\sup \left\{ D^2_\alpha(c_1, \ldots, c_n) \mid (c_1, \ldots, c_n) \in \Lambda(\Gamma_1, \ldots, \Gamma_n) \right\}$$

holds. This proves the inequality "$\leq$ ".

On the other hand, let $(c_1, \ldots, c_n) \in \Lambda(\Gamma_1, \ldots, \Gamma_n)$ be given. Choose $A_i \in \mathfrak{A}$ for $i \in \{1, \ldots, n\}$ such that $A_i \subseteq \Gamma_i$ and $P(A_i) = c_i$ holds. Define

$$B \stackrel{d}{=} \bigcup_{i=1}^{n} A_i.$$

Then $B \cap \Gamma_i = A_i$ holds for all $i \in \{1, \ldots, n\}$. It follows:

$$D^2(c_1, \ldots, c_n) = D^2\left(P(B \cap \Gamma_1), \ldots, P(B \cap \Gamma_n)\right) = \text{ Var } U_B \leq \sup B_\alpha.$$

This shows "$\geq$". •

This expression is difficult to calculate. So we restrict ourselves to the case of a finite space $\Omega = \{\omega_1, \ldots, \omega_n\}$, where $p_i \stackrel{d}{=} P(\{\omega_i\})$ for $i \in \{1, \ldots, n\}$, with the σ-algebra $\mathfrak{A} \stackrel{d}{=} \mathfrak{P}(\Omega)$. Under these conditions we have:

Theorem 8.18 *Define for $\alpha \in (0,1)$, $(p_1, \ldots, p_n) \in P_n$ and $A \subseteq \{1, \ldots, n\}$:*

$$D^2_{\alpha, p_1, \ldots, p_n}(A) \stackrel{d}{=} \sum_{i \in A} p_i[\sup(A_i)_\alpha]^2 + \sum_{i \in \{1, \ldots, n\} \setminus A} p_i[\inf(A_i)_\alpha]^2$$

$$- \left\{ \sum_{i \in A} p_i \sup(A_i)_\alpha + \sum_{i \in \{1, \ldots, n\} \setminus A} p_i \inf(A_i)_\alpha \right\}^2$$

$$\Sigma_\alpha[p_1, \ldots, p_n] \stackrel{d}{=} \max \left\{ D^2_{\alpha, p_1, \ldots, p_n}(A) \mid A \subseteq \{1, \ldots, n\} \right\}$$

Then

$$\sup B_\alpha = \Sigma_\alpha[P(\Gamma_1), \ldots, P(\Gamma_n)]$$

is valid.

Proof It holds:

$$\Lambda(\Gamma_1,\ldots,\Gamma_n) =$$

$$\{(c_1,\ldots,c_n) \in [0,1]^n \mid \forall i \in \{1,\ldots,n\}: \; c_i = 0 \;\text{ or }\; c_i = P(\Gamma_i)\}.$$

Let $(c_1,\ldots,c_n) \in \Lambda(\Gamma_1,\ldots,\Gamma_n)$ be arbitrary. Define

$$A \stackrel{\mathrm{d}}{=} \{i \in \{1,\ldots,n\} \mid c_i = P(\Gamma_i)\}.$$

For all $i \in \{1,\ldots,n\} \backslash A$ $c_i = 0$ holds. We conclude:

$$D^2_{\alpha, P(\Gamma_1),\ldots,P(\Gamma_n)}(A) = D^2_\alpha(c_1,\ldots,c_n).$$

This proves the assertion together with the preceding theorem. •

It is possible that $\sup B_\alpha < \sup \tilde{B}_\alpha$ holds for an $\alpha \in (0,1)$. As an example, consider the case $n = 1$. Let $\Omega = \{\omega\}$ and $\mathfrak{A} = \{\emptyset, \{\omega\}\}$. Let $\Omega' = [0,1]$ and $\mathfrak{A}' = \{A \in \mathfrak{B}_1 \mid A \subseteq [0,1]\}$, where $\mathfrak{B}_1$ denotes the Lebesgue's σ-algebra. Define $X_\omega \stackrel{\mathrm{d}}{=} \mathrm{I}_{[-1,+1]}$. It holds $P_1 = \{(1)\}$ (P_n is defined on page 83).

For $c \in [0,1]$ and $\alpha \in (0,1)$ we have (compare Theorem 8.10):

$$D^2_{\alpha,1}(c) = c \cdot 1^2 + (1-c)\cdot(-1)^2 - \{c \cdot 1 + (1-c)\cdot(-1)\}^2 = 1 - (2c-1)^2.$$

It follows then that

$$\sup \tilde{B}_\alpha = \sup\left\{1 - (2c-1)^2 \mid c \in [0,1]\right\} = 1.$$

But $D^2_\alpha(\emptyset) = D^2_\alpha(\{1\}) = 0$ holds, and therefore $\sup B_\alpha = 0$ is valid. We conclude:

$$\mathrm{Var}\, X = \mathrm{I}_{\{0\}} \quad\text{and}\quad \widetilde{\mathrm{Var}}\, X = \mathrm{I}_{[0,1]}.$$

In the sequel we try to determine $\inf B_\alpha$ and $\inf \tilde{B}_\alpha$. We discover that we have $\inf \tilde{B}_\alpha = \inf B_\alpha$. In order to construct algorithms for their calculation, we must define the following real numbers.

Definition 8.19 *For $t \in \mathbb{R}$ we define:*

$$A[t] \stackrel{\mathrm{d}}{=} \{i \in \{1,\ldots,n\} \mid \inf(A_i)_\alpha > t\},$$

$$B[t] \stackrel{\mathrm{d}}{=} \{i \in \{1,\ldots,n\} \mid \sup(A_i)_\alpha < t\},$$

$$C[t] \stackrel{\mathrm{d}}{=} \{i \in \{1,\ldots,n\} \mid t \in (A_i)_\alpha\},$$

$$D_j[t] \stackrel{\mathrm{d}}{=} \left\{ i \in \{1,\ldots,n\} \;\middle|\; \begin{array}{l} j \in \{1,\ldots,(N_i)_\alpha - 1\} \text{ and} \\ t \in \left((b_j)_\alpha^{(i)}, \frac{1}{2}\left[(b_j)_\alpha^{(i)} + (a_{j+1})_\alpha^{(i)}\right] \right) \end{array} \right\} (j \in \mathbb{N})\},$$

and

$$E_j[t] \stackrel{\mathrm{d}}{=} \left\{ i \in \{1,\ldots,n\} \;\middle|\; \begin{array}{l} i \in \{2,\ldots,(N_i)_\alpha\} \text{ and} \\ t \in \left(\frac{1}{2}\left[(b_{j-1})_\alpha^{(i)} + (a_j)_\alpha^{(i)}\right], (a_j)_\alpha^{(i)} \right) \end{array} \right\} (j \in \mathbb{N}).$$

A vector $(V_1[t],\ldots,V_n[t]) \in \mathbb{R}^n$ is defined by

$$V_i[t] \stackrel{\mathrm{d}}{=} \begin{cases} \inf(A_i)_\alpha & , \text{if } i \in A[t] \\ \sup(A_i)_\alpha & , \text{if } i \in B[t] \\ t & , \text{if } i \in C[t] \\ (b_j)_\alpha^{(i)} & , \text{if } i \in D_j[t] \; (j \in \mathbb{N}) \\ (a_j)_\alpha^{(i)} & , \text{if } i \in E_j[t] \; (j \in \mathbb{N}) \,. \end{cases}$$

These sets and vectors do not depend on $(P(\Gamma_1),\ldots,P(\Gamma_n))$, but of course on $\alpha \in (0,1)$. For the sake of simplicity we omit the index α.

The following lemma shows why we have chosen to define it this way.

Lemma 8.20 *For all $t \in \mathbb{R}$ and $\alpha \in (0,1)$*

$$V_i[t] \in (A_i)_\alpha \quad \text{and} \quad |V_i[t] - t| = \min\{|x-t| \mid x \in (A_i)_\alpha\}$$

is valid.

Proof We have to distinguish the cases "$i \in A[t]$", "$i \in B[t]$", "$i \in C[t]$", "$i \in D_j[t]$", and "$i \in E_j[t]$". •

For all $i \in \{1,\ldots,n\}$ $(A_i)_\alpha$ is a finite union of compact intervals. The set of all "edges" of these intervals as well as of the midpoints of the gaps plays an important role. Let $\alpha \in (0,1)$ be fixed. Define

$$\{c_0,\ldots,c_M\}$$

$$= \bigcup_{i=1}^{n} \left(\{\inf(A_i)_\alpha, \sup(A_i)_\alpha\} \cup \bigcup_{j=1}^{(N_i)_\alpha} \left\{ (a_j)_\alpha^{(i)}, (b_{j-1})_\alpha^{(i)}, \frac{1}{2}\left[(a_j)_\alpha^{(i)} + (b_{j-1})_\alpha^{(i)}\right] \right\} \right)$$

such that $c_{k-1} < c_k$ holds for $k \in \{1, \ldots, M\}$.
Especially we have

$$c_0 = \min_{i \in \{1,\ldots,n\}} \inf(A_i)_\alpha \text{ and } c_M = \max_{i \in \{1,\ldots,n\}} \sup(A_i)_\alpha .$$

We also define $d_k \stackrel{\mathrm{d}}{=} \frac{1}{2}[c_{k-1} + c_k]$ for $k \in \{1, \ldots, M\}$.
It is easy to show:

Lemma 8.21

(i) *For all* $k \in \{1, \ldots, M\}$ *and* $t \in (c_{k-1}, c_k)$
$A[t] = A[d_k]$ *and* $B[t] = B[d_k]$ *and* $C[t] = C[d_k]$ *and* $D_j[t] = D_j[d_k]$ *and* $E_j[t] = E_j[d_k]$ $(j \in \mathbb{N}\ )$.

(ii) *For all* $k \in \{1, \ldots, M\}$, $t \in (c_{k-1}, c_k)$, *and for all* $i \in \{1, \ldots, n\} \backslash C[t]$: $V_i[t] = V_i[d_k]$.

(iii) *For all* $t \leq c_0$, *and* $i \in \{1, \ldots, n\}$: $V_i[t] = V_i[c_0]$ *and for all* $t \geq c_M$, *and* $i \in \{1, \ldots, n\}$: $V_i[t] = V_i[c_M]$ *is valid.*

Definition 8.22 *For* $t \in \mathbb{R}$ *define* $V[t] \in Y_\alpha$ *by*

$$V[t](\omega) \stackrel{\mathrm{d}}{=} \begin{cases} V_i[t] & \text{, if } \omega \in \Gamma_i \text{ for an } i \in \{1,\ldots,n\} \text{ and } \omega' \in \Omega', \\ Z_\alpha(\omega) & \text{, if } \omega \in \Omega \backslash \bigcup_{i=1}^{n} \Gamma_i \end{cases}$$

Lemma 8.23 $\inf \tilde{B}_\alpha = \inf B_\alpha = \inf\{\ \mathit{Var}\ V[t] \mid t \in [c_0, c_M]\}$

Proof For all $t \in [c_0, c_M]$ $V[t] \in Y_\alpha$ is valid. Each mapping $U \in Y_\alpha$ can be considered as a mapping $U \in \tilde{Y}_\alpha$. So we have the inclusions

$$\{\ \mathrm{Var}\ V[t] \mid t \in [c_0, c_M]\} \subseteq B_\alpha \subseteq \tilde{B}_\alpha.$$

This yields the inequalities

$$\inf\{\ \mathrm{Var}\ V[t] \mid t \in [c_0, c_M]\} \geq \inf B_\alpha \geq \inf \tilde{B}_\alpha.$$

On the other hand , let $U \in \tilde{Y}_\alpha$ be arbitrary. Lemma 8.20 shows that

$$|U(\omega, \omega') - t| \geq |V_i[t] - t| = |V[t](\omega, \omega') - t|$$

is valid for $t \in \mathbb{R}$,$i \in \{1, \ldots, n\}$, $\omega \in \Gamma_i$, $\omega' \in \Omega'$. Define $t \stackrel{\mathrm{d}}{=} E\ U \in [c_0, c_M]$. We obtain

$$[U(\omega,\omega') - E\,U]^2 \geq [V[E\,U](\omega,\omega') - E\,U]^2$$

for $\omega \in \bigcup_{i=1}^{n} \Gamma_i$, $\omega' \in \Omega'$.

As $P\left(\bigcup_{i=1}^{n} \Gamma_i\right) = 1$ is valid, we conclude by integration about $\Omega \times \Omega'$:

$$\operatorname{Var} U \geq \int_{\Omega\times\Omega'} [V[E\,U] - E\,U]^2 dP \otimes P' \geq \operatorname{Var} V[E\,U]$$

$$\geq \inf\{\operatorname{Var} V[t] \mid t \in [c_0, c_M]\},$$

as for all random variables $V \in \tilde{\chi}$ with existing variance

$$\operatorname{Var} V = \min_{x\in\mathbb{R}} \int_{\Omega\times\Omega'} [V - x]^2 dP \otimes P'$$

is valid. We conclude:

$$\inf \tilde{B}_\alpha \geq \inf\{\operatorname{Var} V[t] \mid t \in [c_0, c_M]\}. \bullet$$

Definition 8.24 *Define for* $\alpha \in (0,1)$, $(p_1,\ldots,p_n) \in P_n$, $t \in \mathbb{R}$:

(i) $E_{\alpha,p_1,\ldots,p_n}[t] \stackrel{\text{d}}{=} \sum_{i=1}^{n} p_i V_i[t]$.

(ii) $E^2_{\alpha,p_1,\ldots,p_n} \stackrel{\text{d}}{=} \sum_{i=1}^{n} p_i \{V_i[t]\}^2$.

(iii) $Var_{\alpha,p_1,\ldots,p_n}[t] \stackrel{\text{d}}{=} \sum_{i=1}^{n} p_i \{V_i[t] - E_{\alpha,p_1,\ldots,p_n}[t]\}^2$.

(iv) $\Phi_\alpha[p_1,\ldots,p_n] \stackrel{\text{d}}{=} \inf\{Var_{\alpha,p_1,\ldots,p_n}[t] \mid t \in [c_0, c_M]\}$.

It follows immediately:

Lemma 8.25 *For all* $t \in \mathbb{R}$

(i) $$E\,V[t] = E_{\alpha, P(\Gamma_1),\ldots,P(\Gamma_n)}[t]$$

(ii) $$E\,\{V[t]\}^2 = E^2_{\alpha, P(\Gamma_1),\ldots,P(\Gamma_n)}[t]$$

(iii) $$\operatorname{Var} V[t] = Var_{\alpha, P(\Gamma_1),\ldots,P(\Gamma_n)}[t]$$

(iv) $$\inf B_\alpha = \Phi_\alpha[P(\Gamma_1),\ldots,P(\Gamma_n)]$$

(v) $$Var_{\alpha,p_1,\ldots,p_n}[t] = E_{\alpha,p_1,\ldots,p_n}[t] - \{E_{\alpha,p_1,\ldots,p_n}[t]\}^2$$

is valid.

Theorem 8.26 *For all* $\alpha \in (0,1)$ *and* $(p_1,\ldots,p_n) \in P_n$

$$\Phi_\alpha[p_1,\ldots,p_n] = \inf\left\{ \mathit{Var}_{\alpha,p_1,\ldots,p_n}[t] \;\middle|\; \begin{array}{l} t \in [c_0, c_M] \text{ and} \\ t = E_{\alpha,p_1,\ldots,p_n}[t] \end{array} \right\}$$

is valid.

Proof

(a) Let $t \in [c_0, c_M]$ be arbitrary. Define two sequences $\{E_j[t]\}_{j\in\mathbb{N}}$ and $\left\{\left(V_1^{(j)}[t],\ldots,V_n^{(j)}[t]\right)\right\}_{j\in\mathbb{N}}$ recurrently by

(i) $V_i^{(1)}[t] \stackrel{\mathrm{d}}{=} V_i[t]$ for $i \in \{1,\ldots,n\}$ and
$E_1[t] \stackrel{\mathrm{d}}{=} E_{\alpha,p_1,\ldots,p_n}[t]$.

(ii) $V_i^{(j+1)}[t] \stackrel{\mathrm{d}}{=} V_i[E_j[t]]$ for $i \in \{1,\ldots,n\}$ and $n \in \mathbb{N}$ and
$E_{j+1}[t] \stackrel{\mathrm{d}}{=} \sum\limits_{i=1}^{n} p_i V_i^{(j+1)}[t]$.

By distinguishing several cases, we can show:

$$c_0 \le t_1 \le t_2 \le c_M \text{ implies } V_i[t_1] \le V_i[t_2] \text{ for } i \in \{1,\ldots,n\}.$$

Applying the mathematical induction with respect to j, we obtain:

$$E_1[t] \ge t \text{ implies } V_j^{(j+1)}[t] \ge V_i^{(j)}[t] \text{ and } E_{j+1}[t] \ge E_j[t]$$

for $i \in \{1,\ldots,n\}$, $j \in \mathbb{N}$.
If $j = 1$ and $E_1[t] \ge t$, then we have for all $i \in \{1,\ldots,n\}$:

$$V_i^{(2)}[t] = V_i[E_1[t]] \ge V_i[t] = V_i^{(1)}[t].$$

It follows:

$$E_2[t] \ge E_1[t].$$

If $j \in \mathbb{N}$, we have by induction hypothesis

$$E_{j+1}[t] \ge E_j[t].$$

It follows for $i \in \{1,\ldots,n\}$:

$$V_i^{(j+2)}[t] = V_i[E_{j+1}[t]] \ge V_i[E_j[t]] = V_i^{(j+1)}[t].$$

This yields:

$$E_{j+2}[t] \ge E_{j+1}[t].$$

In a similar way we can show that $E_1[t] \leq t$ implies

$$V_i^{(j+1)}[t] \leq V_i^{(j)}[t] \text{ and } E_{j+1}[t] \leq E_j[t].$$

(b) Let $t \in [c_0, c_M]$ be arbitrary. From part (a) follows
$E_1[t] \geq t$ implies $\left\{V_i^{(j)}\right\}_{j \in \mathbb{N}}$ (for $i \in \{1, \ldots, n\}$) and $\{E_j[t]\}_{j \in \mathbb{N}}$ are monotonously non-decreasing.
$E_1[t] \geq t$ implies $\left\{V_i^{(j)}\right\}_{j \in \mathbb{N}}$ (for $i \in \{1, \ldots, n\}$) and $\{E_j[t]\}_{j \in \mathbb{N}}$ are monotonously non-increasing.

As $V_i^{(j)}[t] \in (A_i)_\alpha \subseteq [c_0, c_M] \subseteq [-D, D]$ holds for all $t \in [c_0, c_M]$, $i \in \{1, \ldots, n\}$, $j \in \mathbb{N}$, all these sequences are convergent.
Define

$$\Phi_i[t] \stackrel{\mathrm{d}}{=} \lim_{r \to \infty} V_i^{(j)}[t] \quad (i \in \{1, \ldots, n\}),$$

$$E\ \Phi[t] \stackrel{\mathrm{d}}{=} \lim_{j \to \infty} E_j[t],$$

$$\mathrm{Var}\ \Phi[t] \stackrel{\mathrm{d}}{=} \sum_{i=1}^{n} p_i \{\Phi_i[t] - E\ \Phi[t]\}^2.$$

It follows:

$$E\ \Phi[t] = \sum_{i=1}^{n} p_i \Phi_i[t].$$

Lemma 8.20 (with $x = V_i^{(j)}[t]$) shows that

$$\left|V_i^{(j)}[t] - E_j[t]\right| \geq |V_i[E_j[t]] - E_j[t]| = \left|V_i^{(j+1)}[t] - E_j[t]\right|$$

holds for all $i \in \{1, \ldots, n\}$, $j \in \mathbb{N}$, and this implies

$$\sum_{i=1}^{n} p_i \left\{V_i^{(j)}[t] - E_j[t]\right\}^2 \geq \sum_{i=1}^{n} p_i \left\{V_i^{(j+1)}[t] - E_j[t]\right\}^2.$$

It is easy to show that

$$\sum_{i=1}^{n} p_i \left\{V_i^{(j)}[t] - E_j[t]\right\}^2 = \min_{x \in \mathbb{R}} \sum_{i=1}^{n} p_i \left\{V_i^{(j)}[t] - x\right\}^2$$

holds, so

$$\left\{\sum_{i=1}^{n} p_i \left\{V_i^{(j)}[t] - E_j[t]\right\}^2\right\}_{j \in \mathbb{N}}$$

is monotonously non-increasing against Var $\Phi[t]$.

$$\forall\, t \in [c_0, c_M]: \quad \text{Var}\ \Phi[t] \leq \text{Var}_{\alpha, p_1, \ldots, p_n}[t].$$

(c) Let $t \in [c_0, c_M]$ be arbitrary.
By distinguishing between several cases, we can show that

$$V_i[E\ \Phi[t]] = \Phi_i[t]$$

is valid for $i \in \{1, \ldots, n\}$.
From this and Definition 8.24(i) we can conclude

$$\begin{aligned} E\ \Phi[t] &= \sum_{i=1}^{n} p_i \cdot \Phi_i[t] = \sum_{i=1}^{n} p_i \cdot V_i[E\ \Phi[t]] = \\ &= E_{\alpha, p_1, \ldots, p_n}[E\ \Phi[t]]. \end{aligned}$$

(d) Because of the final equation in step (c) we know that the set

$$B \stackrel{\text{d}}{=} \{\ \text{Var}_{\alpha, p_1, \ldots, p_n}[t] \,|\, t \in [c_0, c_M] \ \text{and}\ \ t = E_{\alpha, p_1, \ldots, p_n}[t]\}$$

is not empty. From Definition 8.24(iv) we know

$$\Phi_\alpha[p_1, \ldots, p_n] \leq \inf B.$$

On the other hand, let $t \in [c_0, c_M]$ arbitrary.
From the equation

$$\Phi[t] = V_i[E\ \Phi[t]]$$

which is valid for all $i \in \{1, \ldots, n\}$ and by step (c) follows:

$$\text{Var}_{\alpha, p_1, \ldots, p_n}[E\ \Phi[t]] = \text{Var}\ \Phi[t].$$

Step (b) shows that

$$\text{Var}_{\alpha, p_1, \ldots, p_n}[t] \geq \text{Var}\ \Phi[t]$$

is valid.
All together we obtain

$$\text{Var}_{\alpha, p_1, \ldots, p_n}[t] \geq \text{Var}_{\alpha, p_1, \ldots, p_n}[E\ \Phi[t]] \ \text{and}$$

$$E\ \Phi[t] = E_{\alpha, p_1, \ldots, p_n}[E\ \Phi[t]] \in [c_0, c_M].$$

It follows:

$$\mathrm{Var}\ \alpha, p_1, \ldots, p_n[E\ \Phi[t]] \in B \quad \text{and} \quad \Phi_\alpha[p_1, \ldots, p_n] \geq \inf B. \bullet$$

Lemma 8.27 *Let $\alpha \in (0,1)$ and $(p_1, \ldots, p_n) \in P_n$. Let $t_0 \in [c_0, c_M]$ arbitrary such that $t_0 = E\alpha, p_1, \ldots, p_n[t_0]$ is valid. Then*

(i) $$\sum_{i \in \{1,\ldots,n\} \setminus C[t_0]} p_i V_i[t_0] = \left(1 - \sum_{i \in C[t_0]} p_i\right) E\alpha, p_1, \ldots, p_n[t_0] \text{ holds.}$$

(ii) Define for $t \in [c_0, c_M]$

$$D\alpha, p_1, \ldots, p_n[t]^2 \stackrel{\mathrm{d}}{=}$$

$$\begin{cases} \sum_{i \in \{1,\ldots,n\} \setminus C[t]} p_i \{V_i[t]\}^2 - \frac{1}{1 - \sum_{i \in C[t]} p_i} \left\{ \sum_{i \in \{1,\ldots,n\} \setminus C[t]} p_i V_i[t] \right\}^2 \\ \quad , \text{if } \sum_{i \in C[t]} p_i < 1 \\ 0 \quad , \text{if } \sum_{i \in C[t]} p_i = 1 \end{cases}$$

Then $\mathrm{Var}\ \alpha, p_1, \ldots, p_n[t_0] = D^2_{\alpha}, p_1, \ldots, p_n[t_0]$ *is valid.*

Proof

(i) follows from Definition 8.24(i) and from the fact that

$$V_i[t_0] = t_0 = E\alpha, p_1, \ldots, p_n[t_0]$$

holds for $i \in C[t_0]$.

(ii) For all $i \in C[t_0]$ $\{V_i[t_0] - E\alpha, p_1, \ldots, p_n[t_0]\}^2 = 0$ holds.
If $\sum_{i \in C[t_0]} p_i = 1$, then $p_i = 0$ for all $i \in \{1, \ldots, n\} \setminus C[t_0]$. It follows
$\mathrm{Var}\ \alpha, p_1, \ldots, p_n[t_0] = 0 = D^2_{\alpha}, p_1, \ldots, p_n[t_0]$.

Let $\sum_{i \in C[t_0]} p_i < 1$ be valid. Then

$$\begin{aligned}
& \text{Var}\, \alpha, p_1, \ldots, p_n[t_0] \\
= & \sum_{i \in \{1,\ldots,n\} \setminus C[t_0]} p_i \{V_i[t_0] - E\alpha, p_1, \ldots, p_n[t_0]\}^2 \\
= & \sum_{i \in \{1,\ldots,n\} \setminus C[t_0]} p_i \{V_i[t_0]\}^2 \\
& - 2E\alpha, p_1, \ldots, p_n[t_0] \sum_{i \in \{1,\ldots,n\} \setminus C[t_0]} p_i V_i[t_0] + \\
& + \{E\alpha, p_1, \ldots, p_n[t_0]\}^2 \sum_{i \in \{1,\ldots,n\} \setminus C[t_0]} p_i \\
= & \sum_{i \in \{1,\ldots,n\} \setminus C[t]} p_i \{V_i[t_0]\}^2 \\
& - \left(1 - \sum_{i \in C[t_0]} p_i\right) \{E\alpha, p_1, \ldots, p_n[t_0]\}^2 \\
= & D^2_{\alpha, p_1, \ldots, p_n}[t_0]
\end{aligned}$$

holds. •

Theorem 8.28 *Let $\alpha \in (0,1)$ and $(p_1, \ldots, p_n) \in P_n$. Then*

$$\Phi_\alpha[p_1, \ldots, p_n] = \beta \stackrel{\mathrm{d}}{=}$$

$$\min \left[\begin{array}{l} \min \left\{ D^2_{\alpha, p_1, \ldots, p_n}[c_k] \;\middle|\; \begin{array}{l} k \in \{1, \ldots, M-1\} \text{ and} \\ \sum\limits_{i \in \{1,\ldots,n\} \setminus C[c_k]} p_i V_i[c_k] = \\ c_k \left(1 - \sum\limits_{i \in C[c_k]} p_i\right) \end{array} \right\}, \\ \min \left\{ D^2_{\alpha, p_1, \ldots, p_n}[d_k] \;\middle|\; \begin{array}{l} k \in \{1, \ldots, M\} \text{ and} \\ c_{k-1} \left(1 - \sum\limits_{i \in C[d_k]} p_i\right) < \\ \sum\limits_{i \in \{1,\ldots,n\} \setminus C[d_k]} p_i V_i[d_k] < \\ c_k \left(1 - \sum\limits_{i \in C[d_k]} p_i\right) \end{array} \right\}, \end{array} \right]$$

is valid; hereby we define the sum of the empty set as 0 and the minimum of the empty set as $+\infty$.

Proof For all $k \in \{1, \ldots, M\}$ we have the inclusion

$$C[d_k] \subseteq C[d_k] \cap C[c_{k-1}].$$

Let us assume that there is a $t_0 \in [c_0, c_M]$ with $\sum\limits_{i \in C[t_0]} p_i = 1$.

If there is a $k \in \{1, \ldots, M\}$ with $t_0 \in (c_{k-1}, c_k)$, then

$$\sum_{i \in C[c_{k-1}]} p_i = \sum_{i \in C[c_k]} p_i = 1.$$

So we can conclude: There exists a $k \in \{1, \ldots, M-1\}$ with $\sum\limits_{i \in C[c_k]} p_i = 1$.

On the one hand we have

$$t_0 = E_{\alpha, p_1, \dots, p_n}[t_0] \quad \text{and} \quad \mathrm{Var}_{\alpha, p_1, \dots, p_n}[t_0] = 0.$$

This follows from Definition 8.24(i) and 8.24(iii). So $\Phi_\alpha[p_1, \dots, p_n] = 0$ is valid.
On the other hand $p_i = 0$ holds for all $i \in \{1, \dots, n\} \backslash C[c_k]$. It follows:

$$D^2_{\alpha, p_1, \dots, p_n}[c_k] = 0 \quad \text{and}$$

$$\sum_{i \in \{1,\dots,n\} \backslash C[c_k]} p_i V_i[c_k] = 0 = c_k \left(1 - \sum_{i \in C[c_k]} p_i\right)$$

and therefore $\beta = 0$ is valid.

Let $\sum_{i \in C[t]} p_i < 1$ be valid for all $t \in [c_0, c_M]$. Define

$$f(t) \overset{\mathrm{d}}{=} E_{\alpha, p_1, \dots, p_n}[t] - t.$$

for $t \in [c_0, c_M]$.
From Definition 8.24(i) we obtain

$$\forall k \in \{0, 1, \dots, M\}: \; f(c_k) = \sum_{i \in \{1,\dots,n\} \backslash C[c_k]} p_i V_i[c_k] - c_k \left(1 - \sum_{i \in C[c_k]} p_i\right),$$

and from Definition 8.24(i) and Lemma 8.21(i) and (ii):

$$\forall k \in \{1, \dots, M\}, \; \forall t \in (c_{k-1}, c_k):$$

$$f(t) = \sum_{i \in \{1,\dots,n\} \backslash C[d_k]} p_i V_i[d_k] - t \left(1 - \sum_{i \in C[d_k]} p_i\right)$$

This yields for $k \in \{1, \dots, M\}$:

$$\lim_{\substack{h \to 0 \\ h > 0}} f(c_{k-1} + h) = \sum_{i \in \{1,\dots,n\} \backslash C[d_k]} p_i V_i[d_k] - c_{k-1} \left(1 - \sum_{i \in C[d_k]} p_i\right)$$

$$\lim_{\substack{h \to 0 \\ h > 0}} f(c_k - h) = \sum_{i \in \{1,\dots,n\} \backslash C[d_k]} p_i V_i[d_k] - c_k \left(1 - \sum_{i \in C[d_k]} p_i\right).$$

(a) Let $t \in [c_0, c_M]$ such that $t = E_{\alpha, p_1, \ldots, p_n}[t]$ holds. There exists an $i \in \{1, \ldots, n\}$ with $\inf(A_i)_\alpha > c_0$ and $p_i > 0$, as otherwise $\sum\limits_{i \in C[c_0]} p_i = 1$ would be valid. With the help of Definition 8.24(i) we obtain:

$$E_{\alpha, p_1, \ldots, p_n}[c_0] > c_0.$$

In a similar way we can show

$$E_{\alpha, p_1, \ldots, p_n}[c_M] < c_M.$$

So $t \in (c_0, c_M)$ is valid.
By assertion we have $f(t) = E_{\alpha, p_1, \ldots, p_n}[t] - t = 0$.
If there is a $k \in \{1, \ldots, M-1\}$ with $t = c_k$, then

$$\sum_{i \in \{1,\ldots,n\} \setminus C[c_k]} p_i V_i[c_k] - c_k \left(1 - \sum_{i \in C[c_k]} p_i\right)$$

follows. By Lemma 8.27(ii) we know that

$$\operatorname{Var}_{\alpha, p_1, \ldots, p_n}[t] = D^2_{\alpha, p_1, \ldots, p_n}[t] \geq \beta$$

is valid. If $t \in (c_{k-1}, c_k)$ for a $k \in \{1, \ldots, M\}$, then we have

$$\sum_{i \in \{1,\ldots,n\} \setminus C[d_k]} p_i V_i[d_k] = t \left(1 - \sum_{i \in C[d_k]} p_i\right)$$

As $1 - \sum\limits_{i \in C[d_k]} p_i > 0$ holds, we obtain

$$\sum_{i \in \{1,\ldots,n\} \setminus C[d_k]} p_i V_i[d_k] > c_{k-1} \left(1 - \sum_{i \in C[d_k]} p_i\right) \quad \text{and}$$

$$\sum_{i \in \{1,\ldots,n\} \setminus C[d_k]} p_i V_i[d_k] < c_k \left(1 - \sum_{i \in C[d_k]} p_i\right)$$

We conclude with help of Lemma 8.21 (ii):

$$\operatorname{Var}_{\alpha, p_1, \ldots, p_n}[t] = D^2_{\alpha, p_1, \ldots, p_n}[t] = D^2_{\alpha, p_1, \ldots, p_n}[d_k] \geq \beta.$$

Finally we obtain with Theorem 8.26:

$$\Phi_\alpha[p_1,\ldots,p_n] \geq \beta.$$

(b) Let $k \in \{1,\ldots,M-1\}$ such that

$$\sum_{i\in\{1,\ldots,n\}\setminus C[c_k]} p_i V_i[c_k] = c_k \left(1 - \sum_{i\in C[c_k]} p_i\right)$$

holds. It follows: $f(c_k) = 0$, i.e. $c_k = E_{\alpha,p_1,\ldots,p_n}[c_k]$.
Applying Lemma 8.27(ii), we obtain:

$$D^2_{\alpha,p_1,\ldots,p_n}[c_k] = \operatorname{Var}_{\alpha,p_1,\ldots,p_n}[c_k] \geq \Phi_\alpha[p_1,\ldots,p_n].$$

Let $k \in \{1,\ldots,M\}$ such that

$$c_{k-1}\left(1 - \sum_{i\in C[d_k]} p_i\right) < \sum_{i\in\{1,\ldots,n\}\setminus C[d_k]} p_i V_i[d_k] < c_k \left(1 - \sum_{i\in C[d_k]} p_i\right)$$

is valid, i.e.

$$\lim_{\substack{h\to 0\\ h>0}} f(c_{k-1}+h) > 0 \quad \text{and} \quad \lim_{\substack{h\to 0\\ h>0}} f(c_k - h) < 0.$$

f is continuous on (c_{k-1}, c_k). There exist numbers $a \in \mathbb{R}$ and $b \in \mathbb{R}$ with $[a,b] \subseteq (c_{k-1}, c_k)$ such that $f(a) > 0$ and $f(b) < 0$ holds. We can apply the first mean value theorem of the analysis. Therefore there exists a $t \in [a,b] \subseteq (c_{k-1}, c_k)$ with $f(t) = 0$, i.e. $t = E_{\alpha,p_1,\ldots,p_n}[t]$. It follows with Lemma 8.27(ii):

$$D^2_{\alpha,p_1,\ldots,p_n}[d_k] = D^2_{\alpha,p_1,\ldots,p_n}[t] = \operatorname{Var}_{\alpha,p_1,\ldots,p_n}[t] \geq$$
$$\geq \Psi_\alpha[p_1,\ldots,p_n].$$

All together :

$$\Psi_\alpha[p_1,\ldots,p_n] \leq \beta. \bullet$$

We can summarize the results concerning $\inf B_\alpha = \inf \ddot{B}_\alpha$ in the following theorem.

Theorem 8.29 *For all $\alpha \in (0,1)$ we have*

$$\inf B_\alpha = \inf \tilde{B}_\alpha =$$

$$\min\left[\min\left\{ D^2_{\alpha, P(\Gamma_1),\dots,P(\Gamma_n)}{}^{[c_k]} \;\middle|\; \begin{array}{l} k \in \{1,\dots,M-1\} \text{ and} \\ \sum_{i\in\{1,\dots,n\}\setminus C[c_k]} P(\Gamma_i) V_i[c_k] = \\ c_k \left(1 - \sum_{i\in C[c_k]} P(\Gamma_i)\right) \end{array} \right\}\right.$$

$$= \min\left.\min\left\{ D^2_{\alpha, P(\Gamma_1),\dots,P(\Gamma_n)}{}^{[d_k]} \;\middle|\; \begin{array}{l} k \in \{1,\dots,M-1\} \text{ and} \\ c_{k-1}\left(1 - \sum_{i\in C[d_k]} P(\Gamma_i)\right) < \\ \sum_{i\in\{1,\dots,n\}\setminus C[d_k]} P(\Gamma_i) V_i[d_k] < \\ c_k \left(1 - \sum_{i\in C[d_k]} P(\Gamma_i)\right) \end{array} \right\}\right]$$

Moreover $\inf B_\alpha \in B_\alpha$ *and* $\inf \tilde{B}_\alpha \in \tilde{B}_\alpha$ *is valid.*

In the next theorem we show that $\tilde{B}_\alpha$ is a compact interval. From this fact we conclude that $\widetilde{\text{Var}}\, X \in U_C(\mathbb{R})$.

Theorem 8.30 $\widetilde{\text{Var}}\, X \in U_C(\mathbb{R})$.

Proof It has already been shown that for all $\alpha \in (0,1)$ $\inf \tilde{B}_\alpha \in \tilde{B}_\alpha$ (see above) and $\sup \tilde{B}_\alpha \in \tilde{B}_\alpha$ (cf. page 83) holds. We also know from Lemma 8.7(ii) that $\inf\left(\widetilde{\text{Var}}\, X\right)_0 \geq 0$ and $\sup\left(\widetilde{\text{Var}}\, X\right)_0 \leq D^2$ holds. If we can show that

$$\left(\inf \tilde{B}_\alpha, \sup \tilde{B}_\alpha\right) \subseteq \tilde{B}_\alpha$$

holds for all $\alpha \in (0,1)$, the proof is complete.

Let $\alpha \in (0,1)$ and $t \in \left(\inf \tilde{B}_\alpha, \sup \tilde{B}_\alpha\right)$. We have to show that there exists a $U \in \tilde{Y}_\alpha$ with $\text{Var}\, U = t$. We distinguish between several cases in all of which we define a polynom of the degree, of two at the most such that $f(0) \geq t$ and $f(1) \leq t$. $f(\lambda)$ is the variance of a random variable $U_\lambda \in \tilde{Y}_\alpha$ for all $\lambda \in [0,1]$. We can apply the first mean value theorem and conclude:

There exists a $\lambda^* \in [0,1]$ with $\text{Var } U_{\lambda^*} = f(\lambda^*) = t$, i.e. $t \in \tilde{B}_\alpha$.

For all $\lambda \in [0,1]$ A_λ denotes an element of $\mathfrak{A}'$ with $P'(A_\lambda) = \lambda$. For abbreviation define

$$M \stackrel{\mathrm{d}}{=} D^2_{\alpha, P(\Gamma_1), \dots, P(\Gamma_n)}(0, 0, \dots, 0)$$

where $D^2_{\alpha, P(\Gamma_1), \dots, P(\Gamma_n)}$ is defined as in Theorem 8.10.

Let $t \leq M$ be valid. Let us assume that there is an $s \in I\!R$ with

$$\sum_{i \in C[s]} P(\Gamma_i) = 1.$$

Define for $\lambda \in [0,1]$ $U_\lambda \in \tilde{Y}_\alpha$ by

$$U_\lambda(\omega, \omega') \stackrel{\mathrm{d}}{=} \begin{cases} s & \text{, if } \omega \in \Gamma_i \text{ for and } i \in C[s] \text{ and } \omega' \in A_\lambda \\ \inf(A_i)_\alpha & \text{, if } (\omega \in \Gamma_i \text{ for an } i \in \{1, \dots, n\} \backslash C[s] \text{ and } \omega' \in \Omega') \\ & \quad \text{or } (\omega \in \Gamma_i \text{ for an } i \in \{1, \dots, n\} \text{ and } \omega' \in \Omega' \backslash A_\lambda) \\ Z_\alpha(\omega) & \text{, if } \omega \in \Omega \backslash \bigcup_{i=1}^{n} \Gamma_i \text{ and } \omega' \in \Omega'. \end{cases}$$

For all $\lambda \in [0,1]$ we have $U_\lambda \in \tilde{Y}_\alpha$. We obtain:

$$\begin{aligned} f(\lambda) \stackrel{\mathrm{d}}{=} & \text{ Var } U_\lambda = E\, U_\lambda^2 - [E\, U_\lambda]^2 \\ & = \lambda s^2 + (1 - \lambda) E^2_{\alpha, P(\Gamma_1), \dots, P(\Gamma_n)}(0, 0, \dots, 0) \\ & \quad - \left[\lambda s + (1 - \lambda) E_{\alpha, P(\Gamma_1), \dots, P(\Gamma_n)}(0, 0, \dots, 0)\right]^2 \end{aligned}$$

for $\lambda \in [0,1]$. We follow:

$$f(0) = D^2_{\alpha, P(\Gamma_1), \dots, P(\Gamma_n)}(0, 0, \dots, 0) = M \geq t$$

$$f(1) = 0 \leq t.$$

We conclude: $t \in \tilde{B}_\alpha$.

If

$$t \leq M \text{ and } \sum_{i \in C[s]} P(\Gamma_i) < 1$$

holds for all $s \in I\!R$, we can apply Lemma 8.25 and obtain: There exists a $t_0 \in (c_0, c_M)$ with $\operatorname{Var} V[t_0] \leq t$.

$\sum_{i \in C[t_0]} P(\Gamma_i) < 1$ is valid.

Define $V_\lambda \in \tilde{Y}_\alpha$ for $\lambda \in [0,1]$ by

$$V_\lambda(\omega, \omega') \stackrel{\mathrm{d}}{=}$$

$$\begin{cases} \inf(A_i)_\alpha & \text{, if } (\omega \in \Gamma_i \text{ for an } i \in A[t_0] \text{ and } \omega' \in \Omega') \\ & \text{or } (\omega \in \Gamma_i \text{ for an } i \in \{1, \ldots, n\} \text{ and } \omega' \in \Omega' \backslash A_\lambda\ ) \\ \sup(A_i)_\alpha & \text{, if } \omega \in \Gamma_i \text{ for an } i \in B[t_0] \text{ and } \omega' \in A_\lambda \\ (b_j)^{(i)}_\alpha & \text{, if } \omega \in \Gamma_i \text{ for an } i \in C[t_0] \text{ and } \omega' \in A_\lambda \\ (a_j)^{(i)}_\alpha & \text{, if } \omega \in \Gamma_i \text{ for an } i \in D_j[t_0] \\ & \quad \text{with } j \in \{1, \ldots, (N_i)_\alpha\} \text{ and } \omega' \in A_\lambda \\ t_0 & \text{, if } \omega \in \Gamma_i \text{ for an } i \in E_j[t_0] \\ & \quad \text{with } j \in \{1, \ldots, (N_i)_\alpha\} \text{ and } \omega' \in A_\lambda \\ Z_\alpha(\omega) & \text{, if } \omega \in \Omega \backslash \bigcup_{i=1}^{n} \Gamma_i \text{ and } \omega' \in \Omega' . \end{cases}$$

We obtain:

$$f(\lambda) \stackrel{\mathrm{d}}{=} \operatorname{Var} V_\lambda$$
$$= \lambda \cdot E^2_{\alpha, P(\Gamma_1), \ldots, P(\Gamma_n)} + (1 - \lambda) \cdot E^2_{\alpha, P(\Gamma_1), \ldots, P(\Gamma_n)}(0, 0, \ldots, 0)$$
$$- \left\{ \lambda \cdot E_{\alpha, P(\Gamma_1), \ldots, P(\Gamma_n)} + (1 - \lambda) \cdot E_{\alpha, P(\Gamma_1), \ldots, P(\Gamma_n)}(0, 0, \ldots, 0) \right\}^2$$

This yields:

$$f(0) = D^2_{\alpha, P(\Gamma_1), \ldots, P(\Gamma_n)}(0, 0, \ldots, 0) = M \geq t \text{ and}$$
$$f(1) = D_{\alpha, P(\Gamma_1), \ldots, P(\Gamma_n)}[t_0] = \operatorname{Var} V[t_0] \leq t.$$

We can conclude: $t \in \tilde{B}_\alpha$.

Now let us consider the case $t \geq M$.

Choose $(d_1, \ldots, d_n) \in C_\alpha[P(\Gamma_1), \ldots, P(\Gamma_n)]$ (compare Lemma 8.12), i.e.

$$D^2_{\alpha, P(\Gamma_1), \ldots, P(\Gamma_n)}(d_1, \ldots, d_n) =$$
$$= \sup \left\{ D^2_{\alpha, P(\Gamma_1), \ldots, P(\Gamma_n)}(c_1, \ldots, c_n) \,|\, (c_1, \ldots, c_n) \in [0,1]^n \right\} =$$
$$= \sup \tilde{B}_\alpha \geq t.$$

Define for $\lambda \in [0,1]$

$$f(\lambda) \stackrel{\mathrm{d}}{=} D^2_{\alpha, P(\Gamma_1),\ldots,P(\Gamma_n)}[(1-\lambda)d_1,\ldots,(1-\lambda)d_n].$$

Then

$$f(0) = D^2_{\alpha, P(\Gamma_1),\ldots,P(\Gamma_n)}(d_1,\ldots,d_n) \geq t \text{ and}$$
$$f(1) = D^2_{\alpha, P(\Gamma_1),\ldots,P(\Gamma_n)}(0,0,\ldots,0) \leq M \leq t$$

holds. It follows:

$$\exists \lambda^* \in [0,1] \text{ with } f(\lambda^*) = t.$$

Define $B \stackrel{\mathrm{d}}{=} \bigcup_{i=1}^{n} \Gamma_i \times A_{(1-\lambda^*)d_i} \in \mathfrak{A} \otimes \mathfrak{A}'$.
It follows (compare with the proof of Theorem 8.10):

$$\mathrm{Var}\ U_B = D^2_{\alpha, P(\Gamma_1),\ldots,P(\Gamma_n)}[(1-\lambda^*)d_1,\ldots,(1-\lambda^*)d_n] = t,$$

where $U_B \in \tilde{Y}_\alpha$ is defined as in Lemma 8.9. We conclude:

$$t \in \tilde{B}_\alpha. \bullet$$

B_α is not necessarily a closed interval: Consider for example the probability space $\{\{\omega_1,\omega_2\},\mathfrak{P}\{\omega_1,\omega_2\},P\}$, where $P\{\omega_1\} = P\{\omega_2\} = \frac{1}{2}$, and the fuzzy random variable X defined by $X(\omega_1) = \mathrm{I}_{\{0,1\}}$, $X(\omega_2) = \mathrm{I}_{\{0\}}$.

Then $U_1 : \omega_1 \mapsto 0$, $\omega_2 \mapsto 0$ and $U_2 : \omega_1 \mapsto 1$, $\omega_2 \mapsto 0$ are the only random variables with an acceptability degree higher than 0. We have $\mu_X(U_1) = \mu_X(U_2) = 1$, $\mathrm{Var}\ U_1 = 0$, and $\mathrm{Var}\ U_2 = 0.25$. Therefore $B_\alpha = \{0, \frac{1}{4}\}$ follows.

If the random variable X is convex, i.e. $X(\Omega) \subseteq U_C(\mathbb{R})$, then the variance of X is a convex fuzzy set.

Theorem 8.31 *If the codomain $\{\mu_1,\ldots,\mu_n\}$ of the f.r.v. X is a subset of $U_C(\mathbb{R})$, then Var X is convex.*

Proof We already know that $\inf B_\alpha \in B_\alpha$ and $\sup B_\alpha \in B_\alpha$ holds for $\alpha \in (0,1)$. It is also shown that $\inf(\mathrm{Var}\ X)_0 \geq 0$ and $\sup(\mathrm{Var}\ X)_0 \leq D^2$ is valid. So it remains to demonstrate that B_α is convex for all $\alpha \in (0,1)$.

We can assume that $(N_i)_\alpha = 1$, i.e. $(A_i)_\alpha = [\inf(A_i)_\alpha, \sup(A_i)_\alpha]$ is valid for $i \in \{1,\ldots,n\}$ and $\alpha \in (0,1)$.

Let $\alpha \in (0,1)$, $t_1 \in B_\alpha$, $t_2 \in B_\alpha$ with $t_1 \leq t_2$ be arbitrary.
There exist $U_1 \in Y_\alpha$ and $U_2 \in Y_\alpha$ with $\operatorname{Var} U_1 = t_1$ and $\operatorname{Var} U_2 = t_2$. Define

$$U_\lambda \stackrel{\mathrm{d}}{=} \lambda \cdot U_1 + (1-\lambda) \cdot U_2$$

for $\lambda \in [0,1]$. If $\omega \in \Gamma_i$ holds for an $i \in \{1, \ldots, n\}$, then

$$U_1(\omega) \in [\inf(A_i)_\alpha, \sup(A_i)_\alpha]$$

and

$$U_2(\omega) \in [\inf(A_i)_\alpha, \sup(A_i)_\alpha]$$

is valid, i.e. $U_\lambda(\omega) \in (A_i)_\alpha$ holds for $\omega \in \Omega$.
If $\omega \in \Omega \setminus \bigcup_{i=1}^{n} \Gamma_i$ holds, then

$$U_1(\omega) = U_2(\omega) = U_\lambda(\omega) = Z_\alpha(\omega)$$

holds for $\lambda \in [0,1]$. So $U_\lambda \in Y_\alpha$ is valid for $\lambda \in [0,1]$.
Define

$$f(\lambda) \stackrel{\mathrm{d}}{=} \operatorname{Var} U_\lambda =$$

$$\lambda^2 E\, U_1^2 + 2 \cdot \lambda(1-\lambda) E\, U_1 U_2 + (1-\lambda)^2 E\, U_2^2 - \{\lambda E\, U1 + (1-\lambda) E\, U_2\}^2$$

for $\lambda \in [0,1]$.
Then $f(0) = \operatorname{Var} U_2$ and $f(1) = \operatorname{Var} U_1$ follows.
As f is continuous on $[0,1]$, we can conclude:

$$[t_1, t_2] \subseteq [f(1), f(0)] \subseteq B_\alpha.$$

It follows that B_α is convex for all $\alpha \in (0,1)$. Therefore $\operatorname{Var} X$ is convex.
•

Example 8.32 As a tutorial example of a discrete random variable, let us consider an opinion poll. The responses to a question concerning the opinion of an expert interviewed about the age of a sample of persons is summarized in the following table.

No of persons	Relative frequence	Response
5	0.25	very old, but less than 100
10	0.5	exactly 85
5	0.25	approximately 90

Representation as fuzzy sets	Internal representation
$\mu_1 = t^{80,90,100,100}$	$\left[t^{80,90,100,100}\right]_{10}$
$\mu_2 = \mathrm{I}_{\{85\}}$	$\left[\mathrm{I}_{\{85\}}\right]_{10}$
$\mu_3 = t^{85,90,90,95}$	$\left[t^{85,90,90,95}\right]_{10}$

We may describe the (not very realistic) opinion poll by the fuzzy random variable $X : \Omega \to \{\mu_1, \mu_2, \mu_3\}$, where $P(X = \mu_1) = P(X = \mu_3) = 0.25, P(X = \mu_2) = 0.5$. In order to calculate the expected values, we use the equality

$$[E\,X]_{10} = \left[\tilde{E}\,X\right]_{10} = \left[\frac{1}{4}(\mu_1 + 2\mu_2 + \mu_3)\right]_{10} = \left[t^{83.75,87.5,90.0,91.25}\right]_{10}.$$

Var X and $\widetilde{\text{Var}}$ X are calculated by Theorem 8.15, Theorem 8.18, and Theorem 8.29. We obtain:

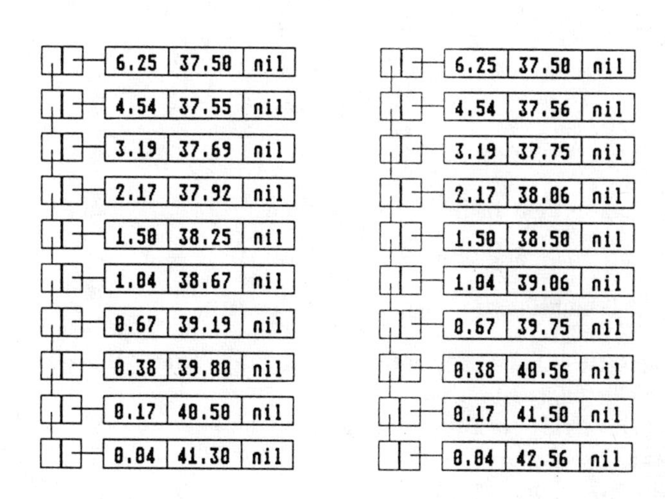

Fig. 8.1. The internal representation of Var X (left) and $\widetilde{\text{Var}}$ X (right).

The fuzzy sets μ_1, μ_2, and μ_3, and the characteristics of X are sketched in the following figure:

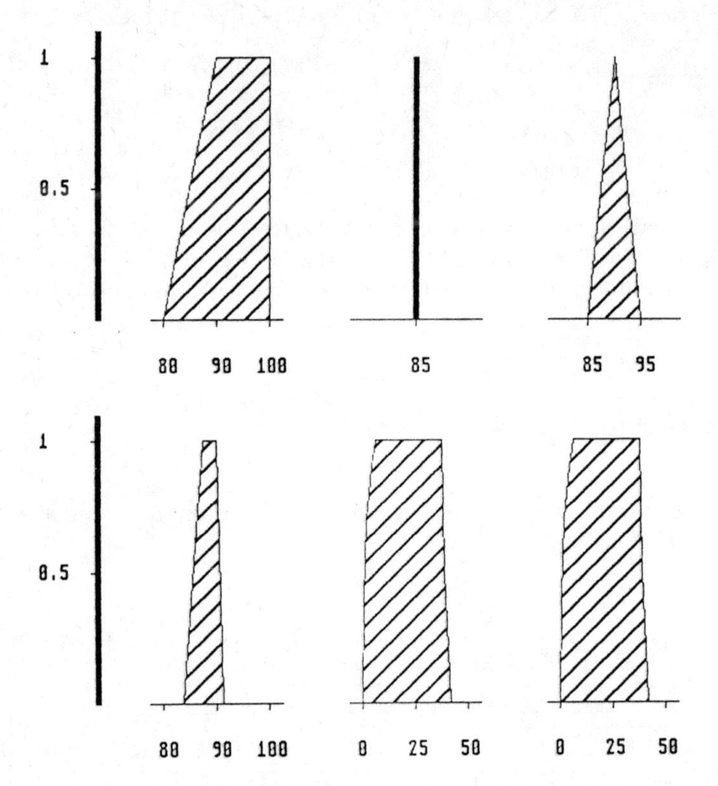

Fig. 8.2. The graphs of $[\mu_1]_{10}$, $[\mu_2]_{10}$, and $[\mu_3]_{10}$ (first line, from left to right), and $[E\ X]_{10}$, $[\text{ Var } X]_{10}$, and $\left[\widetilde{\text{Var}}\ X\right]_{10}$ (second line, from left to right).

These characteristics have to be interpreted by the expert. He could obtain, for example, this information:

- The expected value of X is approximately between 88 – 90.
- The variance is approximately between 0 and 50.
- The data are rather vague.

8.3 EMPIRICAL DISTRIBUTION FUNCTION

It is beyond the scope of this book to give a comprehensive overview of all methods used in descriptive statistics. Here we consider neither robust methods for the description of one dimensional measurements nor correlation considerations of two-dimensional measurements. It should be clear from the preceding considerations how to generalize the results known for sharp data to vague data. We consider only two further concepts of descriptive statistics: quantiles and empirical distribution functions.

If U is a random variable, then a number $x \in I\!R$ satisfying

$$P(U \leq x) \geq p \text{ and}$$
$$P(U \geq x) \geq 1-p, \quad 0 < p < 1$$

is called a quantile of order p for U. The interval of all quantiles of order p for U is denoted by $\Theta_p U$.

Definition 8.33 *Let $X : \Omega \to Q(I\!R)$ be a f.r.v., and let p be a real number such that $0 < p < 1$. We define the following fuzzy sets*

$$(\Theta_p X)(t) \stackrel{\text{d}}{=} \sup\{\mu_x(U) \mid U \in \chi, t \in \Theta_p U\}, \ t \in I\!R,$$

and

$$\left(\tilde{\Theta}_p X\right)(t) \stackrel{\text{d}}{=} \sup\{\tilde{\mu}_x(U) \mid U \in \tilde{\chi}, t \in \Theta_p U\}, \ t \in I\!R.$$

$\Theta_p X$ $\left(\tilde{\Theta}_p X\right)$ is called quantile of order p of X with respect to χ (to $\tilde{\chi}$).

Since the use of quantiles in descriptive statistics should be clear, let us motivate the importance of the empirical distribution function for vague data. Let us consider an opinion poll, where n individuals are questioned and where k different answers $x_1, \ldots, x_k$ are possible. Then

$$H_n(x_j) \stackrel{\text{d}}{=} \text{" number of individuals that chooses answer } x_j\text{"}$$

is called the absolute frequency of x_j and $h_n(x_j) \stackrel{\text{d}}{=} \frac{1}{n} H_n(x_j)$ denotes the relative frequency of x_j for $j \in \{1, \ldots, k\}$. If we have an ordinal scale where the values $x_1, \ldots, x_k$ are ordered, i.e:

$$x_1 < x_2 < \ldots < x_k,$$

then we can describe the opinion poll also by the relative sum frequencies

$$h_n(x_1) + h_n(x_2) + \ldots + h_n(x_j), \ j \in \{1, \ldots, k\}.$$

This treatment of the opinion poll leads to the notion of an empirical distribution function.

For $n \in I\!N$, $(x_1, \ldots, x_n) \in \overline{I\!R}^n \stackrel{\text{d}}{=} (I\!R \cup \{-\infty\} \cup \{+\infty\})^n$, and $t \in I\!R$, we define

$$S_n[x_1, \ldots, x_n](x) \stackrel{\mathrm{d}}{=} \frac{1}{n} \text{card } \{i \in \{1, \ldots, n\} \mid x \geq x_i\}.$$

In this definition, for a finite subset $A \subseteq I\!R$ card A denotes the number of elements of A. The stair function $S_n[x_1, \ldots, x_n]$ is called empirical distribution function. It has the following well-known properties.

Lemma 8.34 *Let $n \in I\!N$ and $(x_1, \ldots, x_n) \in \overline{I\!R}^n$ be arbitrary. Then*

(i) $S_n[x_1, \ldots, x_n]$ is monotonously non-decreasing and is a function continuous on the right.

(ii) $\lim_{x \to -\infty} S_n[x_1, \ldots, x_n](x) = \frac{1}{n} \text{card } \{i \in \{1, \ldots, n\} \mid x_i = -\infty\}$

(iii) $\lim_{x \to +\infty} S_n[x_1, \ldots, x_n](x) = 1 - \frac{1}{n} \text{card } \{i \in \{1, \ldots, n\} \mid x_i = +\infty\}.$

Lemma 8.35 *Let $(x_1, \ldots, x_n) \in \overline{I\!R}^n$ and $(y_1, \ldots, y_n) \in \overline{I\!R}^n$ such that $x_i \leq y_i$ for $i \in \{1, \ldots, n\}$. Then*

$$S_n[x_1, \ldots, x_n](x) \geq S_n[y_1, \ldots, y_n](x).$$

is valid for all $t \in I\!R$.

At this point we generalize and consider the notion of an empirical distribution function for vague data.

Definition 8.36 *Let $n \in I\!N$, define $A_n \stackrel{\mathrm{d}}{=} \{0, \frac{1}{n}, \frac{2}{n}, \ldots, 1\}$. Let $(\mu_1, \ldots, \mu_n) \in [F(I\!R)]^n$. Define $S_n[\mu_1, \ldots, \mu_n] : I\!R \to F(I\!R)$ by*

$$\{S_n[\mu_1, \ldots, \mu_n](x)\}(t) \stackrel{\mathrm{d}}{=}$$

$$\begin{cases} \sup \left\{ \min\limits_{i \in \{1, \ldots, n\}} \mu_i(t_i) \,\middle|\, \begin{array}{l} (t_1, \ldots, t_n) \in I\!R^n, \\ S_n[t_1, \ldots, t_n](x) = t, \end{array} \right\} & , \text{if } t \in A_n \\ 0 & , \text{if } t \in I\!R \backslash A_n \end{cases}$$

for $x \in I\!R$.

It can easily be shown that

$$S_n\left[\mathrm{I}_{\{x_1\}}, \ldots, \mathrm{I}_{\{x_n\}}\right] = \mathrm{I}_{S_n[x_1, \ldots, x_n]}$$

holds for $(x_1, \ldots, x_n) \in I\!R^n$. Therefore $S_n[\mu_1, \ldots, \mu_n]$ is also called empirical distribution function. We have to consider the connections between set representation and empirical distribution function.

Lemma 8.37 *Let $n \in I\!N$ and $(\mu_1, \ldots, \mu_n) \in [Q(I\!R)]^n$. Then we have for all $x \in I\!R$:*

(i) If $\{(A_i)_\alpha \mid \alpha \in (0,1)\}$ *is a normal set representation of* μ_i *for* $i \in \{1,\dots,n\}$ *, then*

$$\left\{ \begin{array}{c} A_n \cap [S_n[\sup(A_1)_\alpha,\dots,\sup(A_n)_\alpha](x), \\ S_n[\inf(A_1)_\alpha,\dots,\inf(A_n)_\alpha](x)] \end{array} \mid \alpha \in (0,1) \right\}$$

is a normal set representation of $S_n[\mu_1,\dots,\mu_n](x)$.

(ii)

$$\left\{ \begin{array}{c} A_n \cap [S_n[(\mu_1)_\alpha,\dots,(\mu_n)_\alpha](x), \\ S_n[(\mu_1)_\alpha,\dots,(\mu_n)_\alpha](x)] \end{array} \mid \alpha \in (0,1) \right\}$$

is a normal set representation of $S_n[\mu_1,\dots,\mu_n](x)$.

Proof Let $x \in I\!R$ be arbitrary.

(i) From Lemma 4.5(i), we know that it is sufficient to show that

$$S_n[(A_1)_\alpha,\dots,(A_n)_\alpha](x) =$$

$$A_n \cap [S_n[\sup(A_1)_\alpha,\dots,\sup(A_n)_\alpha](x), S_n[\inf(A_1)_\alpha,\dots,\inf(A_n)_\alpha](x)]$$

is valid for all $\alpha \in (0,1)$.
Let $\alpha \in (0,1)$.

"$\subseteq$" Let $t \in S_n[(A_1)_\alpha,\dots,(A_n)_\alpha](x)$. Then there exists a $(t_1,\dots,t_n) \in (A_1)_\alpha \times \dots \times (A_n)_\alpha$ with $S_n[t_1,\dots,t_n](x) = t$. On the one hand, this implies: $t \in A_n$.
On the other hand,

$$\inf(A_i)_\alpha \le t_i \le \sup(A_i)_\alpha.$$

follows for all $i \in \{1,\dots,n\}$. Applying Lemma 8.35, we may conclude:

$$S_n[\inf(A_1)_\alpha,\dots\inf(A_n)_\alpha](x) \ge S_n[t_1,\dots,t_n](x)$$

$$= t \ge S_n[\sup(A_i)_\alpha,\dots\sup(A_n)_\alpha](x)$$

"$\supseteq$" Let

$$t \in A_n \cap \begin{bmatrix} S_n[\sup(A_i)_\alpha,\dots,\sup(A_n)_\alpha](x), \\ S_n[\inf(A_i)_\alpha,\dots,\inf(A_n)_\alpha](x) \end{bmatrix}.$$

Then we can find an $r \in \{1,\ldots,n\}$ with $t = \frac{r}{n}$.
For abbreviation define

$$r_1 \stackrel{\mathrm{d}}{=} n \cdot S_n\left[\sup(A_1)_\alpha,\ldots,\sup(A_n)_\alpha\right](x)$$
$$r_2 \stackrel{\mathrm{d}}{=} n \cdot S_n\left[\inf(A_1)_\alpha,\ldots,\inf(A_n)_\alpha\right](x)$$

Then $r_1 \leq r \leq r_2$ is valid.
Define

$$I_1 \stackrel{\mathrm{d}}{=} \{i \in \{1,\ldots,n\} \mid x \geq \sup(A_i)_\alpha\} \quad \text{and}$$
$$I_2 \stackrel{\mathrm{d}}{=} \{i \in \{1,\ldots,n\} \mid x \geq \inf(A_i)_\alpha\} \ .$$

Then

$$I_1 \subseteq I_2 \quad \text{and} \quad \text{card } I_1 = r_1 \quad \text{and} \quad \text{card } I_2 = r_2$$

is valid. We can choose a set $I \subseteq \{1,\ldots,n\}$ such that $I_1 \subseteq I \subseteq I_2$ and card $I = r$ holds.
Define a vector $(t_1,\ldots,t_n) \in \mathbb{R}^n$ by

$$t_i \stackrel{\mathrm{d}}{=} \begin{cases} \inf(A_i)_\alpha & \text{, if } i \in I \text{ and } \inf(A_i)_\alpha > -\infty \\ \min\{x, \sup(A_i)_\alpha\} & \text{, if } i \in I \text{ and } \inf(A_i)_\alpha = -\infty \\ & \quad \text{and } \sup(A_i)_\alpha < +\infty \\ x & \text{, if } i \in I \text{ and } \inf(A_i)_\alpha = -\infty \\ & \quad \text{and } \sup(A_i)_\alpha = +\infty \\ \sup(A_i)_\alpha & \text{, if } i \in \{1,\ldots,n\}\backslash I \text{ and} \\ & \quad \sup(A_i)_\alpha < +\infty \\ \max\{x+1, \inf(A_i)_\alpha\} & \text{, if } i \in \{1,\ldots,n\}\backslash I \text{ and} \\ & \quad \inf(A_i)_\alpha > -\infty \text{ and } \sup(A_i)_\alpha = +\infty \\ x+1 & \text{, if } i \in \{1,\ldots,n\}\backslash I \text{ and} \\ & \quad \inf(A_i)_\alpha = -\infty \text{ and } \sup(A_i)_\alpha = +\infty. \end{cases}$$

As $\{(A_i)_\alpha \mid \alpha \in (0,1)\}$ is a normal set representation of μ_i for $i \in \{1,\ldots,n\}$, $t_i \in (A_i)_\alpha$ holds for all $i \in \{1,\ldots,n\}$. So

$$S_n[t_1,\ldots,t_n](x) \in S_n\left[(A_1)_\alpha,\ldots,(A_n)_\alpha\right](x)$$

follows. The equation $I = \{i \in \{1,\ldots,n\} \mid x \geq t_i\}$ implies

$$S_n[t_1,\ldots,t_n](x) = \frac{1}{n}\text{card } \{i \in \{1,\ldots,n\} \mid x \geq t_i\} = \frac{1}{n}\text{card } I = t.$$

(ii) Because of Theorem 3.5, it is sufficient to show that

$$(S_n[\mu_1,\ldots,\mu_n](x))_\alpha \subseteq$$

$$A_n \cap [S_n [\sup (\mu_1)_\alpha, \ldots, \sup (\mu_n)_\alpha] (x), S_n [\inf (\mu_1)_\alpha, \ldots, \inf (\mu_n)_\alpha] (x)] \subseteq$$
$$\subseteq (S_n[\mu_1, \ldots, \mu_n](x))_{\overline{\alpha}}$$

is valid for all $\alpha \in (0,1)$.
If $t \in (S_n[\mu_1, \ldots, \mu_n](x))_\alpha$, then Lemma 4.6(ii) shows that there exists a $t_i \in (\mu_i)_\alpha$ with $S_n[t_1, \ldots, t_n](x) = t$. On the one hand, this implies $t \in A_n$.
On the other hand,

$$\inf (\mu_i)_\alpha \leq t_i \leq \sup (\mu_i)_\alpha.$$

follows for $i \in \{1, \ldots, n\}$, so we can follow by Lemma 8.35:

$$S_n [\inf (\mu_1)_\alpha, \ldots, \inf (\mu_n)_\alpha] (x) \geq t \geq S_n [\sup (\mu_1)_\alpha, \ldots, \sup (\mu_n)_\alpha] (x).$$

Let $\{(A_i)_\alpha \mid \alpha \in (0,1)\}$ be a normal set representation of μ_i for $i \in \{1, \ldots, n\}$. By Theorem 3.5 we may conclude for all $i \in \{1, \ldots, n\}$:

$$\inf(A_i)_\alpha \leq \inf (\mu_i)_\alpha \quad \text{and} \quad \sup(A_i)_\alpha \geq \sup (\mu_i)_\alpha .$$

Applying Lemma 8.35 and Lemma 8.37(i), we obtain

$$A_n \cap [S_n [\sup (\mu_1)_\alpha, \ldots, \sup (\mu_n)_\alpha] (x), S_n [\inf (\mu_1)_\alpha, \ldots, \inf (\mu_n)_\alpha] (x)] \subseteq$$
$$A_n \cap [S_n [\sup(A_1)_\alpha, \ldots, \sup(A_n)_\alpha] (x), S_n [\inf(A_1)_\alpha, \ldots, \inf(A_n)_\alpha] (x)] \subseteq$$
$$\subseteq (S_n[\mu_1, \ldots, \mu_n](x))_{\overline{\alpha}}. \bullet$$

(iii) Obviously both set representations are normal ones. •

Theorem 8.38 *Let $n \in \mathbb{N}$, $(\mu_1, \ldots, \mu_n) \in [Q(\mathbb{R})]^n$, and let $x \in \mathbb{R}$. Then we have for all $\alpha \in [0,1)$ and $\beta \in (0,1]$:*

(i)
$$(S_n[\mu_1, \ldots, \mu_n](x))_\alpha =$$
$$A_n \cap [S_n [\sup (\mu_1)_\alpha, \ldots, \sup (\mu_n)_\alpha] (x),$$
$$\lim_{r \to \infty} S_n \left[\inf (\mu_1)_{\alpha+(1-\alpha)/(2r)}, \ldots, \inf (\mu_n)_{\alpha+(1-\alpha)/(2r)}\right] (x)]$$

(ii)
$$(S_n[\mu_1, \ldots, \mu_n](x))_{\overline{\beta}} =$$
$$A_n \cap \Big[\lim_{r \to \infty} \inf S_n \left[\sup (\mu_1)_{\beta \cdot (1-1/2r)}, \ldots, \sup (\mu_n)_{\beta \cdot (1-1/2r)}\right](x)$$
$$\lim_{r \to \infty} \sup S_n \left[\inf (\mu_1)_{\beta \cdot (1-1/2r)}, \ldots, \inf (\mu_n)_{\beta \cdot (1-1/2r)}\right](x)\Big]$$

Proof

(i) Let $\alpha \in [0,1)$. For abbreviation define $\alpha_r \stackrel{\text{d}}{=} \alpha + (1-\alpha)/(2r)$ for $r \in \mathbb{N}$.
Lemma 8.37(ii) and Lemma 3.4(i) show that

$$(S_n[\mu_1,\ldots,\mu_n](x))_\alpha =$$

$$A_n \cap \bigcup_{r=1}^{\infty} [S_n\left[\sup(\mu_1)_{\alpha_r},\ldots,\sup(\mu_n)_{\alpha_r}\right](x),$$
$$S_n\left[\inf(\mu_1)_{\alpha_r},\ldots,\inf(\mu_n)_{\alpha_r}\right](x)]$$

holds.
$(\mu_i)_{\alpha_r} \subseteq (\mu_i)_{\alpha_{r+1}}$ is valid for all $i \in \{1,\ldots,n\}$ and $r \in \mathbb{N}$. Lemma 8.38 shows that

$$\left\{S_n\left[\inf(\mu_1)_{\alpha_r},\ldots,\inf(\mu_n)_{\alpha_r}\right](x)\right\}_{r\in\mathbb{N}}$$

is monotonously non-decreasing, and

$$\left\{S_n\left[\sup(\mu_1)_{\alpha_r},\ldots,\sup(\mu_n)_{\alpha_r}\right](x)\right\}_{r\in\mathbb{N}}$$

is monotonously non-increasing.
All elements of these two sequences are in the set A_n, which is a finite set. So both sequences are convergent, and there exists an $R \in \mathbb{N}$ such that

$$\lim_{r\to\infty} S_n\left[\inf(\mu_1)_{\alpha_r},\ldots,\inf(\mu_n)_{\alpha_r}\right](x) =$$
$$S_n\left[\inf(\mu_1)_{\alpha_s},\ldots,\inf(\mu_n)_{\alpha_s}\right](x)$$

and

$$\lim_{r\to\infty} S_n\left[\sup(\mu_1)_{\alpha_r},\ldots,\sup(\mu_n)_{\alpha_r}\right](x) =$$
$$S_n\left[\sup(\mu_1)_{\alpha_s},\ldots,\sup(\mu_n)_{\alpha_s}\right](x)$$

holds for all $s \geq R$. We can conclude

$$\bigcup_{s=1}^{\infty} \left[S_n\left[\sup(\mu_1)_{\alpha_s},\ldots,\sup(\mu_n)_{\alpha_s}\right](x), S_n\left[\inf(\mu_1)_{\alpha_s},\ldots,\inf(\mu_n)_{\alpha_s}\right](x)\right]$$

$$= \left[S_n\left[\sup(\mu_1)_{\alpha_R},\ldots,\sup(\mu_n)_{\alpha_R}\right](x),\right.$$
$$\left. S_n\left[\inf(\mu_1)_{\alpha_R},\ldots,\inf(\mu_n)_{\alpha_R}\right](x)\right] =$$
$$= \left[\lim_{r\to\infty} S_n\left[\sup(\mu_1)_{\alpha_r},\ldots,\sup(\mu_n)_{\alpha_r}\right](x),\right.$$
$$\left.\lim_{r\to\infty} S_n\left[\inf(\mu_1)_{\alpha_r},\ldots,\inf(\mu_n)_{\alpha_r}\right](x)\right].$$

So it remains to prove:

$$\lim_{r\to\infty} S_n\left[\sup(\mu_1)_{\alpha_r},\ldots,\sup(\mu_n)_{\alpha_r}\right](x) = S_n\left[\sup(\mu_1)_{\alpha},\ldots,\sup(\mu_n)_{\alpha}\right](x)$$

On the one hand, $i \in \{1,\ldots,n\}: \quad \sup(\mu_i)_{\alpha_r} \le \sup(\mu_i)_{\alpha}$ is valid for all $i \in \{1,\ldots,n\}$ and $r \in I\!N$, therefore Lemma 8.35 shows that

$$\lim_{r\to\infty} S_n\left[\sup(\mu_1)_{\alpha_r},\ldots,\sup(\mu_n)_{\alpha_r}\right](x) \ge S_n\left[\sup(\mu_1)_{\alpha},\ldots,\sup(\mu_n)_{\alpha}\right](x)$$

holds.
On the other hand, define

$$I_r \stackrel{\mathrm{d}}{=} \left\{i \in \{1,\ldots,n\} \,\middle|\, x \ge \sup(\mu_i)_{\alpha_r}\right\}$$

for $r \in I\!N$. $\{I_r\}_{r\in I\!N}$ is a monotonously non-increasing sequence of subsets of $\{1,\ldots,n\}$. Therefore we can find an $R \in I\!N$ such that $I_r = \bigcap_{s=1}^{\infty} I_s$ is valid for $r \ge R$.
For all $i \in \bigcap_{s=1}^{\infty} I_s$, $x \ge \sup(\mu_i)_{\alpha_r}$ is valid for all $r \ge R$. By Theorem 3.3 and Theorem 3.6, we conclude:

$$x \ge \sup(\mu_i)_{\alpha} \quad \text{for all } i \in \bigcap_{s=1}^{\infty} I_s.$$

It follows:

$$\begin{aligned}
&\lim_{r\to\infty} S_n\left[\sup(\mu_1)_{\alpha_r},\ldots\sup(\mu_n)_{\alpha_r}\right](x)\\
&= \frac{1}{n}\lim_{r\to\infty} \operatorname{card} I_r\\
&= \frac{1}{n}\operatorname{card} \bigcap_{s=1}^{\infty} I_s\\
&\le \frac{1}{n}\operatorname{card}\left\{i \in \{1,\ldots,n\} \mid x \ge \sup(\mu_i)_{\alpha}\right\}\\
&= S_n\left[\sup(\mu_1)_{\alpha},\ldots\sup(\mu_n)_{\alpha}\right](x).
\end{aligned}$$

(ii) Applying Lemma 8.37(ii) and Theorem 3.4(ii), and considering the fact that the intersection of compact subsets of $\mathbb{R}$ is compact, we obtain

$$\begin{aligned}
&(S_n[\mu_1,\ldots,\mu_n](x))_{\overline{\beta}} = \\
&\bigcap_{r=1}^{\infty}\{A_n\cap\Big[S_n\Big[\sup(\mu_1)_{\beta\cdot(1-1/2r)},\ldots,\sup(\mu_n)_{\beta\cdot(1-1/2r)}\Big](x), \\
&\qquad S_n\Big[\inf(\mu_1)_{\beta\cdot(1-1/2r)},\ldots,\inf(\mu_n)_{\beta\cdot(1-1/2r)}\Big](x)\Big]\Big\} = \\
&A_n\cap\bigcap_{r=1}^{\infty}\Big[S_n\Big[\sup(\mu_1)_{\beta\cdot(1-1/2r)},\ldots,\sup(\mu_n)_{\beta\cdot(1-1/2r)}\Big](x), \\
&\qquad S_n\Big[\inf(\mu_1)_{\beta\cdot(1-1/2r)},\ldots,\inf(\mu_n)_{\beta\cdot(1-1/2r)}\Big](x)\Big] = \\
&A_n\cap\Big[\lim_{r\to\infty} S_n\Big[\sup(\mu_1)_{\beta\cdot(1-1/2r)},\ldots,\sup(\mu_n)_{\beta\cdot(1-1/2r)}\Big](x), \\
&\qquad \lim_{r\to\infty} S_n\Big[\inf(\mu_1)_{\beta\cdot(1-1/2r)},\ldots,\inf(\mu_n)_{\beta\cdot(1-1/2r)}\Big](x)\Big]
\end{aligned}$$

In the preceding chapters we saw that the convex fuzzy sets faciliatate the calculations. In the following lemma we show how to get a set representation of the convex hull of the empirical distribution function. For practical application it often suffices to consider the convex hull. We give therefore an upper bound for the Hausdorff distance of the empirical distribution function and its convex hull.

Lemma 8.39 *Let $n \in \mathbb{N}$, $(\mu_1,\ldots,\mu_n) \in [Q(\mathbb{R})]^n$, and let $x \in \mathbb{R}$. Then*

(i)

$$\left\{\begin{array}{l}[S_n[\sup(\mu_1)_\alpha,\ldots,\sup(\mu_n)_\alpha](x), \\ \;S_n[\inf(\mu_1)_\alpha,\ldots,\inf(\mu_n)_\alpha](x)]\end{array} \;\middle|\; \alpha\in(0,1)\right\}$$

is a set representation of co $S_n[\mu_1,\ldots,\mu_n](x)$.

(ii)
$$d_\infty(\text{co } S_n[\mu_1,\ldots,\mu_n](x), S_n[\mu_1,\ldots,\mu_n](x)) \le \frac{1}{2n}.$$

Example 8.40 We continue with our tutorial Example 8.32. In order to determine the empirical distribution function, we have to calculate the α-cuts first. We obtain

$$\begin{aligned}(\mu_1)_\alpha &= (80 + 10 \cdot \alpha, 100], \\ (\mu_2)_\alpha &= \{85\}\ , \qquad\qquad \text{and} \\ (\mu_3)_\alpha &= (85 + 5 \cdot \alpha, 95 - 5 \cdot \alpha)\end{aligned}$$

for $\alpha \in [0,1)$.
Consider the case when the sample is $(\mu_1, \mu_2, \mu_3, \mu_2)$. It follows

$$\begin{aligned}&S_4[\sup(\mu_1)_\alpha, \sup(\mu_2)_\alpha, \sup(\mu_3)_\alpha, \sup(\mu_2)_\alpha](t) \\ &= S_4[100, 85, 95 - 5\alpha, 85](x) = \begin{cases} 0 & , \text{if } x < 85 \\ 0.5 & , \text{if } 85 \le x < 95 - 5\alpha \\ 0.75 & , \text{if } 95 - 5\alpha \le x < 100 \\ 1 & , \text{if } x \ge 100 \end{cases}\end{aligned}$$

and

$$\begin{aligned}&S_4[\inf(\mu_1)_\alpha, \inf(\mu_2)_\alpha, \inf(\mu_3)_\alpha, \inf(\mu_2)_\alpha](x) \\ &= S_4[80 + 10\alpha, 85, 85 + 5\alpha, 85](x) \\ &= \begin{cases} 0 & , \text{if } x < 85,\ \alpha \ge 0.5 \\ 0 & , \text{if } x < 80 + 10\alpha,\ \alpha < 0.5 \\ 0.25 & , \text{if } 80 + 10\alpha \le x < 85,\ \alpha < 0.5 \\ 0.5 & , \text{if } 85 \le x < 80 + 10\alpha,\ \alpha \ge 0.5 \\ 0.75 & , \text{if } 85 \le x < 85 + 5\alpha,\ \alpha < 0.5 \\ 0.75 & , \text{if } 80 + 10\alpha \le x < 85 + 5\alpha,\ \alpha \ge 0.5 \\ 1 & , \text{if } x \ge 85 + 5\alpha\ . \end{cases}\end{aligned}$$

The empirical distribution function can be calculated by Lemma 8.37 and Theorem 8.38. We obtain:

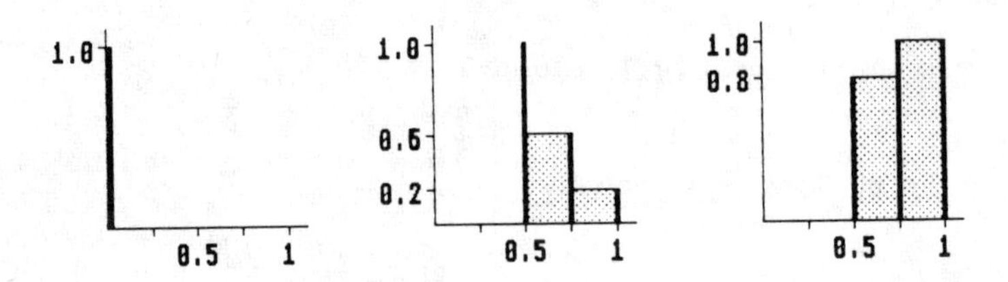

Fig. 8.3. The membership function of $S_4(\mu_1, \mu_2, \mu_3, \mu_2)(x)$.

a) $x = 80$ b) $x = 86$ c) $x = 91$.

The convex hull of $S_4(\mu_1, \mu_2, \mu_3, \mu_2)(x)$ is indicated by the shaded region. The distance between the empirical distribution function and its convex hull measured by d_∞ has the value 0 for $x = 80$ and 0.125 for $x = 86$ or $x = 91$.

If we have a (more realistic) sample with $n >> 10$ elements, then the Hausdorff distance is less than or equal to $1/2n$. For practical purposes it is often sufficient to consider the convex hull. Especially in the case when the fuzzy sets are in $C_n(I\!R)$, we can construct an efficient algorithm for the calculation of co $S_n\{\mu_1, \ldots, \mu_n\}(x)$.

9 Distribution Functions and iid sequences of random variables

In Section 7 we introduced the concept of a f.r.v. and noted that the concept of probability on the sample space was not included in this definition. Let $(\Omega, \mathfrak{A}, P)$ be a probability space, and let X be a f.r.v. defined on our probability space. The r.v. X induces a σ−algebra $\mathfrak{B}$ over $Q(\mathbb{R})$ by setting

$$A \in \mathfrak{B} \overset{\mathrm{d}}{\Leftrightarrow} X^{-1}A \overset{\mathrm{d}}{=} \{\omega \in \Omega \mid X(\omega) \in A\} \in \mathfrak{A}$$

and a probability measure Q on $(Q(\mathbb{R}), \mathfrak{B})$ by means of the correspondence

$$Q(A) \overset{\mathrm{d}}{=} P\left(X^{-1}(A)\right).$$

Since Q is a set function, and set functions are not easy to handle, let us introduce a point function called distribution function. There are several possibilities to define such a function. We restrict ourselves, for simplicity, to the case where the originals are in $\tilde{\chi}$, and define

Definition 9.1 *Let $X : \Omega \to Q(\mathbb{R})$ be a f.r.v. The distribution function (d.f.) of X is a mapping $\tilde{F}_X : \mathbb{R} \to F(\mathbb{R})$ defined by*

$$\left(\tilde{F}_X(x)\right)(p) \overset{\mathrm{d}}{=}$$
$$\begin{cases} \sup\{\tilde{\mu}_X(V) \mid V \in \tilde{\chi},\ (P \otimes P')[V \le x] = p\} & , \text{if } p \in [0,1] \\ 0 & , \text{if } p \in \mathbb{R} \setminus [0,1] \end{cases}$$

for $x \in \mathbb{R}$.

In many experiments an observation is expressible not as a single vague quantity, but as a family of several separate vague quantities. To describe such experiments mathematically, we must study multi-dimensional fuzzy distribution functions.

Definition 9.2 *Let $n \in \mathbb{N}$, let $(X_1, \ldots, X_n) : \Omega \to [Q(\mathbb{R})]^n$ be a f.r.vector. The (common) distribution function (d.f.) of $(X_1, \ldots, X_n)$ is a mapping*

$$\tilde{F}_{(X_1,\ldots,X_n)} : \mathbb{R}^n \to F(\mathbb{R}) \text{ with}$$

$$\left(\tilde{F}_{(X_1,\ldots,X_n)}(x_1, \ldots, x_n)\right)(p) \stackrel{\mathrm{d}}{=}$$

$$\begin{cases} \sup \left\{ \tilde{\mu}_{(X_1,\ldots,X_n)}(V_1, \ldots, V_n) \;\middle|\; \begin{array}{l} (V_1, \ldots, V_n) \in \tilde{\chi}^n, \\ (P \otimes P')(\bigcap_{i=1}^{n} [V_i \le x_i]) = p \end{array} \right\} \\ \qquad\qquad , \text{ if } p \in [0,1] \\ 0 \qquad\qquad , \text{ if } p \in \mathbb{R} \setminus [0,1] \end{cases}$$

for $(x_1, \ldots, x_n) \in \mathbb{R}^n$.

The connections between distribution functions and set representations are clarified in the following two theorems.

Lemma 9.3 *Let $n \in \mathbb{N}$, let $(X_1, \ldots, X_n)$ be a f.r.vector. Let $\{(A_i)_\alpha(\omega) \mid \omega \in \Omega, \alpha \in (0,1)\}$ be a measurable set representation of X_i for $i \in \{1, \ldots, n\}$. Let $(x_1, \ldots, x_n) \in \mathbb{R}^n$.*
Define $(\underline{A_i})_\alpha(\omega) \stackrel{\mathrm{d}}{=} \inf\, (A_i)_\alpha(\omega)$ and $(\overline{A_i})_\alpha(\omega) \stackrel{\mathrm{d}}{=} \sup\, (A_i)_\alpha(\omega)$ for $\omega \in \Omega, \alpha \in (0,1)$, $i \in \{1, \ldots, n\}$. Then

(i) $\left\{ \left[P\left(\bigcap_{i=1}^{n} [(\overline{A_i})_\alpha \le x_i] \right), P\left(\bigcap_{i=1}^{n} [(\underline{A_i})_\alpha \le x_i] \right) \right] \mid \alpha \in (0,1) \right\}$ *is set representation of $\tilde{F}_{(X_1,\ldots,X_n)}(x_1, \ldots, x_n)$.*

(ii) $\tilde{F}_{(X_1,\ldots,X_n)}(x_1, \ldots, x_n) \in U_C(\mathbb{R})$.

Proof

(i) Theorem 3.5 shows that it is sufficient to prove for all $\alpha \in (0,1)$

$$\left(\tilde{F}_{(X_1,\ldots,X_n)}(x_1, \ldots, x_n)\right)_\alpha$$
$$\subseteq \left[P\left(\bigcap_{i=1}^{n} \left[\overline{(A_i)_\alpha} \le x_i\right] \right), P\left(\bigcap_{i=1}^{n} \left[\underline{(A_i)_\alpha} \le x_i\right] \right) \right]$$
$$\subseteq \left(\tilde{F}_{(X_1,\ldots,X_n)}(x_1, \ldots, x_n)\right)_{\overline{\alpha}} .$$

Let $\alpha \in (0,1)$. We know that

$$((X_i)_\omega)_\alpha \subseteq (A_i)_\alpha(\omega) \subseteq ((X_i)_\omega)_{\overline{\alpha}}$$

is valid for $i \in \{1,\ldots,n\}$, $\omega \in \Omega$.
If $p \in \left(\tilde{F}_{(X_1,\ldots,X_n)}(x_1,\ldots,x_n)\right)_\alpha$, there exists a random vector

$$(V_1,\ldots,V_n)\epsilon\tilde{\chi}^n \quad \text{with} \quad \tilde{\mu}_{X_i}(V_i) > \alpha$$

for $i \in \{1,\ldots,n\}$ such that

$$(P \otimes P')\left(\bigcap_{i=1}^{n}[V_i \leq x_i]\right) = p\ .$$

For all $i \in \{1,\ldots,n\}$, $\omega \in \Omega$, $\omega' \in \Omega'$

$$V_i(\omega,\omega') \in ((X_i)_\omega)_\alpha \subseteq (A_i)_\alpha(\omega) \text{ and}$$
$$(\underline{A}_i)_\alpha(\omega) \leq V_i(\omega,\omega') \leq \left(\overline{A}_i\right)_\alpha(\omega) \text{ is valid.}$$

As $P \otimes P'$ is monotonous,

$$\begin{aligned} P\left(\bigcap_{i=1}^{n}[(\overline{A}_i)_\alpha \leq x_i]\right) &\leq p \\ &= (P \otimes P')\left(\bigcap_{i=1}^{n}[V_i \leq x_i]\right) \\ &\leq P\left(\bigcap_{i=1}^{n}[(\underline{A}_i)_\alpha \leq x_i]\right) \end{aligned}$$

holds.
Let

$$P\left(\bigcap_{i=1}^{n}[(\overline{A}_i)_\alpha \leq x_i]\right) \leq p \leq P\left(\bigcap_{i=1}^{n}[(\underline{A}_i)_\alpha \leq x_i]\right).$$

Choose $A_\lambda \in \mathfrak{A}$ with $P'(A_\lambda) = \lambda$ for $\lambda \in [0,1]$.

Define for $i \in \{1,\ldots,n\}$ and $\lambda \in [0,1]$ a r.v. $(V_i)_\lambda \in \tilde{\chi}$ by

$$(V_i)_\lambda(\omega,\omega') \stackrel{\mathrm{d}}{=}$$

$$= \begin{cases} (\overline{A}_i)_\alpha(\omega) & \text{, if } (\overline{A}_i)_\alpha(\omega) < +\infty \text{ and } \omega' \in A_\lambda \\ \max\{(\underline{A}_i)_\alpha(\omega), x_i+1\} & \text{, if } (\overline{A}_i)_\alpha(\omega) = +\infty \text{ and } (\underline{A}_i)_\alpha(\omega) > -\infty \\ & \quad \text{and } \omega' \in A_\lambda \\ x_i+1 & \text{, if } (\overline{A}_i)_\alpha(\omega) = +\infty \text{ and } (\underline{A}_i)_\alpha(\omega) = -\infty \\ & \quad \text{and } \omega' \in A_\lambda \\ (\underline{A}_i)_\alpha(\omega) & \text{, if } (\underline{A}_i)_\alpha(\omega) > -\infty \text{ and } \omega' \in \Omega'\backslash A_\lambda \\ \min\{(\overline{A}_i)_\alpha(\omega), x_i\} & \text{, if } (\underline{A}_i)_\alpha(\omega) = -\infty \text{ and } (\overline{A}_i)_\alpha(\omega) < +\infty \\ & \quad \text{and } \omega' \in \Omega'\backslash A_\lambda \\ x_i & \text{, if } (\underline{A}_i)_\alpha(\omega) = -\infty \text{ and } (\overline{A}_i)_\alpha(\omega) = +\infty \\ & \quad \text{and } \omega' \in \Omega'\backslash A_\lambda\,. \end{cases}$$

For all $i \in \{1,\ldots,n\}$, $\lambda \in [0,1]$, $\omega \in \Omega$, $\omega' \in \Omega'$

$$(V_i)_\lambda(\omega,\omega') \in (A_i)_\alpha(\omega) \subseteq ((X_i)_\omega)_{\overline{\alpha}}$$

is valid. This implies

$$\tilde{\mu}_{(X_1,\ldots,X_n)}((V_1)_\lambda,\ldots,(V_n)_\lambda) \geq \alpha \quad \text{for} \quad \lambda \in [0,1].$$

For $i \in \{1,\ldots,n\}$ and $\lambda \in [0,1]$ the equation

$$\{(\omega,\omega') \in \Omega \times \Omega' \mid (V_i)_\lambda(\omega,\omega') \leq x_i\} =$$

$$\{\omega \in \Omega \mid (\overline{A}_i)_\alpha(\omega) \leq x_i\} \times A_\lambda + \{\omega \in \Omega \mid (\underline{A}_i)_\alpha(\omega) \leq x_i\} \times (\Omega'\backslash A_\lambda)$$

can be obtained after a short calculation. "+" denotes the disjoint union. A continuous function $f : [0,1] \to [0,1]$ is defined by

$$f(\lambda) \stackrel{\mathrm{d}}{=} (P \otimes P')\left(\bigcap_{i=1}^{n} [(V_i)_\lambda \leq x_i]\right) =$$

$$\lambda P\left(\bigcap_{i=1}^{n} [(\overline{A}_i)_\alpha \leq x_i]\right) + (1-\lambda)P\left(\bigcap_{i=1}^{n} [(\underline{A}_i)_\alpha \leq x_i]\right)$$

for $\lambda \in [0,1]$. Obviously $f(0) \geq p$ and $f(1) \leq p$ holds. Applying the first mean value theorem of the analysis, we obtain:
There exists a $\lambda^* \in [0,1]$ with

$$p = f(\lambda^*) = (P \otimes P')\left(\bigcap_{i=1}^{n} [(V_i)_{\lambda^*} \leq x_i]\right).$$

So

$$p \in \left(\tilde{F}_{(X_1,\ldots,X_n)}(x_1,\ldots,x_n)\right)_{\overline{\alpha}}$$

holds.

(ii) follows immediately from (i). •

Theorem 9.4 *Let $n \in \mathbb{N}$, let $(X_1, \dots, X_n) : \Omega \to [Q(\mathbb{R})]^n$ be a f.r.vector, and let $(x_1, \dots, x_n) \in \mathbb{R}^n$.*

$$(\underline{Y}_i)_\alpha(\omega) \stackrel{\mathrm{d}}{=} \inf((X_i)_\omega)_\alpha \quad \text{and} \quad (\overline{Y}_i)_\alpha(\omega) \stackrel{\mathrm{d}}{=} \sup((X_i)_\omega)_\alpha$$

for $i \in \{1, \dots, n\}$, $\omega \in \Omega$, $\alpha \in [0,1)$.

(i) Let $\alpha \in [0,1)$. Then

$$\inf\left(\tilde{F}_{(X_1,\dots,X_n)}(x_1,\dots,x_n)\right)_\alpha = P\left(\bigcap_{i=1}^{n}\left[(\overline{Y_i})_\alpha \le x_i\right]\right)$$

and

$$\sup\left(\tilde{F}_{(X_1,\dots,X_n)}(x_1,\dots,x_n)\right)_\alpha = \lim_{r\to\infty} P\left(\bigcap_{i=1}^{n}\left[(\underline{Y_i})_{\alpha+(1-\alpha)/(2r)} \le x_i\right]\right)$$

holds.

(ii)
$$\left\{\left[P\left(\bigcap_{i=1}^{n}[(\overline{Y}_i)_\alpha \le x_i]\right), P\left(\bigcap_{i=1}^{n}[(\underline{Y}_i)_\alpha \le x_i]\right)\right] \mid \alpha \in (0,1)\right\}$$

is a set representation for $\tilde{F}_{(X_1,\dots,X_n)}(x_1,\dots,x_n)$.

Proof Let $\alpha \in [0,1)$, define $\alpha_r \stackrel{\mathrm{d}}{=} \alpha + (1-\alpha)/(2r)$ for $r \in \mathbb{N}$. Theorem 3.6 and Lemma 9.3(i) as well as Lemma 7.5 show that

$$\inf\left(\tilde{F}_{(X_1,\dots,X_n)}(x_1,\dots,x_n)\right)_\alpha = \lim_{r\to\infty} P\left(\bigcap_{i=1}^{n}\left[(\overline{A_i})_{\alpha_r} \le x_i\right]\right)$$

$$= P\left(\bigcap_{r=1}^{\infty}\bigcap_{i=1}^{n}\left[(\overline{A_i})_{\alpha_r}(\omega) \le x_i\right]\right) = P\left(\bigcap_{i=1}^{n}\left[(\overline{Y_i})_\alpha \le x_i\right]\right) \quad \text{holds.}$$

Since $\left\{\overline{(A_i)}_{\alpha_r}(\omega)\right\}_{r\in\mathbb{N}}$ is monotonously non-decreasing against $(\overline{Y_i})_\alpha(\omega)$ for all $i \in \{1, \dots, n\}$ and $\omega \in \Omega$,

$$\bigcap_{r=1}^{\infty}\bigcap_{i=1}^{n}\left[(\overline{A_i})_{\alpha_r} \le x_i\right] = \bigcap_{i=1}^{n}\bigcap_{r=1}^{\infty}\left[(\overline{A_i})_{\alpha_r} \le x_i\right] = \bigcap_{i=1}^{n}\left[(\overline{Y_i})_\alpha \le x_i\right]$$

is valid.
Applying Theorem 3.6 and Lemma 9.3(i), we obtain:

$$\sup\left(\tilde{F}_{(X_1,\ldots,X_n)}(x_1,\ldots,x_n)\right)_\alpha = \lim_{r\to\infty} P\left(\bigcap_{i=1}^{n}\left[\left(\underline{A_i}\right)_{\alpha_r} \leq x_i\right]\right).$$

Moreover,

$$\left(\underline{A_i}\right)_{\alpha_r}(\omega) \geq \left(\underline{Y_i}\right)_{\alpha_{r+1}}(\omega) \geq \left(\underline{A_i}\right)_{\alpha_{r+1}}(\omega)$$

holds for $\omega \in \Omega$, $i \in \{1,\ldots,n\}$, $r \in I\!N$, it follows:

$$\bigcap_{i=1}^{n}\left[\left(\underline{A_i}\right)_{\alpha_r} \leq x_i\right] \subseteq \bigcap_{i=1}^{n}\left[\left(\underline{Y_i}\right)_{\alpha_{r+1}} \leq x_i\right] \subseteq \bigcap_{i=1}^{n}\left[\left(\underline{A_i}\right)_{\alpha_r} \leq x_i\right]$$

for $r \in I\!N$. Therefore

$$\sup\left(\tilde{F}_{(X_1,\ldots,X_n)}(x_1,\ldots,x_n)\right)_\alpha = \lim_{r\to\infty} P\left(\bigcap_{i=1}^{n}\left[\left(\underline{A_i}\right)_{\alpha_r} \leq x_i\right]\right) \leq$$

$$\lim_{r\to\infty} P\left(\bigcap_{i=1}^{n}\left[\left(\underline{Y_i}\right)_{\alpha_r} \leq x_i\right]\right) \leq \lim_{r\to\infty} P\left(\bigcap_{i=1}^{n}\left[\left(\underline{A_i}\right)_{\alpha_r} \leq x_i\right]\right)$$

is valid, which shows assertion (i).
By Theorem 9.4(i), Theorem 3.4(i), Theorem 3.5, and Lemma 9.3(i), we know that

$$\begin{aligned}
&\left(\tilde{F}_{(X_1,\ldots,X_n)}(x_1,\ldots,x_n)\right)_\alpha \\
&\subseteq \left[P\left(\bigcap_{i=1}^{n}\left[\left(\overline{Y_i}\right)_{\alpha_r} \leq x_i\right]\right), \lim_{r\to\infty} P\left(\bigcap_{i=1}^{n}\left[\left(\underline{Y_i}\right)_{\alpha_r} \leq x_i\right]\right)\right] \\
&\subseteq \left[P\left(\bigcap_{i=1}^{n}\left[\left(\overline{Y_i}\right)_{\alpha_r} \leq x_i\right]\right), \lim_{r\to\infty} P\left(\bigcap_{i=1}^{n}\left[\left(\underline{Y_i}\right)_{\alpha_r} \leq x_i\right]\right)\right] \\
&\subseteq \left[P\left(\bigcap_{i=1}^{n}\left[\left(\overline{A_i}\right)_{\alpha_r} \leq x_i\right]\right), \lim_{r\to\infty} P\left(\bigcap_{i=1}^{n}\left[\left(\underline{A_i}\right)_{\alpha_r} \leq x_i\right]\right)\right] \\
&\subseteq \left(\tilde{F}_{(X_1,\ldots,X_n)}(x_1,\ldots,x_n)\right)_{\overline{\alpha}}
\end{aligned}$$

is valid for $\alpha \in (0,1)$, and therefore assertion (ii) follows by Theorem 3.5. ∎

Recall that the joint distribution function of a random vector uniquely determines the marginal distributions of the component random variables, but in general, knowledge of marginal distribution is not enough to determine the joint distribution. In the following we deal with a very special class of distribution in which the marginal distribution determine the joint distribution of a fuzzy random vector.

Definition 9.5 *Let $n \in \mathbb{N}$, let $X_i : \Omega \to Q(\mathbb{R})$ be a f.r.v. for all $i \in \{1,\dots,n\}$. $X_1,\dots,X_n$ is called (completely) independent if and only if*

$$\tilde{F}_{(X_1,\dots,X_n)}(x_1,\dots,x_n) = \prod_{i=1}^{n} \tilde{F}_{X_i}(x_i).$$

is valid for all $(x_1,\dots,x_n) \in \mathbb{R}^n$.

Definition 9.6 *Let $X,Y : \Omega \to Q(\mathbb{R})$ be two f.r.v.'s. X,Y are called (pairwise) independent if and only if*

$$\tilde{F}_{(X,Y)}(x,y) = \tilde{F}_X(x) \cdot \tilde{F}_Y(y)$$

is valid for all $(x,y) \in \mathbb{R}^2$.

The notion of independence is one of the most central ideas in stochastics. The independence of f.r.v.'s can be checked by considering ordinary random variables.

Theorem 9.7 *Let $n \in \mathbb{N}$ and let $X_i : \Omega \to Q(\mathbb{R})$ be a f.r.v. for all $i \in \{1,\dots,n\}$. Define*

$$(\underline{Y_i})_\alpha(\omega) \overset{\mathrm{d}}{=} \inf((X_i)_\omega)_\alpha \quad \text{and} \quad (\overline{Y_i})_\alpha(\omega) \overset{\mathrm{d}}{=} \sup((X_i)_\omega)_\alpha \ .$$

for $i \in \{1,\dots,n\}$, $\omega \in \Omega$, and $\alpha \in [0,1)$.
Then $X_1,\dots,X_n$ are (completely) independent if and only if on the one hand $(\underline{Y}_1)_\alpha,\dots,(\underline{Y}_n)_\alpha$, and on the other hand $(\overline{Y}_1)_\alpha,\dots,(\overline{Y}_n)_\alpha$ are completely independent for $\alpha \in [0,1)$.

Proof
(a) Let $(x_1,\dots,x_n) \in \mathbb{R}^n$. By Lemma 9.4(ii) we know that

$$\{[P((\overline{Y_i})_\alpha \le x_i), P((\underline{Y_i})_\alpha \le x_i)] \mid \alpha \in (0,1)\}$$

is a set representation of $\tilde{F}_{X_i}(x_i)$ for $i \in \{1,\dots,n\}$. As

$$0 \le P((\overline{Y_i})_\alpha \le x_i) \le P((\underline{Y_i})_\alpha \le x_i) \le 1$$

holds for $i \in \{1, \ldots, n\}$,

$$\left\{ \left[\prod_{i=1}^{n} P\left(\left(\overline{Y_i}\right)_\alpha \le x_i \right), \prod_{i=1}^{n} P\left(\left(\underline{Y_i}\right)_\alpha \le x_i \right) \right] \mid \alpha \in (0,1) \right\}$$

is a set representation of $\prod_{i=1}^{n} \tilde{F}_{X_i}(x_i)$, and $\prod_{i=1}^{n} \tilde{F}_{X_i}(x_i) \in U_C(I\!R)$ holds. Applying Theorem 3.6 we obtain for $\alpha \in [0,1)$:

$$\begin{aligned} \inf \left(\prod_{i=1}^{n} \tilde{F}_{X_i}(x_i) \right)_\alpha &= \lim_{r\to\infty} \left(\prod_{i=1}^{n} P\left((\overline{Y}_i)_{\alpha+(1-\alpha)/(2r)} \le x_i \right) \right) \\ &= \prod_{i=1}^{n} \lim_{r\to\infty} P\left((\overline{Y}_i)_{\alpha+(1-\alpha)/(2r)} \le x_i \right) \\ &= \prod_{i=1}^{n} P\left(\left(\overline{Y_i}\right)_\alpha \le x_i \right) \end{aligned}$$

as well as

$$\begin{aligned} \sup \left(\prod_{i=1}^{n} \tilde{F}_{X_i}(x_i) \right)_\alpha &= \lim_{r\to\infty} \left(\prod_{i=1}^{n} P\left((\underline{Y}_i)_{\alpha+(1-\alpha)/(2r)} \le x_i \right) \right) = \\ &= \prod_{i=1}^{n} \lim_{r\to\infty} P\left((\underline{Y}_i)_{\alpha+(1-\alpha)/(2r)} \le x_i \right). \end{aligned}$$

(b) Definition 9.5, Lemma 3.12 and step (a) of this proof as well as Theorem 9.4 show the following equivalences:

$$X_1, \ldots, X_n \text{ are completely independent}$$

$$\Leftrightarrow \left\{ \begin{array}{l} \forall (x_1, \ldots, x_n) \epsilon I\!R^n,\ \forall \alpha \in [0,1) : \\ \inf \left(\tilde{F}_{(X_1,\ldots,X_n)}(x_1, \ldots, x_n) \right)_\alpha = \inf \left(\prod_{i=1}^{n} \tilde{F}_{X_i} \right)_\alpha \quad \text{and} \\ \sup \left(\tilde{F}_{(X_1,\ldots,X_n)}(x_1, \ldots, x_n) \right)_\alpha = \sup \left(\prod_{i=1}^{n} \tilde{F}_{X_i} \right)_\alpha \end{array} \right.$$

$$
\Leftrightarrow \begin{cases} \forall (x_1,\ldots,x_n) \epsilon \mathbb{R}^n,\ \forall \alpha \in [0,1): \\ P\left(\bigcap_{i=1}^{n} [(\overline{Y}_i)_\alpha \leq x_i]\right) = \prod_{i=1}^{n} P\left([(\overline{Y}_i)_\alpha \leq x_i]\right) \quad \text{and} \\ \forall (x_1,\ldots,x_n) \epsilon \mathbb{R}^n,\ \forall \alpha \in [0,1) : \\ \lim_{r\to\infty} P\left(\bigcap_{i=1}^{n} \left[(\underline{Y_i})_{\alpha+(1-\alpha)/(2r)} \leq x_i\right]\right) = \\ = \lim_{r\to\infty} \prod_{i=1}^{n} P\left(\left[(\underline{Y_i})_{\alpha+(1-\alpha)/(2r)} \leq x_i\right]\right). \end{cases}
$$

The first proposition is obviously equivalent to the complete independence of $(\overline{Y}_1)_\alpha, \ldots, (\overline{Y}_n)_\alpha$ for all $\alpha \in [0,1)$. If $(\underline{Y}_1)_\alpha, \ldots, (\underline{Y}_n)_\alpha$ are completely independent for all $\alpha \in [0,1)$, the second proposition follows.

Let, on the other hand, the second proposition be valid. We want to show that

$$
P\left(\bigcap_{i=1}^{n} [(\underline{Y}_i)_\alpha < x_i]\right) = \prod_{i=1}^{n} P[(\underline{Y}_i)_\alpha < x_i]
$$

is valid for all $\alpha \in [0,1)$ and $(x_1,\ldots,x_n) \in \mathbb{R}^n$.

Let $\alpha \in [0,1)$ be arbitrary. Define $\alpha_r \stackrel{\mathrm{d}}{=} \alpha + (1-\alpha)/(2r)$.

$\left\{(\underline{Y_i})_{\alpha_r}(\omega)\right\}_{r\in\mathbb{N}}$ is monotonously non-increasing for $i \in \{1,\ldots,n\}$ and $\omega \in \Omega$. By a short calculation we can follow:

$$
\begin{aligned} &\bigcap_{i=1}^{n} \bigcup_{r=1}^{\infty} \bigcup_{k=1}^{\infty} \left[(\underline{Y_i})_{\alpha_r} \leq x_i - \frac{1}{k}\right] \\ = &\bigcup_{k=1}^{\infty} \bigcup_{r=1}^{\infty} \bigcap_{i=1}^{n} \left[(\underline{Y_i})_{\alpha_r} \leq x_i - \frac{1}{k}\right]. \end{aligned}
$$

By Lemma 7.5 and Theorem 3.6 we can conclude:

$$\bigcap_{i=1}^{n} \left[\left(\underline{Y_i}\right)_\alpha < x_i \right] = \Omega \setminus \bigcup_{i=1}^{n} \bigcap_{r=1}^{\infty} \left[\left(\underline{Y_i}\right)_{\alpha_r} \geq x_i \right]$$

$$= \Omega \setminus \bigcup_{i=1}^{n} \bigcap_{r=1}^{\infty} \bigcap_{k=1}^{\infty} \left[\left(\underline{Y_i}\right)_{\alpha_r} > x_i - \frac{1}{k} \right]$$

$$= \bigcap_{i=1}^{n} \bigcup_{r=1}^{\infty} \bigcup_{k=1}^{\infty} \left[\left(\underline{Y_i}\right)_{\alpha_r} \leq x_i - \frac{1}{k} \right]$$

$$= \bigcup_{k=1}^{\infty} \bigcup_{r=1}^{\infty} \bigcap_{i=1}^{n} \left[\left(\underline{Y_i}\right)_{\alpha_r} \leq x_i - \frac{1}{k} \right].$$

If we put $n = 1$, we obtain for all $i \in \{1, \ldots, n\}$:

$$\left[\left(\underline{Y_i}\right)_\alpha < x_i \right] = \bigcup_{r=1}^{\infty} \bigcup_{k=1}^{\infty} \left[\left(\underline{Y_i}\right)_{\alpha_r} \leq x_i - \frac{1}{k} \right].$$

If the second proposition is valid, we can follow:

$$\lim_{r \to \infty} P \left(\bigcap_{i=1}^{n} \left[\left(\underline{Y_i}\right)_{\alpha_r} \leq x_i - \frac{1}{k} \right] \right) = \prod_{i=1}^{n} \lim_{r \to \infty} P \left(\left(\underline{Y_i}\right)_{\alpha_r} \leq x_i - \frac{1}{k} \right).$$

Combining these three equations, we obtain by applying the monotony of the probability measure P:

$$P \left(\bigcap_{i=1}^{n} \left[\left(\underline{Y_i}\right)_\alpha < x_i \right] \right) = P \left(\bigcup_{k=1}^{\infty} \bigcup_{r=1}^{\infty} \bigcap_{i=1}^{n} \left[\left(\underline{Y_i}\right)_{\alpha_r} \leq x_i - \frac{1}{k} \right] \right)$$

$$= \lim_{k \to \infty} \lim_{r \to \infty} P \left(\bigcap_{i=1}^{n} \left[\left(\underline{Y_i}\right)_{\alpha_r} \leq x_i - \frac{1}{k} \right] \right)$$

$$= \lim_{k \to \infty} \prod_{i=1}^{n} \lim_{r \to \infty} P \left[\left(\underline{Y_i}\right)_{\alpha_r} \leq x_i - \frac{1}{k} \right]$$

$$= \prod_{i=1}^{n} \lim_{k \to \infty} \lim_{r \to \infty} P \left[\left(\underline{Y_i}\right)_{\alpha_r} \leq x_i - \frac{1}{k} \right]$$

$$= \prod_{i=1}^{n} P \left[\left(\underline{Y_i}\right)_\alpha \leq x_i \right]. \bullet$$

In a similar way to the proof of Theorem 9.7 we can show

Theorem 9.8 *Let $X_1, X_2 : \Omega \to Q(\mathbb{R})$ be two f.r.v.'s.*
Define

$$\left(\underline{Y_i}\right)_\alpha(\omega) \stackrel{\mathrm{d}}{=} \inf\left(\left(X_i\right)_\omega\right)_\alpha \quad \text{and} \quad \left(\overline{Y_i}\right)_\alpha(\omega) \stackrel{\mathrm{d}}{=} \sup\left(\left(X_i\right)_\omega\right)_\alpha$$

for $i = 1, 2$, $\omega \in \Omega$, and $\alpha \in [0,1)$.
Then X_1, X_2 are (pairwise) independent if and only if on the one hand $\left(\underline{Y}_1\right)_\alpha$ and $\left(\underline{Y}_2\right)_\alpha$, on the other hand $\left(\overline{Y}_1\right)_\alpha$ and $\left(\overline{Y}_2\right)_\alpha$ are pairwise independent for $\alpha \in [0,1)$.

Two f.r.v.'s are, from a stochastical view, identical, if they have the same distribution function.

Definition 9.9 *The two f.r.v.'s $X_1, X_2 : \Omega \to Q(\mathbb{R})$ are called identically distributed if*

$$\tilde{F}_{X_1}(x) = \tilde{F}_{X_2}(x)$$

holds for all $x \in \mathbb{R}$.

According to this definition, X_1 and X_2 are identically distributed, if and only if they have the same distribution function. It does not follow that $X_1 = X_2$ holds with probability 1.

Theorem 9.10 *Let $X_1, X_2 : \Omega \to Q(\mathbb{R})$ be two f.r.v.'s.*
Define $\left(\underline{Y_i}\right)_\alpha(\omega) \stackrel{\mathrm{d}}{=} \inf\left(\left(X_i\right)_\omega\right)_\alpha$ and $\left(\overline{Y_i}\right)_\alpha(\omega) \stackrel{\mathrm{d}}{=} \sup\left(\left(X_i\right)_\omega\right)_\alpha$ for $i = 1, 2$, $\omega \in \Omega$, and $\alpha \in [0,1)$.
Then X_1, X_2 are identically distributed if and only if on the one hand $\left(\underline{Y}_1\right)_\alpha$ and $\left(\underline{Y}_2\right)_\alpha$, on the other hand $\left(\overline{Y}_1\right)_\alpha$ and $\left(\overline{Y}_2\right)_\alpha$ are identically distributed for $\alpha \in [0,1)$.

An important notion in classical stochastics it that of an i.i.d. sequence. We define:

Definition 9.11 *Let $X_i : \Omega \to Q(\mathbb{R})$ be a f.r.v. for all $i \in \{1, \ldots, n\}$. $\{X_i\}_{i \in \mathbb{N}}$ is called i.i.d.-sequence with the common distribution function $\tilde{F}_X : \mathbb{R} \to U_C(\mathbb{R})$ if and only if*

(i) $\tilde{F}_{X_i} = \tilde{F}_X$ holds for all $i \in \mathbb{N}$, and
(ii) $X_1, \ldots, X_n$ are completely independent for all $n \in \mathbb{N}$.

A less severe notion only requires the pairwise indepence.

Definition 9.12 *The sequence $\{X_i\}_{i \in \mathbb{N}}$ of f.r.v.'s is called p.i.i.d.-sequence (pairwise independent and identically distributed sequence) with the common distribution function $\tilde{F}_X : \mathbb{R} \to U_C(\mathbb{R})$ if and only if*
(i) $\tilde{F}_{X_i} = \tilde{F}_X$ holds for all $i \in \mathbb{N}$ and
(ii) X_i and X_j are pairwise independent for all $(i,j) \in \mathbb{N}^2$ with $i \neq j$.

We immediately conclude from the preceding theorems:

Theorem 9.13 *Let $\{X_i\}_{i\in\mathbb{N}}$ be a sequence of f.r.v.'s.*

(i) $\{X_i\}_{i\in\mathbb{N}}$ is an i.i.d. sequence if and only if $\{(\underline{Y}_i)_\alpha\}_{i\in\mathbb{N}}$ and $\{(\overline{Y}_i)_\alpha\}_{i\in\mathbb{N}}$ are two (usual) i.i.d. sequences for all $\alpha \in [0,1)$.

(ii) $\{X_i\}_{i\in\mathbb{N}}$ is a p.i.i.d. sequence if and only if $\{(\underline{Y}_i)_\alpha\}_{i\in\mathbb{N}}$ and $\{(\overline{Y}_i)_\alpha\}_{i\in\mathbb{N}}$ are two (usual) p.i.i.d. sequences for all $\alpha \in [0,1)$.

A (usual) p.i.i.d. sequence is a sequence of pairwise independent and identically distributed (usual) random variables.

We conclude this section by defining the notion of independent and identically shaped sequences of finite f.r.v.s. This notion corresponds to the notion of independent and identically distributed sequences of f.r.v.'s, but it is easier to treat in the case of finite f.r.v.'s.

Definition 9.14 *Let $X_i : \Omega \to Q(\mathbb{R})$ be a finite f.r.v. for all $i \in \{1,\ldots,n\}$. $\{X_i\}_{i\in\mathbb{N}}$ is called p.i.i.s.-sequence (pairwise independent and identically shaped) with the common codomain $\{\mu_1,\ldots,\mu_n\} \subseteq Q(\mathbb{R})$ if and only if*

(i) $\{\mu_1,\ldots,\mu_n\}$ is codomain of X_i for all $i \in \{1,\ldots,n\}$, and

(ii) $\{Z_i\}_{i\in\mathbb{N}}$ is a (usual) p.i.i.d.-sequence where

$$(Z_i)(\omega) = j \overset{\mathrm{d}}{\Leftrightarrow} (X_i)_\omega = \mu_j$$

is defined for $\omega \in \Omega$, $i \in \mathbb{N}$, $j \in \{1,\ldots,n\}$.

10 Limit theorems

In order to make the theory of fuzzy random variables useful for the statistical analysis of vague data, it is necessary to derive limit theorems of the type of the strong law of large numbers, the central limit theorem, and the Gliwenko-Cantelli-Theorem.

10.1 STRONG LAW OF LARGE NUMBERS

This section is devoted to the investigation of sums $S_n \stackrel{\mathrm{d}}{=} \sum_{k=1}^{n} X_k$ of independent f.r.v.'s $X_1, X_2, \ldots$, and especially, of their convergence properties. Consider the simplest case when $P(X_i = 1) = p$ and $P(X_i = 0) = 1 - p$ for all $i \in I\!N$. Then the Borel strong law of large numbers states that S_n/n converges against p almost everywhere, i.e. there exist a set $N \in \mathfrak{B}$, $P(N) = 0$, such that for all $\omega \in \Omega \backslash N$ the property $\lim_{n \to \infty} S_n/n = p$ holds. The strong law of large numbers proves the intuitive fact that the relative frequency of an event converges against its probability. This Borel law is a special case of Komogorov's strong law of large numbers (see [19,31,52,183], e.g.) for usual random variables: If $\{X_i\}_{i \in \boldsymbol{N}}$ is an i.i.d.-sequence with $E|X_1| < \infty$ then S_n/n converges against $E\, X_1$ almost sure. N. Etemati [43] proved that we can replace "i.i.d." by "p.i.i.d.".

We can say that the arithmetic mean is a "reasonable" estimate for the expected value. Z. Artstein and R.A. Vitale [2] extended the strong law of large numbers to random sets (being compact set valued), and M.L. Puri and D.A. Ralescu [172] to normed spaces. We start our considerations by proving a very general result on random sets.

Theorem 10.1 *Let $(\Omega, \mathfrak{A}, P)$ be a probability space. Let $(V_i)(\omega)$ be a non-empty subset of $I\!R$ for $i \in I\!N$ and $\omega \in \Omega$. Define for $i \in I\!N$ and $\omega \in \Omega$:*

$$\left(\underline{V_i}\right)(\omega) \stackrel{\mathrm{d}}{=} \inf (V_i)(\omega) \quad \textit{and} \quad \left(\overline{V_i}\right)(\omega) \stackrel{\mathrm{d}}{=} \sup (V_i)(\omega).$$

If the four properties

(i) $(\underline{V_i})$ *and* $(\overline{V_i})$ *are* $\mathfrak{A} \longrightarrow \mathfrak{B}_1$*-measurable,*

(ii) $E\,|\,(\underline{V_i})\,| < +\infty$ *and* $E\,\left|(\overline{V_i})\right| < +\infty$,

(iii) $\{(\underline{V_i})\}_{i\in I\!N}$ *and* $\{(\overline{V_i})\}_{i\in I\!N}$ *fulfill the strong law of large numbers, i.e. there exist zero sets* N_1 *and* N_2 *such that*

$$\lim_{n\to\infty} \frac{1}{n} \sum_{i=1}^{n} \left[(\underline{V_i})(\omega) - E(\underline{V_i})\right] = 0 \quad \textit{for } \omega \in \Omega\backslash N_1 \quad \textit{and}$$

$$\lim_{n\to\infty} \frac{1}{n} \sum_{i=1}^{n} \left[(\overline{V_i})(\omega) - E(\overline{V_i})\right] = 0 \quad \textit{for } \omega \in \Omega\backslash N_2 \quad \textit{holds,}$$

and

(iv) *The sequence* $\left\{\frac{1}{n}\left[E(\overline{V_n}) - E(\underline{V_n})\right]\right\}_{n\in I\!N}$ *is convergent against 0*

are satisfied,
then there exists a zero set M *such that*

$$\lim_{n\to\infty} d_H \left(\frac{1}{n} \sum_{i=1}^{n} (V_i)(\omega), \left[\frac{1}{n} \sum_{i=1}^{n} E(\underline{V_i}), \frac{1}{n} \sum_{i=1}^{n} E(\overline{V_i}) \right] \right) = 0$$

holds for all $\omega \in \Omega\backslash N$ *where* $\frac{1}{n}\sum_{i=1}^{n}(V_i)(\omega)$ *is defined according to Lemma 4.5* *(see page 30).*

Proof Define $M \stackrel{d}{=} N_1 \cup N_2 \in \mathfrak{A}$, M is a zero set. Let $\omega \in \Omega\backslash M$ be arbitrary. Then $(\underline{V_i})(\omega) > -\infty$ and $(\overline{V_i})(\omega) < +\infty$ is valid for all $i \in I\!N$.

Let $\epsilon > 0$ be arbitrary. Choose $K_0 = K_0(\omega, \epsilon) \in I\!N$ such that for all $k \geq K_0$:

$$(\alpha) \qquad \left| \frac{1}{k} \sum_{i=1}^{k} \left[(\underline{V_i})(\omega) - E(\underline{V_i})\right] \right| < \frac{\epsilon}{6}$$

$$(\beta) \qquad \left| \frac{1}{k} \sum_{i=1}^{k} \left[(\overline{V_i})(\omega) - E(\overline{V_i})\right] \right| < \frac{\epsilon}{6}$$

$$(\gamma) \qquad \left|\frac{1}{k}\left[E\left(\overline{V_k}\right) - E\left(\underline{V_k}\right)\right]\right| < \frac{\epsilon}{6}$$

Then choose $N_0 = N_0(\omega, \epsilon) \in I\!N$ such that $N_0 > K_0$ and

$$(\delta) \qquad \frac{1}{N}\sum_{i=1}^{K_0}\left[E\left(\overline{V_i}\right) - E\left(\underline{V_i}\right)\right] < \frac{\epsilon}{6}$$

holds.
Let $n \in I\!N$ with $n \geq N_0$ be arbitrary. We want to show that

$$d_H\left(\frac{1}{n}\sum_{i=1}^{n}(V_i)(\omega), \left[\frac{1}{n}\sum_{i=1}^{n}E\left(\underline{V_i}\right), \frac{1}{n}\sum_{i=1}^{n}E\left(\overline{V_i}\right)\right]\right) < \epsilon$$

is valid.
Define for $k \in \{0, 1, \ldots, n\}$

$$a_k \stackrel{\mathrm{d}}{=} \frac{1}{n}\left\{\sum_{i=1}^{k}E\left(\overline{V_i}\right) + \sum_{i=k+1}^{n}E\left(\underline{V_i}\right)\right\}.$$

$a_0 = \frac{1}{n}\sum\limits_{i-1}^{n}E\left(\underline{V_i}\right)$ and $a_n = \frac{1}{n}\sum\limits_{i-1}^{n}E\left(\overline{V_i}\right)$ follows as well as

$$0 \leq a_k - a_{k-1} = \frac{1}{n}\left[E\ \left(\overline{V}_k\right) - E\ \left(\underline{V}_k\right)\right] \leq \frac{1}{k}\left[E\ \left(\overline{V}_k\right) - E\ \left(\underline{V}_k\right)\right] \leq \frac{\epsilon}{3}$$

for $k \in \{0, 1, \ldots, n\}$.
(A) We want to show that

$$\sup_{x \in \left[\frac{1}{n}\sum\limits_{i=1}^{n}E(\underline{V_i}), \frac{1}{n}\sum\limits_{i=1}^{n}E(\overline{V_i})\right]} \quad \inf_{y \in \frac{1}{n}\sum\limits_{i=1}^{n}(V_i)(\omega)} \quad |x - y| \ \leq \ \epsilon$$

is valid.
Let

$$x \in \left[\frac{1}{n}\sum_{i=1}^{n}E\left(\underline{V_i}\right), \frac{1}{n}\sum_{i=1}^{n}E\left(\overline{V_i}\right)\right]$$

be arbitrary. There exists a $k \in \{1, \ldots, n\}$ with $x \in [a_{k-1}, a_k]$. If $k \in \{K_0 + 1, \ldots, n\}$, then

$$
\begin{aligned}
&\left| \frac{1}{n} \sum_{i=1}^{k-1} (\overline{V_i})(\omega) + \frac{1}{n} \sum_{k=1}^{n} (\underline{V_i})(\omega) - a_{k-1} \right| \\
&= \left| \frac{1}{n} \sum_{i=1}^{k-1} (\overline{V_i})(\omega) - \frac{1}{n} \sum_{i=1}^{k-1} E(\overline{V_i}) + \frac{1}{n} \sum_{i=k}^{n} (\underline{V_i})(\omega) - \frac{1}{n} \sum_{i=k}^{n} E(\underline{V_i}) \right| \\
&\leq \frac{k-1}{n} \left| \frac{1}{k-1} \sum_{i=1}^{k-1} \left[(\overline{V_i})(\omega) - E(\overline{V_i}) \right] \right| + \\
&\quad + \left| \frac{1}{n} \sum_{i=1}^{n} \left[(\underline{V_i})(\omega) - E(\underline{V_i}) \right] - \frac{1}{n} \sum_{i=1}^{k-1} \left[(\underline{V_i})(\omega) - E(\underline{V_i}) \right] \right| \\
&\leq \left| \frac{1}{k-1} \sum_{i=1}^{k-1} \left[(\overline{V_i})(\omega) - E(\overline{V_i}) \right] \right| + \\
&\quad + \left| \frac{1}{n} \sum_{i=1}^{n} \left[(\underline{V_i})(\omega) - E(\underline{V_i}) \right] \right| + \left| \frac{1}{n} \sum_{i=1}^{k-1} \left[(\underline{V_i})(\omega) - E(\underline{V_i}) \right] \right| \\
&\leq 3 \frac{\epsilon}{6} = \frac{\epsilon}{2}.
\end{aligned}
$$

It follows

$$
\begin{aligned}
&\left| x - \frac{1}{n} \sum_{i=1}^{k-1} (\overline{V_i})(\omega) - \frac{1}{n} \sum_{i=k}^{n} (\underline{V_i})(\omega) \right| \\
&\leq |x - a_{k-1}| + \left| \frac{1}{n} \sum_{i=1}^{k-1} (\overline{V}_i)(\omega) + \frac{1}{n} \sum_{i=k}^{n} (\underline{V}_i)(\omega) \right| \\
&\leq |a_k - a_{k-1}| + \frac{\epsilon}{2} \leq \frac{5}{6} \epsilon.
\end{aligned}
$$

For all $i \in \{1, \ldots, n\}$ choose $u_i \in (V_i)(\omega)$ and $v_i \in (V_i)(\omega)$ such that

$$
u_i \leq (\underline{V_i})(\omega) + \frac{\epsilon}{6}
$$

and

$$
v_i \geq (\overline{V_i})(\omega) - \frac{\epsilon}{6}
$$

holds.

Define

$$
y = y(\omega, \epsilon, n) \overset{\mathrm{d}}{=} \frac{1}{n} \sum_{i=1}^{k-1} v_i + \frac{1}{n} \sum_{i=k}^{n} u_i \in \frac{1}{n} \sum_{i=1}^{n} (V_i)(\omega).
$$

$$\left| y - \frac{1}{n} \sum_{i=1}^{k-1} (\overline{V_i})(\omega) - \frac{1}{n} \sum_{i=k}^{n} (\underline{V_i})(\omega) \right| \leq \frac{\epsilon}{2}$$

is valid. All together we have $|x - y| \leq \epsilon$.

If $x \in [a_{k-1}, a_k]$ for a $k \in \{1, \ldots, K_0\}$, we have

$$\begin{aligned}
&\left| x - \frac{1}{n} \sum_{i=1}^{n} (\underline{V_i})(\omega) \right| \\
\leq &\left| x - \frac{1}{n} \sum_{i=1}^{n} E(\underline{V_i}) \right| + \left| \frac{1}{n} \sum_{i=1}^{n} \left[(\underline{V_i})(\omega) - E(\underline{V_i}) \right] \right| \\
\leq &\left| a_{K_0} - \frac{1}{n} \sum_{i=1}^{n} E(\underline{V_i}) \right| + \frac{\epsilon}{6} \\
= &\left| \frac{1}{n} \sum_{i=1}^{K_0} E(\overline{V_i}) - \frac{1}{n} \sum_{i=1}^{K_0} E(\underline{V_i}) \right| + \frac{\epsilon}{6} \leq \frac{\epsilon}{3}.
\end{aligned}$$

Choose for $i \in \{1, \ldots, n\}$ $u_i \in (V_i)(\omega)$ such that $u_i \leq (\underline{V}_i)(\omega) + \frac{\epsilon}{6}$ is valid. Define $y = y(\omega, \epsilon, n) \stackrel{\text{d}}{=} \frac{1}{n} \sum_{i=1}^{n} u_i$. It follows

$$|x - y| \leq \epsilon.$$

(B) Now we want to prove

$$\sup_{y \in \frac{1}{n} \sum_{i=1}^{n} (V_i)(\omega)} \quad \inf_{x \in \left[\frac{1}{n} \sum_{i=1}^{n} E(\underline{V_i}), \frac{1}{n} \sum_{i=1}^{n} E(\overline{V_i}) \right]} |x - y| \leq \epsilon.$$

Let $y \in \frac{1}{n} \sum_{i=1}^{n} (V_i)(\omega)$ be arbitrary. Then

$$y \in \left[\frac{1}{n} \sum_{i=1}^{n} (\underline{V_i})(\omega), \frac{1}{n} \sum_{i=1}^{n} (\overline{V_i})(\omega) \right]$$

follows.
Because of the inequations (α) and (β) of the beginning, we can follow:

$$y \in \left[\frac{1}{n} \sum_{i=1}^{n} E(\underline{V_i}) - \frac{\epsilon}{6}, \frac{1}{n} \sum_{i=1}^{n} E(\overline{V_i}) + \frac{\epsilon}{6} \right]$$

So there exists an

$$x \in \left[\frac{1}{n}\sum_{i=1}^{n} E\left(\underline{V_i}\right), \frac{1}{n}\sum_{i=1}^{n} E\left(\overline{V_i}\right)\right]$$

with

$$|x - y| \leq \epsilon.$$

(C) If we summarize part (A) and (B), we obtain

$$\forall\, \omega \in \Omega \backslash N,\ \forall\, \epsilon > 0\ \ \exists\, N = N(\omega, \epsilon)\ \forall\, n \geq N:$$

$$d_H\left(\frac{1}{n}\sum_{i=1}^{n}(V_i)(\omega), \left[\frac{1}{n}\sum_{i=1}^{n} E\left(\underline{V_i}\right), \frac{1}{n}\sum_{i=1}^{n} E\left(\overline{V_i}\right)\right]\right) \quad \leq \quad \epsilon. \bullet$$

The assumption (iv) in the Theorem 10.1 cannot be omitted. Otherwise the theorem may be wrong.
As an example define for $i \in \mathbb{N}$, $\omega \in \Omega$:

$$(V_i)(\omega) \stackrel{\mathrm{d}}{=} \begin{cases} \{0, i^2\} & \text{, if there is a } k \in \mathbb{N} \text{ with } i = 5^k \\ \{0\} & \text{, otherwise .} \end{cases}$$

Then $\left(\underline{V_i}\right)$ and $\left(\overline{V_i}\right)$ are one-point random variables for all $i \in \mathbb{N}$, i.e. $\left\{\left(\underline{V_i}\right)\right\}_{i \in \mathbb{N}}$ and $\left\{\left(\overline{V_i}\right)\right\}_{i \in \mathbb{N}}$ fulfil the strong law of large numbers.
For $i \in \mathbb{N}$

$$E\left(\underline{V_i}\right) = 0 \text{ and}$$

$$E\left(\overline{V_i}\right) = \begin{cases} i^2 & \text{, if } k \in \mathbb{N} \text{ with } i = 5^k \\ 0 & \text{, otherwise} \end{cases}$$

is valid. The assertion (iv) is broken, as

$$\left\{\frac{1}{5^k}\left[E\left(\overline{V_{5^k}}\right) - E\left(\underline{V_{5^k}}\right)\right]\right\}_{k \in \mathbb{N}} = \left\{5^k\right\}_{k \in \mathbb{N}} \stackrel{k \to \infty}{\to} \infty \text{ holds,}$$

i.e. we have a divergent subsequence. For all $\omega \in \Omega$ and $k \in \mathbb{N}$ the set

$$\frac{1}{5^k}\sum_{j=1}^{5^k}(V_j)(\omega) =$$

$$\left\{x \in \mathbb{R} \,\middle|\, \forall j \in \{1, \ldots, k\}\ \exists\, x_j \in \left\{0, 25^j\right\} \text{ with } x = \frac{1}{5^k}\sum_{j=1}^{k} x_j\right\}$$

has at the most 2^k elements, as the mapping

$$\begin{array}{llcl} \Phi : & \{0,1\}^k & \rightarrow & \frac{1}{5^k} \sum_{i=1}^{5^k} (V_j)(\omega) \\ & (l_1, \dots, l_k) & \mapsto & \frac{1}{5^k} \sum_{j=1}^{5^k} \left(25^i\right)^{l_i} \end{array}$$

is a surjection.

Let $\frac{1}{5^k} \sum_{j=1}^{5^k} (V_j)(\omega) = \left\{ x_0^{(k)}, \dots, x_{M_k}^{(k)} \right\}$ be valid such that $x_{l-1}^{(k)} < x_l^{(k)}$ is valid for all $k \in I\!N$ and $l \in \{1, \dots, M_k^{(k)}\}$. Then

$$x_0^{(k)} = 0 \text{ and } x_{M_k}^{(k)} = \frac{1}{5^k} \sum_{j=1}^{k} 25^j \geq 5^k$$

holds. Therefore for all $k \in I\!N$ there exists an $l_0 \in \{1, \dots, M_k^{(k)}\}$ with $x_{l_0}^{(k)} - x_{l_0}^{(k)} \geq \left(\frac{5}{2}\right)^k$.

$$\frac{1}{2} \left(x_{l_0}^{(k)} + x_{l_0}^{(k)} \right) \in \left[\frac{1}{5^k} \sum_{i=1}^{5^k} E\left(\underline{V_i}\right), \frac{1}{5^k} \sum_{i=1}^{5^k} E\left(\overline{V_i}\right) \right]$$

is valid as well as

$$\left| \frac{1}{2} \left(x_j^{(k)} + x_{j-1}^{(k)} \right) - y \right| \geq \frac{1}{2} \left(\frac{5}{2} \right)^k .$$

for all $y \in \frac{1}{5^k} \sum_{i=1}^{5^k} (V_i)(\omega)$ It follows

$$d_H \left[\frac{1}{n} \sum_{i=1}^{n} (V_i)(\omega), \left[\frac{1}{n} \sum_{i=1}^{n} E\left(\underline{V_i}\right), \frac{1}{n} \sum_{i=1}^{n} E\left(\overline{V_i}\right) \right] \right] \quad \overset{n \to \infty}{\rightarrow} \quad \infty.$$

This result on random sets can be generalized to fuzzy random variables. We distinguish between k odd and k even.

Theorem 10.2 *Let $k \in I\!N$ be odd. Let $X_i : \Omega \to Q(I\!R)$ be a f.r.v. for $i \in I\!N$. Define*

$$\left(\underline{Y_i}\right)_\alpha(\omega) \stackrel{\mathrm{d}}{=} \inf\left(\left(X_i\right)_\omega\right)_\alpha \text{ and } \left(\overline{Y_i}\right)_\alpha(\omega) \stackrel{\mathrm{d}}{=} \sup\left(\left(X_i\right)_\omega\right)_\alpha$$

for $\omega \in \Omega$, $\alpha \in [0,1)$, and $i \in I\!N$. If

(i) $E\left|\left(\underline{Y_i}\right)_0^k\right| < \infty$ and $E\left|\left(\overline{Y_i}\right)_0^k\right| < \infty$ holds for $i \in I\!N$, and

(ii) the sequences $\left\{\left(\underline{Y_i}\right)_\alpha{}^k\right\}_{i\in I\!N}$ and $\left\{\left(\overline{Y_i}\right)_\alpha{}^k\right\}_{i\in I\!N}$ fulfill the strong law of large numbers for all $\alpha \in [0,1)$, and

(iii) the sequence $\left\{\frac{1}{n}\left[E\left\{\left(\overline{Y_n}\right)_0^k\right\} - E\left\{\left(\underline{Y_n}\right)_0^k\right\}\right]\right\}_{n\in I\!N}$ is convergent against 0,

then there exists a zero set M such that

$$\lim_{n\to\infty} d_H\left(\left(\frac{1}{n}\sum_{i=1}^{n}\left(X_i\right)_\omega{}^k\right)_\alpha, \left(\frac{1}{n}\sum_{i=1}^{n} E\left\{\mathrm{co}\ \left(X_i\right)^k\right\}\right)_\alpha\right) = 0$$

is valid for all $\omega \in \Omega \backslash M$ and $\alpha \in [0,1) \cap Q\!\!\!\!I$.

Proof

(A) Theorem 3.3 and Lemma 4.5(ii) show that

$$\left[\left(\underline{Y_i}\right)_\alpha(\omega)\right]^k = \inf\left(\left(X_i\right)_\omega{}^k\right)_\alpha \text{ and } \left[\left(\overline{Y_i}\right)_\alpha(\omega)\right]^k = \sup\left(\left(X_i\right)_\omega{}^k\right)_\alpha$$

holds for $\omega \in \Omega$, $\alpha \in [0,1)$, and $i \in I\!N$.
For all $\omega \in \Omega$, $\alpha \in [0,1)$, and $i \in I\!N$

$$\left[\left(\underline{Y}_i\right)_0(\omega)\right]^k \le \left[\left(\underline{Y}_i\right)_\alpha(\omega)\right]^k \le \left[\left(\overline{Y}_i\right)_\alpha(\omega)\right]^k \le \left[\left(\overline{Y}_i\right)_0(\omega)\right]^k$$

is valid. This implies

$$E\left|\left(\underline{Y_i}\right)_\alpha{}^k\right| < \infty \text{ and } E\left|\left(\overline{Y_i}\right)_\alpha{}^k\right| < \infty$$

for $i \in I\!N$ and $\alpha \in [0,1)$ as well as

$$\left\{\frac{1}{n}\left[E\left\{\left(\overline{Y_n}\right)_\alpha{}^k\right\} - E\left\{\left(\underline{Y_n}\right)_\alpha{}^k\right\}\right]\right\}_{n\in I\!N} \stackrel{n\to\infty}{\longrightarrow} 0.$$

In order to apply the Theorem 10.1 for a fixed $\alpha \in [0,1)$, we define $(V_i)(\omega) \stackrel{\mathrm{d}}{=} \left((X_i)_\omega{}^k\right)_\alpha$ for $\omega \in \Omega$, $i \in I\!N$. We obtain the following: There exists a zero set $M_\alpha \in \mathfrak{A}$ such that

$$\lim_{n\to\infty} d_H\left(\left(\frac{1}{n}\sum_{i=1}^{n}(X_i)_\omega{}^k\right)_\alpha, \left[\frac{1}{n}\sum_{i=1}^{n}E\left(\underline{Y_i}\right)_\alpha{}^k, \frac{1}{n}\sum_{i=1}^{n}E\left(\overline{Y_i}\right)_\alpha{}^k\right]\right) = 0$$

holds for all $\omega \in \Omega \backslash M_\alpha$.

(B) Lemma 4.5(ii) shows that

$$\left(\frac{1}{n}\sum_{i=1}^{n}(X_i)_\omega{}^k\right)_\alpha = \frac{1}{n}\sum_{i=1}^{n}\left((X_i)_\omega{}^k\right)_\alpha$$

is valid for all $\omega \in \Omega$, $\alpha \in [0,1)$, and $n \in I\!N$.
Define the zero set

$$M \stackrel{\mathrm{d}}{=} \bigcup_{\alpha\in[0,1)\cap Q} M_\alpha \in \mathfrak{A}.$$

For all $\omega \in \Omega \backslash M$ and $\alpha \in [0,1) \cap Q$:

$$\lim_{n\to\infty} d_H\left(\left(\frac{1}{n}\sum_{i=1}^{n}(X_i)_\omega{}^k\right)_\alpha, \left[\frac{1}{n}\sum_{i=1}^{n}E\left(\underline{Y_i}\right)_\alpha{}^k, \frac{1}{n}\sum_{i=1}^{n}E\left(\overline{Y_i}\right)_\alpha{}^k\right]\right) = 0$$

holds.

(C) Define

$$Z_n \stackrel{\mathrm{d}}{=} \frac{1}{n}\sum_{i=1}^{n}(X_i)^k$$

for $n \in I\!N$. Theorem 7.4 shows that Z_n is a f.r.v. for all $n \in I\!N$. By Theorem 8.3(iii) and Theorem 8.4, we know that $(E \text{ co } Z_n)_\alpha$ is convex, and that

$$\inf(E \text{ co } Z_n)_\alpha = E\left(\underline{Z_n}\right)_\alpha \quad \text{and} \quad \sup(E \text{ co } Z_n)_\alpha = E\left(\overline{Z_n}\right)_\alpha$$

holds for $\alpha \in (0,1)$. Furthermore we conclude with the help of

Lemma 4.5(ii):

$$\left(\underline{Z_n}\right)_\alpha(\omega) \stackrel{\mathrm{d}}{=} \inf\left(\left(Z_n\right)_\omega\right)_\alpha = \inf\left(\frac{1}{n}\sum_{i=1}^{n}\left(X_i\right)_\omega{}^k\right)_\alpha$$

$$= \inf\left(\frac{1}{n}\sum_{i=1}^{n}\left(\left(X_i\right)_\omega{}^k\right)_\alpha\right)$$

$$= \frac{1}{n}\sum_{i=1}^{n}\inf\left(\left(X_i\right)_\omega{}^k\right)_\alpha = \frac{1}{n}\sum_{i=1}^{n}\left(\underline{Y_i}\right)_\alpha{}^k(\omega)$$

and

$$\left(\overline{Z_n}\right)_\alpha(\omega) \stackrel{\mathrm{d}}{=} \sup\left(\left(Z_n\right)_\omega\right)_\alpha = \frac{1}{n}\sum_{i=1}^{n}\left(\overline{Y_i}\right)_\alpha{}^k(\omega)$$

for $\omega \in \Omega$, $\alpha \in [0,1)$, and $n \in I\!N$. By Lemma 6.2 we conclude:

$$0 = d_H\left(\left(E\,\mathrm{co}\;Z_n\right)_\alpha, \left[\inf\left(E\,\mathrm{co}\;Z_n\right)_\alpha, \sup\left(E\,\mathrm{co}\;Z_n\right)_\alpha\right]\right) =$$

$$= d_H\left(\left(E\left\{\frac{1}{n}\sum_{i=1}^{n}\left(X_i\right)^k\right\}\right)_\alpha, \left[\frac{1}{n}\sum_{i=1}^{n}E\left(\underline{Y_i}\right)_\alpha{}^k, \frac{1}{n}\sum_{i=1}^{n}E\left(\overline{Y_i}\right)_\alpha{}^k\right]\right)$$

(D) Part (B) and the last equation yield:
For all $\omega \in \Omega\backslash M$ and $\alpha \in [0,1) \cap Q$

$$\lim_{n\to\infty} d_H\left(\left(E\left\{\frac{1}{n}\sum_{i=1}^{n}\left(X_i\right)^k\right\}\right)_\alpha, \left[\frac{1}{n}\sum_{i=1}^{n}E\left(\underline{Y_i}\right)_\alpha{}^k, \frac{1}{n}\sum_{i=1}^{n}E\left(\overline{Y_i}\right)_\alpha{}^k\right]\right)$$

$$= 0. \bullet$$

The analogous theorem for the case "k even" states:

Theorem 10.3 *Let $k \in I\!N$ be even. Let $X_i : \Omega \to F(I\!R)$ be a mapping for all $i \in I\!N$ such that $|X_i|$ is a f.r.v. Define*

$$\left(\underline{Z_i}\right)_\alpha(\omega) \stackrel{\mathrm{d}}{=} \inf\left(\left|X_i\right|_\omega\right)_\alpha \quad \text{and} \quad \left(\overline{Z_i}\right)_\alpha(\omega) \stackrel{\mathrm{d}}{=} \sup\left(\left|X_i\right|_\omega\right)_\alpha \ .$$

for $\omega \in \Omega$, $\alpha \in [0,1)$, and $i \in I\!N$. If *(i)* $E\left(\overline{Z_i}\right)_0^k < \infty$ *holds for all* $i \in \mathbb{N}$,

(ii) $\left\{\left(\underline{Z_i}\right)_\alpha{}^k\right\}_{i\in\mathbb{N}}$ *and* $\left\{\left(\underline{Z_i}\right)_\alpha{}^k\right\}_{i\in\mathbb{N}}$ *fulfil the strong law of large numbers for all* $\alpha \in [0,1)$,

and

(iii) The sequence $\left\{\frac{1}{n}\left[E\left\{\left(\overline{Z_n}\right)_0^k\right\} - E\left\{\left(\underline{Z_n}\right)_0^k\right\}\right]\right\}_{n\in\mathbb{N}}$ *is convergent against* 0,

then there is a zero set M *such that*

$$\lim_{n\to\infty} d_H\left(\left(\frac{1}{n}\sum_{i=1}^{n}(X_i)_\omega{}^k\right)_\alpha, \left(\frac{1}{n}\sum_{i=1}^{n}E\left\{(X_i)^k\right\}\right)_\alpha\right) = 0$$

is valid for all $\omega \in \Omega\backslash M$ *and* $\alpha \in [0,1) \cap \mathbb{Q}$.

Proof For all $i \in \mathbb{N}$, $\omega \in \Omega$, $\alpha \in [0,1)$

$$0 \le \left(\underline{Z_i}\right)_0(\omega) \le \left(\underline{Z_i}\right)_\alpha(\omega) \le \left(\overline{Z_i}\right)_\alpha(\omega) \le \left(\overline{Z_i}\right)_0(\omega)$$

holds. This implies

$$E\left(\underline{Z_i}\right)_\alpha{}^k < \infty \text{ and } E\left(\overline{Z_i}\right)_\alpha{}^k < \infty$$

for $i \in \mathbb{N}$ and $\alpha \in [0,1)$. We can apply Theorem 10.1 on $\{|X_i|^k\}_{i\in\mathbb{N}}$ and obtain the following:
For all $\alpha \in (0,1)$ there exists a zero set M_α such that for all $\omega \in \Omega\backslash M_\alpha$:

$$\lim_{n\to\infty} d_H\left(\left(\frac{1}{n}\sum_{i=1}^{n}(X_i)_\omega{}^k\right)_\alpha, \left[\frac{1}{n}\sum_{i=1}^{n}E\left\{\left(\underline{Z_i}\right)_\alpha{}^k\right\}, \frac{1}{n}\sum_{i=1}^{n}E\left\{\left(\overline{Z_i}\right)_\alpha{}^k\right\}\right]\right) = 0.$$

Define

$$M \stackrel{\text{d}}{=} \bigcup_{\alpha\in[0,1)\cap\mathbb{Q}} M_\alpha .$$

Similar to the proof of Theorem 10.2, we can show:

$$\lim_{n\to\infty} d_H\left(\left(\frac{1}{n}\sum_{i=1}^{n}|X_i|_\omega^k\right)_\alpha, \left(\frac{1}{n}\sum_{i=1}^{n}E\left\{\text{co } |X_i|^k\right\}\right)_\alpha\right) = 0.$$

is valid for $\omega \in \Omega\backslash M$ and $\alpha \in [0,1) \cap \mathbb{Q}$.
For all $i \in \mathbb{N}$ and $\omega \in \Omega$

$$\left(|X_i|_\omega\right)^k = \left((X_i)_\omega\right)^k ,$$

is valid, this implies

$$E\left\{\text{co } |X_i|^k\right\} = E\left\{\text{co } (X_i)^k\right\}$$

So the proof is completed. •

For the most interesting case of p.i.i.d.-sequences, we have shown the following result:

Theorem 10.4 *Let $(\Omega, \mathfrak{A}, P)$ be a probability space and $X_i : \Omega \to F(\mathbb{R})$ be a mapping for all $i \in \mathbb{N}$. Let $k \in \mathbb{N}$.*

(i) If k is odd, $\{X_i\}_{i\in\mathbb{N}}$ is a p.i.i.d. sequence of f.r.v.'s, and $E\left|\left(\underline{Y_1}\right)_0\right| < \infty$ and $E\left|\left(\overline{Y_1}\right)_0\right| < \infty$, then there exists a zero set M such that

$$\left\{\frac{1}{n}\sum_{i=1}^{n} (X_i)_\omega{}^k\right\}_{n\in\mathbb{N}} \overset{d_H,Q}{\longrightarrow} E\left(\text{co } X_1^k\right)$$

is valid for all $\omega \in \Omega\backslash M$.

(ii) If k is even, $\{X_i\}_{i\in\mathbb{N}}$ is a p.i.i.d. sequence of f.r.v.'s, and $E\left|\left(\overline{Z_1}\right)_0\right| < \infty$, then there exists a zero set M such that

$$\left\{\frac{1}{n}\sum_{i=1}^{n} (X_i)_\omega\right\}_{n\in\mathbb{N}} \overset{d_H,Q}{\longrightarrow} E\left(\text{co } X_1^k\right)$$

is valid for all $\omega \in \Omega\backslash M$.

Proof
A usual p.i.i.d. sequence $\{Y_i\}_{i\in\mathbb{N}}$ with existing expected value fulfills the strong law of large numbers as N. Etemadi [35] has shown. We can apply Theorem 10.2 and Theorem 10.3. Note that we obtain an estimate for $E\left(\text{co } X_1^k\right)$ as well as for $\tilde{E}\, X_1^k = E\left(\text{co } X_1^k\right)$.
For the special case $k = 1$, we prove that the arithmetic mean converges also in the stronger convergence with respect to d_∞. •

Theorem 10.5 *Let $(\Omega, \mathfrak{A}, P)$ be a probability space. Let $\{X_i\}_{i\in\mathbb{N}}$ be a p.i.i.d. sequence on $(\Omega, \mathfrak{A}, P)$. Define*

$$\left(\underline{Y_i}\right)_\alpha(\omega) \overset{\text{d}}{=} \inf\left(\left(X_i\right)_\omega\right)_\alpha \quad \text{and} \quad \left(\overline{Y_i}\right)_\alpha(\omega) \overset{\text{d}}{=} \sup\left(\left(X_i\right)_\omega\right)_\alpha$$

for $\omega \in \Omega$, $\alpha \in [0,1)$, and $i \in \mathbb{N}$. If

(i) $E\left|\left(\underline{Y_1}\right)_0\right| < \infty$ and $E\left|\left(\overline{Y_1}\right)_0\right| < \infty$ is valid, and

(ii) for all $i \in \mathbb{N}$ there exists a zero set M_i such that $(X_i)_\omega \in F_C(\mathbb{R})$ holds for all $\omega \in \Omega \setminus M_i$,

then we can find a zero set N such that

$$\left\{ \frac{1}{n} \sum_{i=1}^{n} (X_i)_\omega \right\}_{n \in \mathbb{N}} \overset{d_\infty}{\rightarrow} EX_1$$

for all $\omega \in \Omega \setminus N$.

Proof

(a) Let A, A_1, A_2 and B, B_1, B_2 be six nonempty subsets of $\mathbb{R}$. We want to show that $A_1 \subseteq A \subseteq A_2$ and $B_1 \subseteq B \subseteq B_2$ implies:

$$d_H(A, B) \leq \max\{d_H[A_1, B_1], d_H[A_2, B_2]\} + d_H[B_1, B_2].$$

Let $a \in A$ be arbitrary. Then

$$\inf_{b \in B} |a - b| \leq \inf_{b_1 \in B_1} |a - b_1|$$

is valid. For all $b_1 \in B_1$, $b_2 \in B_2$, we have

$$|a - b_1| \leq |a - b_2| + |b_1 - b_2|.$$

It follows

$$\begin{aligned} &\inf_{b_1 \in B_1} |a - b_1| \\ &\leq |a - b_2| + \inf_{b_1 \in B_1} |b_1 - b_2| \\ &\leq |a - b_2| + \sup_{b_2 \in B_2} \inf_{b_1 \in B_1} |b_1 - b_2|. \end{aligned}$$

As this inequality holds for all $b_2 \in B_2$, we obtain:

$$\inf_{b_1 \in B_1} |a - b_1| \leq \inf_{b_2 \in B_2} |a - b_2| + \sup_{b_2 \in B_2} \inf_{b_1 \in B_1} |b_1 - b_2|.$$

We can conclude:

$$\inf_{b \in B} |a - b| \leq \inf_{b_1 \in B_1} |a - b_1| \leq \inf_{b_2 \in B_2} |a - b_2| + d_H[B_1, B_2].$$

This is valid for all $a \in A$, so it follows:

$$\begin{aligned} \sup_{a \in A} \inf_{b \in B} |a - b| &\leq \sup_{a \in A} \inf_{b_2 \in B_2} |a - b_2| + d_H[B_1, B_2] \\ &\leq \sup_{a_2 \in A_2} \inf_{b_2 \in B_2} |a_2 - b_2| + d_H[B_1, B_2] \\ &\leq d_H[A_2, B_2] + d_H[B_1, B_2] \\ &\leq \max\{d_H[A_1, B_1], d_H[A_2, B_2]\} + d_H[B_1, B_2]. \end{aligned}$$

In a similar way, it is possible to show:

$$\sup_{b\in B}\inf_{a\in A}|a-b|$$
$$\leq d_H[A_1,B_1]+d_H[B_1,B_2]\leq$$
$$\max\{d_H[A_1,B_1],d_H[A_2,B_2]\}+d_H[B_1,B_2].$$

Combining these two inequalities, we obtain the assertion.

(b) Define $M_0 \overset{\text{d}}{=} \bigcup_{i=1}^{\infty} M_i$; M_0 is a zero set. Define

$$\left(\underline{X_i}\right)_\alpha(\omega) \overset{\text{d}}{=} \inf\left(\left(X_i\right)_\omega\right)_{\overline{\alpha}} \quad \text{and} \quad \left(\overline{X_i}\right)_\alpha(\omega) \overset{\text{d}}{=} \sup\left(\left(X_i\right)_\omega\right)_{\overline{\alpha}}.$$

for $\omega \in \Omega$, $\alpha \in (0,1]$,$i \in \mathbb{N}$.
Let $\beta \in (0,1]$ be arbitrary, define $\beta_r \overset{\text{d}}{=} \beta\cdot(1-1/2r)$ for $r \in \mathbb{N}$.
Lemma 3.17 shows that for all $\omega \in \Omega$, $i \in \mathbb{N}$ $\left\{(\underline{Y}_i)_{\beta_r}\right\}_{r\in\mathbb{N}}$ is monotonously non-decreasing against $(\underline{X}_i)_\beta(\omega)$, and $\left\{\left(\overline{Y}_i\right)_{\beta_r}\right\}_{r\in\mathbb{N}}$ is monotonously non-increasing against $\left(\overline{X}_i\right)_\beta(\omega)$.
By Lemma 7.5 we obtain for $i \in \mathbb{N}$, $x \in \mathbb{R}$:

$$\left\{\omega\in\Omega\backslash M_0 \,\middle|\, \left(\underline{X_i}\right)_\beta(\omega)\leq x\right\} = \bigcap_{r=1}^{\infty}\left\{\omega\in\Omega\backslash M_0 \,\middle|\, \left(\underline{X_i}\right)_{\beta_r}(\omega)\leq x\right\}$$

and

$$\left\{\omega\in\Omega\backslash M_0 \,\middle|\, \left(\overline{X_i}\right)_\beta(\omega)\geq x\right\} = \bigcap_{r=1}^{\infty}\left\{\omega\in\Omega\backslash M_0 \,\middle|\, \left(\overline{X_i}\right)_{\beta_r}(\omega)\geq x\right\}.$$

Therefore $\left(\overline{X_i}\right)_\beta$ and $\left(\underline{X_i}\right)_\beta$ are measurable for all $i \in \mathbb{N}$.
We also obtain from the monotony mentioned above that $\left\{\left(\underline{X_i}\right)_\beta\right\}_{i\in\mathbb{N}}$ and $\left\{\left(\overline{X_i}\right)_\beta\right\}_{i\in\mathbb{N}}$ are i.i.d. sequences.

(c) Let $\alpha \in [0,1)$ be arbitrary.
$\left\{\left(\underline{Y_i}\right)_\alpha\right\}_{i\in\mathbb{N}}$ and $\left\{\left(\overline{Y_i}\right)_\alpha\right\}_{i\in\mathbb{N}}$ are p.i.i.d. sequences with existing expected values. We can apply N. Etemadi's [43] strong law of large numbers without assuming the existence of the variances of i.i.d. sequences (see [35]), and can conclude that $\left\{\left(\underline{Y_i}\right)_\alpha\right\}_{i\in\mathbb{N}}$ and $\left\{\left(\overline{Y_i}\right)_\alpha\right\}_{i\in\mathbb{N}}$ fulfill

the strong law of large numbers. Therefore Theorem 10.1 shows that there exists a zero set $M_1(\alpha)$ such that

$$\lim_{n\to\infty} d_H\left[\left(\frac{1}{n}\sum_{i=1}^{n}(X_i)_\omega\right)_\alpha, (E \text{ co } X_1)_\alpha\right] = 0$$

is valid for all $\omega \in \Omega\backslash M_1(\alpha)$ as

$$\text{cl } (E \text{ co } X_1)_\alpha = \left[E \left(\underline{Y_i}\right)_\alpha, E \left(\overline{Y_i}\right)_\alpha\right]$$

holds.
Let $\beta \in (0,1]$ be arbitrary.
We also can apply Theorem 10.1 to the sequence $\{(V_i)\}_{i\in I\!N}$ with $(V_i)(\omega) \stackrel{\text{d}}{=} ((X_i)_\omega)_{\overline{\alpha}}$ for $\omega \in \Omega$, $i \in I\!N$. As $\left\{\left(\underline{X_i}\right)_\beta\right\}_{i\in I\!N}$ and $\left\{\left(\overline{X_i}\right)_\beta\right\}_{i\in I\!N}$ are p.i.i.d. sequences with existing expected values, there exists a zero set $M_2(\beta)$ with $M_0 \subseteq M_2(\beta)$ such that

$$\lim_{n\to\infty} d_H\left[\frac{1}{n}\sum_{i=1}^{n}((X_i)_\omega)_{\overline{\beta}}, \left[E\left(\underline{X_i}\right)_\beta, E\left(\overline{X_i}\right)_\beta\right]\right] = 0$$

is valid for all $\omega \in \Omega\backslash M_2(\beta)$.
Since $(E \text{ co } X)_{\overline{\beta}}$ is closed and convex, by applying Lebesgue's theorem of the dominated convergence (see [81,82,84],e.g.) on Lemma 3.17(ii), and utilizing Theorem 3.3 and Theorem 8.3, we obtain

$$\inf (E \text{ co } X_1)_{\overline{\beta}} = \lim_{r\to\infty} E \left(\underline{Y}_1\right)_{\beta\cdot(1-1/2r)}$$
$$= E\left[\lim_{r\to\infty} E \left(\underline{Y}_1\right)_{\beta\cdot(1-1/2r)}\right] = E \left(\underline{X}_1\right)_\beta$$

and

$$\sup (E \text{ co } X_1)_{\overline{\beta}} = E \left(\overline{X_1}\right)_\beta.$$

This implies

$$(E \text{ co } X_1)_{\overline{\beta}} = \left[E \left(\underline{X}_1\right)_\beta, E \left(\overline{X_1}\right)_\beta\right].$$

Lemma 3.15 shows that for all $\omega \in \Omega\backslash M_0$ and $n \in I\!N$

$$\left(\frac{1}{n}\sum_{i=1}^{n}(X_i)_\omega\right)_{\overline{\beta}} = \frac{1}{n}\sum_{i=1}^{n}((X_i)_\omega)_{\overline{\beta}}$$

holds. Therefore we have for all $\omega \in \Omega \backslash M_2(\beta)$:

$$\lim_{n\to\infty} d_H \left[\left(\frac{1}{n} \sum_{i=1}^{n} (X_i)_\omega \right)_{\overline{\beta}}, (E \text{ co } X)_{\overline{\beta}} \right] = 0.$$

(d) Let $R \in \mathbb{N}$ be arbitrary. Define

$$u_j^{(R)} \stackrel{\text{d}}{=} E\left(\underline{Y_1}\right)_0 + \frac{j}{R}\left[E\left(\underline{X_1}\right)_{\overline{1}} - E\left(\underline{Y_1}\right)_0\right] \quad \text{and}$$
$$v_j^{(R)} \stackrel{\text{d}}{=} E\left(\overline{X_1}\right)_1 + \frac{j}{R}\left[E\left(\overline{Y_1}\right)_0 - E\left(\overline{X_1}\right)_1\right] \quad \text{for } j \in \{0, 1, \ldots, R\} .$$

The mapping f with $f(\alpha) \stackrel{\text{d}}{=} E\left(\underline{Y_1}\right)_\alpha$ is mononously non-decreasing on $[0, 1)$. Applying the theorem of the dominated convergence, we obtain:

$$\lim_{\substack{\alpha\to 1\\ \alpha<1}} E\left(\underline{Y_1}\right)_\alpha = E\left(\underline{X_1}\right)_1 .$$

Therefore the set

$$B_j^{(R)} \stackrel{\text{d}}{=} \left\{ \alpha \in [0,1) \,\middle|\, \lim_{\substack{h\to 0\\ h>0}} E\left(\underline{Y_1}\right)_{\alpha-h} \leq u_j^{(R)} \leq \lim_{\substack{h\to 0\\ h>0}} E\left(\underline{Y_1}\right)_{\alpha+h} \right\}$$

is not empty for all $j \in \{1, \ldots, R-1\}$. Choose $x_j^{(R)} \in B_j^{(R)}$ for all $j \in \{1, \ldots, R-1\}$ arbitrary such that

$$x_1^{(R)} \leq x_2^{(R)} \leq \ldots \leq x_{R-1}^{(R)}$$

is valid.
In the same way, we can find points

$$z_j^{(R)} \in \left\{ \alpha \in [0,1) \,\middle|\, \lim_{\substack{h\to 0\\ h>0}} E\left(\overline{Y_1}\right)_{\alpha-h} \leq v_j^{(R)} \leq \lim_{\substack{h\to 0\\ h>0}} E\left(\overline{Y_1}\right)_{\alpha+h} \right\}$$

for $j \in \{1, \ldots, R-1\}$ with

$$z_1^{(R)} \leq z_2^{(R)} \leq \ldots \leq z_{R-1}^{(R)} .$$

Define

$$\left\{ \alpha_j^{(R)} \mid i \in \{1, 2, \ldots, 2R-2\} \right\}$$

such that

(i) $\left\{\alpha_j^{(R)} \mid j \in \{1, \ldots, 2R-2\}\right\} = \bigcup_{j=1}^{R-1} \left\{x_j^{(R)}, z_j^{(R)}\right\}$ is valid, and

(ii) $\alpha_j^{(R)} \leq \alpha_{j+1}^{(R)}$ holds for all all $j \in \{1, \ldots, 2R-3\}$.

Define $\alpha_0^{(R)} \stackrel{\mathrm{d}}{=} 0$ and $\alpha_{2R-1}^{(R)} \stackrel{\mathrm{d}}{=} 1$.
Now we want to show that

$$d_H\left[(E \text{ co } X_1)_{\alpha_j^{(R)}}, (E \text{ co } X_1)_{\overline{\alpha_{j+1}^{(R)}}}\right]$$
$$\leq \frac{1}{R}\, d_H\left[(E \text{ co } X_1)_0, (E \text{ co } X_1)_{\overline{1}}\right].$$

is valid for all $j \in \{0, 1, \ldots, 2R-2\}$ if $\alpha_j^{(R)} < \alpha_{j+1}^{(R)}$ holds.
From Lemma 3.4 and part (c) of this proof, we know that

$$\lim_{\substack{h \to 0 \\ h > 0}} E\left(\underline{Y_1}\right)_{\alpha+h} = E\left(\underline{Y_1}\right)_\alpha \quad \text{and}$$
$$\lim_{\substack{h \to 0 \\ h > 0}} E\left(\overline{Y_1}\right)_{\alpha+h} = E\left(\overline{Y_1}\right)_\alpha$$

is valid for $\alpha \in [0, 1)$ as well as

$$\lim_{\substack{h \to 0 \\ h > 0}} E\left(\underline{Y_1}\right)_\alpha = E\left(\underline{X_1}\right)_\alpha \quad \text{and}$$
$$\lim_{\substack{h \to 0 \\ h > 0}} E\left(\overline{Y_1}\right)_\alpha = E\left(\overline{X_1}\right)_\alpha$$

for $\alpha \in (0, 1]$.
Let $j \in \{1, \ldots, 2R-3\}$ such that $\alpha_j^{(R)} < \alpha_{j+1}^{(R)}$. Then Lemma 6.2 shows that

$$d_H\left[(E \text{ co } X_1)_{\alpha_j^{(R)}}, (E \text{ co } X_1)_{\overline{\alpha_{j+1}^{(R)}}}\right] =$$

$$d_H\left[\left[E\left(\underline{Y_1}\right)_{\alpha_j^{(R)}}, E\left(\overline{Y_1}\right)_{\alpha_j^{(R)}}\right], \left[E\left(\underline{X_1}\right)_{\alpha_{j+1}^{(R)}}, E\left(\overline{X_1}\right)_{\alpha_{j+1}^{(R)}}\right]\right]$$

$$\max\left\{E\left(\underline{X_1}\right)_{\alpha_{j+1}^{(R)}} - E\left(\underline{Y_1}\right)_{\alpha_j^{(R)}}, E\left(\overline{X_1}\right)_{\alpha_{j+1}^{(R)}} - E\left(\overline{Y_1}\right)_{\alpha_j^{(R)}}\right\}$$

is valid. We can choose a $k \in \{1, \ldots, R-1\}$ with

$$x_k^{(R)} \leq \alpha_j^{(R)} < \alpha_{j+1}^{(R)} \leq x_{k+1}^{(R)}.$$

Then

$$\begin{aligned} & E\left(\underline{X_1}\right)_{\alpha_{j+1}^{(R)}} - E\left(\underline{Y_1}\right)_{\alpha_j^{(R)}} \leq E\left(\underline{X_1}\right)_{x_{k+1}^{(R)}} - E\left(\underline{Y_1}\right)_{x_k^{(R)}} = \\ = & \lim_{\substack{h \to 0 \\ h > 0}} E\left(\underline{Y_1}\right)_{x_{k+1}^{(R)} - h} - \lim_{\substack{h \to 0 \\ h > 0}} E\left(\underline{Y_1}\right)_{x_k^{(R)} + h} \leq \\ \leq & u_{k+1}^{(R)} - u_k^{(R)} = \\ = & \frac{1}{R}\left[E\left(\underline{X_1}\right)_1 - E\left(\underline{Y_1}\right)_0\right] \\ \leq & \frac{1}{R} d_H\left[(E \text{ co } X_1)_0, (E \text{ co } X_1)_{\bar{1}}\right]. \end{aligned}$$

is valid. We can also choose an $l \in \{1, \ldots, R-1\}$ with

$$z_l^{(R)} \leq \alpha_j^{(R)} < \alpha_{j+1}^{(R)} \leq z_{l+1}^{(R)},$$

and we can follow:

$$\begin{aligned} & E\left(\overline{Y_1}\right)_{\alpha_j^{(R)}} - E\left(\overline{X_1}\right)_{\alpha_{j+1}^{(R)}} \leq \\ \leq & \frac{1}{R}\left[E\left(\underline{X_1}\right)_1 - E\left(\underline{Y_1}\right)_0\right] \\ \leq & \frac{1}{R} d_H\left[(E \text{ co } X_1)_0, (E \text{ co } X_1)_{\bar{1}}\right]. \end{aligned}$$

If $j = 0$ or $j = 2R - 2$, the proof can be done in a similar way.

(e) Define for $R \in I\!N$, $n \in I\!N$, $\omega \in \Omega$:

$$S_n^{(R)}(\omega) \stackrel{d}{=} \max \left\{ \begin{array}{l} d_H\left[\left(\frac{1}{n}\sum_{i=1}^{n}(X_i)_\omega\right)_{\overline{1}}, (E \text{ co } X_1)_{\overline{1}}\right], \\ d_H\left[\left(\frac{1}{n}\sum_{i=1}^{n}(X_i)_\omega\right)_{0}, (E \text{ co } X_1)_{0}\right], \\ \max\limits_{j \in \{1,\ldots,2R-2\}} d_H\left[\begin{array}{r}\left(\frac{1}{n}\sum_{i=1}^{n}(X_i)_\omega\right)_{\alpha_j^{(R)}}, \\ (E \text{ co } X_1)_{\alpha_j^{(R)}}\end{array}\right] \\ \max\limits_{j \in \{1,\ldots,2R-2\}} d_H\left[\begin{array}{r}\left(\frac{1}{n}\sum_{i=1}^{n}(X_i)_\omega\right)_{\overline{\alpha_j^{(R)}}}, \\ (EX_1)_{\overline{\alpha_j^{(R)}}}\end{array}\right] \end{array} \right\}.$$

We want to show that for all $R \in I\!N$, $n \in I\!N$, $\omega \in \Omega$

$$\forall\ \alpha \in [0,1) : d_H\left[\left(\frac{1}{n}\sum_{i=1}^{n}(X_i)_\omega\right)_{\alpha}, (E \text{ co } X_1)_{\alpha}\right] \leq S_n^{(R)}(\omega) + \frac{1}{R}D$$

and

$$\forall\ \alpha \in (0,1] : d_H\left[\left(\frac{1}{n}\sum_{i=1}^{n}(X_i)_\omega\right)_{\overline{\alpha}}, (E \text{ co } X_1)_{\overline{\alpha}}\right] \leq S_n^{(R)}(\omega) + \frac{1}{R}D\ .$$

is valid where, for abbreviation,

$$D \stackrel{d}{=} d_H\left[(E \text{ co } X_1)_0, (E \text{ co } X_1)_{\overline{1}}\right].$$

Let $\alpha \in (0,1]$ be arbitrary.
If there is a $j \in \{0,1,\ldots,2R-1\}$ with $\alpha = \alpha_j^{(R)}$, the assertion is obvious.
If $\alpha \in \left(\alpha_j^{(R)}, \alpha_{j+1}^{(R)}\right)$ for a $j \in \{0,1,\ldots,2R-2\}$, then $\alpha_j^{(R)} < \alpha_{j+1}^{(R)}$ follows.

$$\left(\frac{1}{n}\sum_{i=1}^{n}(X_i)_\omega\right)_{\overline{\alpha_{j+1}^{(R)}}} \subseteq \left(\frac{1}{n}\sum_{i=1}^{n}(X_i)_\omega\right)_{\alpha}$$

$$\subseteq \left(\tfrac{1}{n}\sum_{i=1}^{n}(X_i)_\omega\right)_{\overline{\alpha}} \subseteq \left(\frac{1}{n}\sum_{i=1}^{n}(X_i)_\omega\right)_{\alpha_j^{(R)}}$$

and

$$(E \text{ co } X_1)_{\overline{\alpha_{j+1}^{(R)}}} \subseteq (E \text{ co } X_1)_\alpha \subseteq (E \text{ co } X_1)_{\overline{\alpha}} \subseteq (E \text{ co } X_1)_{\alpha_j^{(R)}}$$

holds.

We can apply part (a) and (d) and obtain:

$$d_H\left[\left(\frac{1}{n}\sum_{i=1}^n (X_i)_\omega\right)_\alpha, (E \text{ co } X_1)_\alpha\right]$$

$$\leq \max\left\{\begin{array}{l} d_H\left[\left(\frac{1}{n}\sum_{i=1}^n (X_i)_\omega\right)_{\overline{\alpha_{j+1}^{(R)}}}, (E \text{ co } X_1)_{\overline{\alpha_{j+1}^{(R)}}}\right], \\ d_H\left[\left(\frac{1}{n}\sum_{i=1}^n (X_i)_\omega\right)_{\alpha_j^{(R)}}, (E \text{ co } X_1)_{\alpha_j^{(R)}}\right] \end{array}\right\} +$$

$$+ d_H\left[(E \text{ co } X_1)_{\overline{\alpha_{j+1}^{(R)}}}, (E \text{ co } X_1)_{\alpha_j^{(R)}}\right]$$

$$\leq S_n^{(R)}(\omega) + \frac{1}{R}D \quad \text{for } \omega \in \Omega,\ R \in I\!N,\ n \in I\!N,$$

and in a similar way the other assertion can be shown.

(f) Define

$$M \stackrel{\mathrm{d}}{=} \bigcup_{R=1}^{\infty} \bigcup_{j=0}^{2R-1} M_1\left(\alpha_j^{(R)}\right) \cup M_2\left(\alpha_j^{(R)}\right) \in \mathfrak{A},$$

M is a zero set.

$\left\{S_n^{(R)}(\omega)\right\}_{n \in I\!N}$ (for $\omega \in \Omega \backslash M$, $R \in I\!N$) is the maximum of a finite number of zero sequences, as being shown in part (c) and (e). Therefore

$$\lim_{n\to\infty} S_n^{(R)}(\omega) = 0 \text{ is valid for all } R \in I\!N,\ \omega \in \Omega \backslash N.$$

As it is shown in part (e),

$$d_\infty\left(\frac{1}{n}\sum_{i=1}^n (X_i)_\omega, E \text{ co } X_1\right) \leq S_n^{(R)}(\omega) + \frac{1}{R}D.$$

Let $\omega \in \Omega \backslash M$ and $\epsilon > 0$ be arbitrary. Define $R = R(\epsilon) \stackrel{\mathrm{d}}{=} \left[\frac{2D}{\epsilon}\right] + 1$. There exists an $N_0 = N_0(\omega, \epsilon) \in I\!N$ such that

$$\left| S_n^{(R)}(\omega) \right| \leq \frac{\epsilon}{2}$$

is valid for $n \geq N_0$. This implies

$$d_\infty \left(\frac{1}{n} \sum_{i=1}^{n} (X_i)_\omega, E \text{ co } X_1 \right) \leq \frac{\epsilon}{2} + \frac{D}{R(\epsilon)} \leq \epsilon. \bullet$$

Example 10.6 Let $\{R_i\}_{i \in I\!N}$ and $\{X_i\}_{i \in I\!N}$ be two sequences of (usual) random variables such that $\left\{ \binom{R_i}{S_i} \right\}_{i \in I\!N}$ is a p.i.i.d. sequence of two dimensional random vectors. Define

$$(X_i)_\omega \stackrel{\mathrm{d}}{=} g^{R_i(\omega), S_i(\omega)^2} \mathrm{I}_{[-D,D]}$$

for $i \in I\!N$,$\omega \in \Omega$ where $D > 0$, and $g^{a,b}$ is defined as on page 12 for $a \in I\!R$, $b \in I\!R$.

(i) Then $\{X_i\}_{i \in I\!N}$ is an i.i.d. sequence,

(ii)
$$\frac{1}{n} \sum_{i=1}^{n} (X_i)_\omega = \frac{1}{n} \sum_{i=1}^{n} g^{R_i(\omega), S_i(\omega)^2} \mathrm{I}_{[-D,+D]}$$
$$= g^{\frac{1}{n} \sum_{i=1}^{n} R_i(\omega), \frac{1}{n} \sum_{i=1}^{n} S_i(\omega)^2} \mathrm{I}_{[-D,+D]},$$

and

(iii)
$$\left\{ \frac{1}{n} \sum_{i=1}^{n} (X_i)_\omega \right\}_{n \in I\!N} \xrightarrow{d_\infty} g^{ER, ES^2} \cdot \mathrm{I}_{[-D,+D]}$$

for all $\omega \in \Omega \backslash M$ where $P(M) = 0$.

An important theorem in the probability theory is the central limit theorem. We define the notion "asymptotically normally distributed" and formulate a central limit theorem for sequences of fuzzy random variables. It holds for i.i.d. sequences.

Definition 10.7 *Let $\{X_i\}_{i\in\mathbb{N}}$ be a sequence of f.r.v.'s on the probability space $(\Omega, \mathfrak{A}, P)$.*
$\{X_i\}_{i\in\mathbb{N}}$ is called asymptotically convex if there exists a sequence $\{\Gamma_i\}_{i\in\mathbb{N}}$ of convex f.r.v.'s (i.e. $\Gamma_i : \Omega \to U(\mathbb{R})$ for all $i \in \mathbb{N}$) such that

$$\lim_{i\to\infty} d_\infty\left((X_i)_\omega, (\Gamma_i)_\omega\right) = 0$$

is valid for all $\omega \in \Omega\backslash N$ where N is a zero set.

Definition 10.8 *Let $\{X_i\}_{i\in\mathbb{N}}$ be a sequence of f.r.v.'s .*
$\{X_i\}_{i\in\mathbb{N}}$ is called asymptotically normally distributed if
(i) $\{X_i\}_{i\in\mathbb{N}}$ is asymptotically convex, and
(ii) $\left\{(\underline{Y_i})_\alpha\right\}_{i\in\mathbb{N}}$ and $\left\{(\overline{Y_i})_\alpha\right\}_{i\in\mathbb{N}}$ are asymptotically normally distributed where we define for $i \in \mathbb{N}$, $\omega \in \Omega$, $\alpha \in [0,1)$:

$$(\underline{Y_i})_\alpha(\omega) \overset{\text{d}}{=} \inf\left(\left(X_i\right)_\omega\right)_\alpha$$

and

$$(\overline{Y_i})_\alpha(\omega) \overset{\text{d}}{=} \sup\left(\left(X_i\right)_\omega\right)_\alpha .$$

Theorem 10.9 *Let $(\Omega, \mathfrak{A}, P)$ be a probability space, and let $\{X_i\}_{i\in\mathbb{N}}$ be an i.i.d. sequence on $(\Omega, \mathfrak{A}, P)$. Let M_1 be a zero set such that $(X_i)_\omega \in F_C(\mathbb{R})$ holds for all $\omega \in \Omega\backslash M_1$, let $E\ (\underline{Y}_i)_0^2 < +\infty$ and $E\ \left(\overline{Y}_i\right)_0^2 < +\infty$ be valid. Then*

$$\left\{\frac{1}{n}\sum_{i=1}^{n} X_i\right\}_{n\in\mathbb{N}}$$

is asymptotically normally distributed.

Proof From Theorem 9.13(i) we know that $\left\{(\underline{Y_i})_\alpha\right\}_{i\in\mathbb{N}}$ and $\left\{(\overline{Y_i})_\alpha\right\}_{i\in\mathbb{N}}$ are two i.i.d. sequences for all $\alpha \in [0,1)$.

(i) As $E\ |(\underline{Y}_1)_0| < +\infty$ and $E\ \left|\left(\overline{Y}_1\right)_0\right| < +\infty$ holds, we can find a zero set M_2 (because of Theorem 10.5) such that

$$\left\{\frac{1}{n}\sum_{i=1}^{n}(X_i)_\omega\right\}_{n\in\mathbb{N}} \xrightarrow{d_\infty} E \text{ co } X_1$$

for all $\omega \in \Omega\backslash M_2$. So $\left\{\frac{1}{n}\sum\limits_{i=1}^{n}(X_i)\right\}_{n\in\mathbb{N}}$ is asymptotically convex. Define

$$(\Gamma_i)_\omega \overset{\text{d}}{=} E \text{ co } X_1$$

for $i \in \mathbb{N}$, $\omega \in \Omega$.

(ii) As

$$\left(\underline{Y_i}\right)_\alpha^2(\omega) \leq \left(\underline{Y}_i\right)_0^2(\omega) + \left(\overline{Y}_i\right)_0^2(\omega)$$

and

$$\left(\overline{Y_i}\right)_\alpha^2(\omega) \leq \left(\underline{Y}_i\right)_0^2(\omega) + \left(\overline{Y}_i\right)_0^2(\omega)$$

holds for all $i \in \mathbb{N}$, $\alpha \in [0,1)$, and $\omega \in \Omega$, $E\left(\underline{Y_i}\right)_\alpha$, $E\left(\overline{Y_i}\right)_\alpha$, and $\operatorname{Var}\left(\underline{Y_i}\right)_\alpha$, $\operatorname{Var}\left(\overline{Y_i}\right)_\alpha$ exist for all $\alpha \in [0,1)$. Therefore $\left\{\frac{1}{n}\sum_{i=1}^{n}\left(\overline{Y_i}\right)_\alpha\right\}_{n\in\mathbb{N}}$ and $\left\{\frac{1}{n}\sum_{i=1}^{n}\left(\overline{Y_i}\right)_\alpha\right\}_{n\in\mathbb{N}}$ are asymptotically normally distributed for all $\alpha \in [0,1)$ (compare M. Fisz [44], or [12,24,173], e.g.). As

$$\inf\left(\frac{1}{n}\sum_{i=1}^{n}(X_i)_\omega\right)_\alpha = \frac{1}{n}\sum_{i=1}^{n}\left(\underline{Y_i}\right)_\alpha(\omega)$$

and

$$\sup\left(\frac{1}{n}\sum_{i=1}^{n}(X_i)_\omega\right)_\alpha = \frac{1}{n}\sum_{i=1}^{n}\left(\overline{Y_i}\right)_\alpha(\omega)$$

holds for $n \in \mathbb{N}$, $\omega \in \Omega$, and $\alpha \in [0,1)$, the proof is complete.

10.2 CONSISTENT ESTIMATORS IN THE FINITE CASE

In Section 8 we considered algorithms for the calculation of the expected value and the variance. We assumed that the probabilities of the different vague data are known. In practical application, this assumption is not realistic, because the probabilities are often unknown. How can we get good estimates for the characteristics of a population, if we know that the observed values are in a finite set $\{\mu_1, \ldots, \mu_n\}$ of vague data?
A natural way of solving this problem is to draw a random sample of the population, to calculate the relative frequencies $\tilde{p}_i$ of the μ_i's, and to use the results of Section 8 to approximate the unknown characteristic. For the case of the expectation and the variance we will demonstrate in the following that this procedure does indeed lead to consistent estimates.
Throughout this section we use the notions of Section 8.

Theorem 10.10 *Let $\{X_j\}_{j\in\mathbb{N}}$ be a p.i.i.d. sequence of finite discrete f.r.v.'s on the probability space $(\Omega, \mathfrak{A}, P)$. Let $\{\mu_1, \ldots, \mu_n\} \subseteq F_C(\mathbb{R})$ be a set containing the (common) codomain of $\{X_j\}_{j\in\mathbb{N}}$.*
Define for $\omega \in \Omega$, $N \in \mathbb{N}$, $i \in \{1, \ldots, n\}$

$$P_i^N(\omega) \overset{\text{d}}{=} \frac{1}{n} \text{card } \{j \in \{1, \ldots, N\} \mid (X_j)_\omega = \mu_i\};$$

$P_i^{(N)}$ is the relative frequence of μ_i. Define for $\omega \in \Omega$, $N \in \mathbb{N}$ $E_\omega^{(N)} \in U(\mathbb{R})$ by

$$E_\omega^{(N)}(t) \overset{\text{d}}{=}$$

$$\sup \left\{ \alpha \mathrm{I}_{\left[\sum_{i=1}^{n} P_i^{(N)}(\omega) \cdot \inf(A_i)_\alpha, \sum_{i=1}^{n} P_i^{(N)}(\omega) \cdot \sup(A_i)_\alpha \right]}(t) \mid \omega \in \Omega \right\}$$

for $t \in \mathbb{R}$, where $\{(A_i)_\alpha \mid \alpha \in (0,1)\}$ is a set representation of μ_i for $i \in \{1, \ldots, n\}$. Then there exists a zero set M such that

$$\left\{ E_\omega^{(N)} \right\}_{N \in \mathbb{N}} \xrightarrow{d_\infty} E \text{ co } X_1 = \tilde{E}\, X_1$$

holds for $\omega \in \Omega \backslash M$.

Proof

(a) Define

$$D \overset{\text{d}}{=} \max_{i \in \{1, \ldots, n\}} \max \{|\inf (\mu_i)_0|, |\sup (\mu_i)_0|\}$$

and

$$\Gamma \overset{\text{d}}{=} \{\omega \in \Omega \mid (X_1)_\omega \in \{\mu_1, \ldots, \mu_n\}\}.$$

For all $\omega \in \Gamma$ $(X_i)_\omega \in [-D, +D]$ is valid, this implies $E\,|(\underline{Y}_1)_0| \leq D$ and $E\,\left|(\overline{Y}_1)_0\right| \leq D$ where

$$(\underline{Y}_1)_\alpha(\omega) \overset{\text{d}}{=} \inf((X_1)_\omega)_\alpha$$

and

$$(\overline{Y}_1)_\alpha(\omega) \overset{\text{d}}{=} \sup((X_1)_\omega)_\alpha$$

for $\alpha \in [0,1)$.

(b) Let $\alpha \in [0,1)$ be arbitrary, define $\alpha_r \overset{\text{d}}{=} \alpha + 1/2r(1-\alpha)$ for $r \in \mathbb{N}$. Theorem 3.6 shows that for all $\omega \in \Omega$, $N \in \mathbb{N}$

$$\inf \left(E_\omega^{(N)} \right)_\alpha = \lim_{r \to \infty} \sum_{i=1}^{n} P_i^{(N)}(\omega) \cdot \inf(A_i)_{\alpha_r} = \sum_{i=1}^{n} P_i^{(N)}(\omega) \cdot \inf(\mu_i)_\alpha$$

and

$$\sup\left(E_\omega^{(N)}\right)_\alpha = \lim_{r\to\infty}\sum_{i=1}^{n} P_i^{(N)}(\omega)\cdot\sup(A_i)_{\alpha_r} = \sum_{i=1}^{n} P_i^{(N)}(\omega)\cdot\sup(\mu_i)_\alpha$$

holds.
By Theorem 8.3 we know that $\left(E_\omega^{(N)}\right)_\alpha$ is convex for $\omega \in \Omega$, $N \in I\!N$.

(c) Combining Lemma 3.15(i) and part (b), we obtain that

$$\left\{\left[\sum_{i=1}^{n} P_i^{(N)}(\omega)\inf(\mu_i)_\alpha, \sum_{i=1}^{n} P_i^{(N)}(\omega)\sup(\mu_i)_\alpha\right] \mid \alpha\in(0,1)\right\}$$

is a set representation of $E_\omega^{(N)}$ for all $\omega \in \Omega$, $N \in I\!N$. As

$$\{[E(\underline{Y}_1)_\alpha, E(\overline{Y}_1)_\alpha] \mid \alpha\in(0,1)\}$$

is a set representation for E co X_1 (compare Theorem 8.3(iv)), we can follow with the help of Lemma 6.8:

$$\begin{aligned} & d_\infty\left(E\ \text{co}\ X_1, E_\omega^{(N)}\right) \\ \le & \sup_{\alpha\in(0,1)} d_H\left[\begin{array}{l}\left[\sum_{i=1}^{n} P_i^{(N)}(\omega)\inf(\mu_i)_\alpha, \sum_{i=1}^{n} P_i^{(N)}(\omega)\sup(\mu_i)_\alpha\right], \\ \left[E(\underline{Y}_1)_\alpha, E(\overline{Y}_1)_\alpha\right]\end{array}\right] \\ = & \sup_{\alpha\in(0,1)} \max\left\{\begin{array}{l}\left|\sum_{i=1}^{n} P_i^{(N)}(\omega)\inf(\mu_i)_\alpha - E(\underline{Y}_1)_\alpha\right|, \\ \left|\sum_{i=1}^{n} P_i^{(N)}(\omega)\sup(\mu_i)_\alpha - E(\overline{Y}_1)_\alpha\right|\end{array}\right\} \end{aligned}$$

(d) Define

$$\{\nu_1,\ldots,\nu_n\} \stackrel{\text{d}}{=}$$

$$\{\nu\in U(I\!R) \mid \exists i\in\{1,\ldots,n\} \text{ with } \nu = \text{co } \mu_i \text{ and } P(X_1=\mu_i)>0\}$$

(notice that co μ_i = co μ_j may be possible for $i \neq j$).
Define

$$M_j \stackrel{\text{d}}{=} \{i\in\{1,\ldots,n\} \mid \text{co } \mu_i = \nu_i\}$$

for $j \in \{1,\ldots,m\}$ and $k_j \stackrel{d}{=} \min M_j$. For all $j \in \{1,\ldots,m\}$, $k \in M_j$, and $\alpha \in [0,1)$ the equations $\inf(\mu_k)_\alpha = \inf\left(\mu_{k_j}\right)_\alpha$ and $\sup(\mu_k)_\alpha = \sup\left(\mu_{k_j}\right)_\alpha$ are valid.

Define sequences $\left\{Y_i^{(j)}\right\}_{i\in\mathbb{N}}$ of (usual) random variables by

$$Y_i^{(j)}(\omega) \stackrel{d}{=} \begin{cases} 1 & \text{, if co } (X_j)_\omega = \nu_i \\ 0 & \text{, if co } (X_j)_\omega \neq \nu_i \end{cases}$$

for $\omega \in \Omega$, $j \in \mathbb{N}$, $i \in \{1,\ldots,m\}$.

As $\{X_i\}_{i\in\mathbb{N}}$ is a p.i.i.d. sequence, $\{\text{co } X_i\}_{i\in\mathbb{N}}$ is a p.i.i.d. sequence because of Lemma 3.9 and Theorem 9.13(i).

With the help of Lemma 3.12, we conclude:

$\left\{Y_i^{(j)}\right\}_{j\in\mathbb{N}}$ is an i.i.d. sequence for all $i \in \{1,\ldots,m\}$. Due to the strong law of large numbers, there exist zero sets $N_1,\ldots,N_n$ such that

$$\lim_{N\to\infty} \frac{1}{N} \sum_{j=1}^{N} Y_i^{(j)}(\omega) = E\left[Y_i^{(1)}\right] = P(\text{co } X_1 = \nu_i) = \sum_{k\in M_i} P(X_1 = \mu_k)$$

holds for $i \in \{1,\ldots,m\}$ and $\omega \in \Omega\backslash N_i$.

Define the zero set $M \stackrel{d}{=} \bigcup_{i=1}^{m} M_i$.

For all $\omega \in \Omega\backslash M$ and $i \in \{1,\ldots,m\}$

$$\lim_{N\to\infty} \sum_{k\in M_i} P_k^{(N)}(\omega) = \lim_{N\to\infty} \frac{1}{N} \sum_{j=1}^{N} Y_i^{(j)}(\omega) = \sum_{k\in M_i} P(X_1 = \mu_k)$$

is valid.

(e) Let $\omega \in \Omega\backslash M$ and $\epsilon > 0$ be arbitrary.
Choose $N_0 = N_0(\omega,\epsilon) \in \mathbb{N}$ such that

$$\sum_{i=1}^{m} \left| \sum_{k\in M_i} P_k^{(N)}(\omega) - \sum_{k\in M_i} P(X_1 = \mu_k) \right| \leq \frac{\epsilon}{D}$$

is valid for all $N \geq N_0$.

Let $\alpha \in (0,1)$ and $N \geq N_0$. Then

$$\begin{aligned}
&\left|\sum_{i=1}^{n} P_i^{(N)}(\omega) \inf (\mu_i)_\alpha - E\,(\underline{Y}_1)_\alpha\right| \\
&= \left|\sum_{i=1}^{n} \left\{P_i^{(N)}(\omega) - P(X_1 = \mu_i)\right\} \cdot \inf (\mu_i)_\alpha\right| \\
&= \left|\sum_{i=1}^{m} \sum_{k \in M_i} \left\{P_k^{(N)}(\omega) - P\,(X_1 = \mu_k)\right\} \cdot \inf (\mu_k)_\alpha\right| \\
&\leq \sum_{i=1}^{m} \left|\sum_{k \in M_i} \left\{P_k^{(N)}(\omega) - P\,(X_1 = \mu_\alpha)\right\}\right| \cdot \left|\inf (\mu_{k_i})_\alpha\right| \\
&\leq D \cdot \sum_{i=1}^{m} \left|\sum_{k \in M_i} \left\{P_k^{(N)}(\omega) - P\,(X_1 = \mu_k)\right\}\right| \leq \epsilon
\end{aligned}$$

follows, and in the same way, we conclude:

$$\left|\sum_{i=1}^{n} P_i^{(N)}(\omega) \cdot \sup (\mu_i)_\alpha - E\,\left(\overline{Y}_1\right)_\alpha\right| \leq \epsilon$$

for all $\alpha \in (0,1)$ and $N \geq N_0$.
Together with part (c), this completes the proof. •

Remark We say that

$$\left\{E^{(N)}\right\}_{N \in I\!N} \quad \text{with} \quad E^{(N)} : \Omega \to F_C\,(I\!R),\ \omega \mapsto E_\omega^{(N)},\ N \in I\!N$$

is a consistent estimate (with respect to d_∞) for $\tilde{E}\,X_1 = E\,(\text{co } X_1)$. $\left\{E^{(N)}\right\}_{N \in I\!N}$ is also an unbiased estimate in the sense that $\tilde{E}\,\left(E^{(N)}\right) = E\,\left(\text{co } E^{(N)}\right) = E \text{ co } X_1 = \tilde{E}\,X_1$ for all $N \in I\!N$ holds.

The case of the variance is complicated. We will prove three technical theorems first.

Lemma 10.11 *Let $n \in I\!N$ and $D \subseteq I\!R^n$. Let $f, g : D \to I\!R$ be two mappings. Then*

$$\left|\sup_{x \in D} f(x) - \sup_{x \in D} g(x)\right| \leq \sup_{x \in D} |f(x) - g(x)|$$

and

$$\left| \inf_{x \in D} f(x) - \inf_{x \in D} g(x) \right| \le \sup_{x \in D} |f(x) - g(x)|$$

is valid, if

$$\sup_{x \in D} |f(x) - g(x)| < +\infty$$

holds.

Proof For all $y \in D$

$$f(y) - \sup_{x \in D} g(x) \le f(y) - g(y) \le \sup_{x \in D} |f(x) - g(x)|$$

holds, which implies

$$\sup_{x \in D} f(x) - \sup_{x \in D} g(x) = \sup_{y \in D} \left\{ f(y) - \sup_{x \in D} g(x) \right\} \le \sup_{x \in D} |f(x) - g(x)|.$$

In the same way

$$\sup_{x \in D} g(x) - \sup_{x \in D} f(x) \le \sup_{x \in D} |f(x) - g(x)|$$

can be shown.
In addition, for all $y \in D$

$$\inf_{x \in D} f(x) - g(y) \le f(y) - g(y) \le \sup_{x \in D} |f(x) - g(x)|$$

holds, which implies

$$\inf_{x \in D} f(x) - \inf_{x \in D} g(x) = \sup_{y \in D} \left\{ \inf_{x \in D} f(x) - g(y) \right\} \le \sup_{x \in D} |f(x) - g(x)|. \bullet$$

Lemma 10.12 *For all $\epsilon > 0$ there exists a $\delta = \delta(\epsilon) > 0$, such that $\sum_{i=1}^{n} |p_i - \tilde{p}_i| \le \delta$ implies*

$$|\Psi_\alpha[p_1, \ldots, p_n] - \Psi_\alpha[\tilde{p}_1, \ldots, \tilde{p}_n]| \le \epsilon$$

is valid for all $\alpha \in (0,1)$, $(p_1, \ldots, p_n) \in P_n$, $(\tilde{p}_1, \ldots, \tilde{p}_n) \in P_n$. Hereby

$$P_n \stackrel{\mathrm{d}}{=} \left\{ (p_1, \ldots, p_n) \in [0,1]^n \,\middle|\, \sum_{i=1}^{n} p_i = 1 \right\},$$

and Ψ is defined as in Theorem 8.10.

Proof Define

$$D \stackrel{\mathrm{d}}{=} \max_{i \in \{1,\ldots,n\}} \max\{|\inf(\mu_i)_0|, |\sup(\mu_i)_0|\}.$$

(a) Let $\alpha \in (0,1)$, $(p_1,\ldots,p_n) \in P_n$, $(\tilde{p}_1,\ldots,\tilde{p}_n) \in P_n$ and $(c_1,\ldots,c_n) \in [0,1]^n$. Due to the notions of Theorem 8.10, it follows:

$$\left|D^2_{\alpha,p_1,\ldots,p_n}(c_1,\ldots,c_n) - D^2_{\alpha,\tilde{p}_1,\ldots,\tilde{p}_n}(c_1,\ldots,c_n)\right|$$

$$\leq \left|E^2_{\alpha,p_1,\ldots,p_n}(c_1,\ldots,c_n) - E^2_{\alpha,\tilde{p}_1,\ldots,\tilde{p}_n}(c_1,\ldots,c_n)\right|$$

$$+ \left|E_{\alpha,p_1,\ldots,p_n}(c_1,\ldots,c_n) + E_{\alpha,\tilde{p}_1,\ldots,\tilde{p}_n}(c_1,\ldots,c_n)\right| \cdot$$

$$\cdot \left|E_{\alpha,p_1,\ldots,p_n}(c_1,\ldots,c_n) - E_{\alpha,\tilde{p}_1,\ldots,\tilde{p}_n}(c_1,\ldots,c_n)\right|$$

$$\leq \sum_{i=1}^n |p_i - \tilde{p}_i| c_i [\sup(A_i)_\alpha]^2 + \sum_{i=1}^n |p_i - \tilde{p}_i| \cdot (1 - c_i)[\inf(A_i)_\alpha]^2$$

$$+ \left\{\sum_{i=1}^n (p_i + \tilde{p}_i) c_i |\sup(A_i)_\alpha| + \sum_{i=1}^n (p_i + \tilde{p}_i)(1 - c_i)|\inf(A_i)_\alpha|\right\}$$

$$\cdot \left\{\sum_{i=1}^n |p_i - \tilde{p}_i| c_i |\sup(A_i)_\alpha| + \sum_{i=1}^n |p_i - \tilde{p}_i|(1 - c_i)|\inf(A_i)_\alpha|\right\}$$

$$\leq D^2 \sum_{i=1}^n |p_i - \tilde{p}_i| [c_i + (1 - c_i)]$$

$$+ D \sum_{i=1}^n (p_i + \tilde{p}_i)[c_i + (1 - c_i)]\, D \sum_{i=1}^n |p_i - \tilde{p}_i|$$

$$= 3D^2 \sum_{i=1}^n |p_i - \tilde{p}_i|,$$

since $|\inf(A_i)_\alpha| \leq D$ and $|\sup(A_i)_\alpha| \leq D$ holds for $\alpha \in (0,1)$ and $i \in \{1,\ldots,n\}$.

(b) In Theorem 8.10 we define

$$\Psi_\alpha[p_1,\ldots,p_n] = \sup\left\{D^2_{\alpha,p_1,\ldots,p_n}(c_1,\ldots,c_n) \mid (c_1,\ldots,c_n) \in [0,1]^n\right\}$$

for $\alpha \in (0,1)$ and $(p_1,\ldots,p_n) \in P_n$.
Let $\epsilon > 0$, define $\delta(\epsilon) \stackrel{\mathrm{d}}{=} \epsilon/(3D^2)$.

Let $\alpha \in (0,1)$, $(p_1,\dots,p_n) \in P_n$, and $(\tilde{p}_1,\dots,\tilde{p}_n) \in P_n$ be arbitrary. Then we conclude with Lemma 10.11:

$$\begin{aligned}
&|\Psi_\alpha[p_1,\dots,p_n] - \Psi_\alpha[\tilde{p}_1,\dots,\tilde{p}_n]| \\
&= |\sup\left\{ D^2_{\alpha,p_1,\dots,p_n}(c_1,\dots,c_n) \,|\, (c_1,\dots,c_n) \in [0,1]^n \right\} \\
&\quad - \sup\left\{ D^2_{\alpha,\tilde{p}_1,\dots,\tilde{p}_n}(c_1,\dots,c_n) \,|\, (c_1,\dots,c_n) \in [0,1]^n \right\}| \\
&\le \sup\left\{ \left| D^2_{\alpha,p_1,\dots,p_n}(c_1,\dots,c_n) - D^2_{\alpha,\tilde{p}_1,\dots,\tilde{p}_n}(c_1,\dots,c_n) \right| \right. \\
&\qquad \left. |(c_1,\dots,c_n) \in [0,1]^n \right\} \\
&\le 4\, D^2 \sum_{i=1}^n |p_i - \tilde{p}_i|.
\end{aligned}$$

Therefore $\sum_{i=1}^n |p_i - \tilde{p}_i| \le \delta(\epsilon)$ implies

$$|\Psi_\alpha[p_1,\dots,p_n] - \Psi_\alpha[\tilde{p}_1,\dots,\tilde{p}_n]| \le \epsilon. \quad \bullet$$

Lemma 10.13 *For all $\epsilon > 0$ exists a $\delta(\epsilon) > 0$ such that $\sum_{i=1}^n |p_i - \tilde{p}_i| \le \delta$ implies:*

$$|\Phi_\alpha[p_1,\dots,p_n] - \Phi_\alpha[\tilde{p}_1,\dots,\tilde{p}_n]| \le \epsilon,$$

for all $\alpha \in (0,1)$, where Φ_α is defined as in Definition 8.17(iv).

Proof Let D be defined as in the preceding lemma.

(a) Let $\alpha \in (0,1)$, $(p_1,\dots,p_n) \in P_n$, $(\tilde{p}_1,\dots,\tilde{p}_n) \in P_n$ and $t \in \mathbb{R}$ be arbitrary.

Due to the notion of Definition 8.24, we have

$$\left| \text{Var}_{\alpha,p_1,\ldots,p_n}[t] - \text{Var}_{\alpha,\tilde{p_1},\ldots,\tilde{p_n}}[t] \right| \leq$$

$$\leq \left| E^2_{\alpha,p_1,\ldots,p_n}[t] - E^2_{\alpha,\tilde{p}_1,\ldots,\tilde{p}_n}[t] \right|$$
$$+ \left| E_{\alpha,p_1,\ldots,p_n}[t] + E_{\alpha,\tilde{p}_1,\ldots,\tilde{p}_n}[t] \right|$$
$$\cdot \left| E_{\alpha,p_1,\ldots,p_n}[t] - E_{\alpha,\tilde{p}_1,\ldots,\tilde{p}_n}[t] \right|$$

$$\leq \sum_{i=1}^{n} |p_i - \tilde{p}_i| \{(V_i)_\alpha[t]\}^2$$
$$+ \left\{ \sum_{i=1}^{n} (p_i + \tilde{p}_i)\, |(V_i)_\alpha[t]| \right\} \left\{ \sum_{i=1}^{n} |p_i - \tilde{p}_i| \cdot |(V_i)_\alpha[t]| \right\}$$

$$\leq \; D^2 \sum_{i=1}^{n} |p_i - \tilde{p}_i| + 2DD \sum_{i=1}^{n} |p_i - \tilde{p}_i| = 3D^2 \sum_{i=1}^{n} |p_i - \tilde{p}_i|.$$

(b) According to Definition 8.24(iv),

$$\Phi_\alpha[p_1,\ldots,p_n] = \inf \left\{ \text{Var}_{\alpha,p_1,\ldots,p_n}[t] \mid t \in [c_0, c_M] \right\}.$$

holds for $\alpha \in (0,1)$ and $(p_1,\ldots,p_n) \in P_n$.
Let $\epsilon > 0$ be arbitrary, define $\delta(\epsilon) \stackrel{\text{d}}{=} \epsilon / \left(3D^2\right)$. Let $\alpha \in (0,1)$, $(p_1,\ldots,p_n) \in P_n$, and $(\tilde{p}_1,\ldots,\tilde{p}_n) \in P_n$.
Then we can apply Lemma 10.11 with $D = I\!R$ and obtain:

$$|\Phi_\alpha[p_1,\ldots,p_n] - \Phi_\alpha[\tilde{p}_1,\ldots,\tilde{p}_n]|$$
$$= |\inf \left\{ \text{Var}_{\alpha,p_1,\ldots,p_n}[t] \mid t \in I\!R \right\}$$
$$- \inf \left\{ \text{Var}_{\alpha,\tilde{p}_1,\ldots,\tilde{p_n}}[t] \mid t \in I\!R \right\} |$$
$$\leq \sup \left\{ \left| \text{Var}_{\alpha,p_1,\ldots,p_n}[t] - \text{Var}_{\alpha,\tilde{p}_1,\ldots,\tilde{p_n}}[t] \right| \mid t \in I\!R \right\}$$
$$\leq \; 3D^2 \sum_{i=1}^{n} |p_i - \tilde{p}_i| \leq \epsilon,$$

if

$$\sum_{i=1}^{n} |p_i - \tilde{p}_i| \leq \delta(\epsilon). \bullet$$

The following two theorems demonstrate how to get a consistent estimate for $\widetilde{\text{Var}}\, X$ and $\text{Var}\,(\text{co}\ X)$ in the case of finite discrete f.r.v.'s with values in $N(\mathbb{R})$.

Theorem 10.14 *Let $\{X_j\}_{j\in\mathbb{N}}$ be a p.i.i.s.-sequence of finite f.r.v.'s on the probability space $(\Omega, \mathfrak{A}, P)$. Let $\{\mu_1, \ldots, \mu_n\} \subseteq N(\mathbb{R})$ containing the (common) codomain of $\{X_j\}_{j\in\mathbb{N}}$.*
Define for $\omega \in \Omega$, $n \in \mathbb{N}$, $i \in \{1, \ldots, n\}$:

$$P_i^{(N)}(\omega) \stackrel{\text{d}}{=} \frac{1}{N} \text{card}\ \left\{ j \in \{1, \ldots, N\} \,\middle|\, (X_j)_\omega = \mu_i \right\};$$

$P_i^{(N)}$ is the relative frequency of μ_i.
Define for $\omega \in \Omega$, $N \in \mathbb{N}$ $B_\omega^{(N)} \in U(\mathbb{R})$ by

$$B_\omega^{(N)}(t) \stackrel{\text{d}}{=}$$

$$\sup\left\{ \alpha \mathrm{I}_{\left[\Phi_\alpha\left[P_1^{(N)}(\omega), \ldots, P_n^{(N)}(\omega)\right], \Psi_\alpha\left[P_1^{(N)}(\omega), \ldots, P_n^{(N)}(\omega)\right]\right]}(t) \,\middle|\, \alpha \in (0,1) \right\}$$

for $t \in \mathbb{R}$, where Φ_α is defined as in Definition 8.24(iv) and Ψ_α as in Theorem 8.10. Then there exists a zero set $M \in \mathfrak{A}$ such that for all $\omega \in \Omega \backslash M$

$$\left\{ B_\omega^{(N)} \right\}_{N\in\mathbb{N}} \stackrel{d_\infty}{\to} \widetilde{\text{Var}}\, X_1$$

holds.

Proof
(a) Define for $i \in \{1, \ldots, n\}$ a sequence $\left\{ Y_i^{(j)} \right\}_{j\in\mathbb{N}}$ by

$$Y_i^{(j)}(\omega) \stackrel{\text{d}}{=} \begin{cases} 1 & , \text{if } (X_j)_\omega = \mu_i \\ 0 & , \text{if } (X_j)_\omega \neq \mu_i \end{cases} \quad \text{for } \omega \in \Omega,\ j \in \mathbb{N}.$$

As $\{X_j\}_{j\in\mathbb{N}}$ is an i.i.s.-sequence, $\left\{ Y_i^{(j)} \right\}_{j\in\mathbb{N}}$ is a (usual) i.i.d.-sequence for all $i \in \{1, \ldots, n\}$.

$$P_i^{(N)}(\omega) = \frac{1}{N} \sum_{j=1}^{N} Y_i^{(j)}(\omega)$$

is valid for $N \in \mathbb{N}$, $\omega \in \Omega$, $i \in \{1, \ldots, n\}$.
Applying Etemadi's strong law of large numbers (compare [43]), we can

find zero sets $M_1, \ldots, M_n$ such that

$$\lim_{N \to \infty} \frac{1}{N} \sum_{j=1}^{n} Y_i^{(j)}(\omega) = E\left[Y_i^{(1)}\right] = P(\Gamma_i)$$

is valid for $i \in \{1, \ldots, n\}$ and $\omega \in \Omega \backslash M_i$. Define $M \stackrel{\mathrm{d}}{=} \bigcup_{i=1}^{n} M_i$, M is a zero set.

$$\lim_{N \to \infty} P_i^{(N)}(\omega) = P(\Gamma_i)$$

is valid for all $\omega \in \Omega \backslash M$ and $i \in \{1, \ldots, n\}$.

(b) Let $\epsilon > 0$ and $\omega \in \Omega \backslash M$ be arbitrary. Because of Lemma 10.12 and Lemma 10.13, we can choose $\delta = \delta(\epsilon)$ such that

$$|\Phi_\alpha[p_1, \ldots, p_n] - \Phi_\alpha[P(\Gamma_1), \ldots, P(\Gamma_n)]| \le \epsilon \text{ and}$$
$$|\Psi_\alpha[p_1, \ldots, p_n] - \Psi_\alpha[P(\Gamma_1), \ldots, P(\Gamma_n)]| \le \epsilon$$

is valid for all $\alpha \in (0,1)$, $(p_1, \ldots, p_n) \in P_n$, $(\tilde{p}_1, \ldots, \tilde{p}_n) \in P_n$ if $\sum_{i=1}^{n} |p_i - \tilde{p}_i| \le \delta(\epsilon)$ holds.

Choose $N_0 = N_0(\epsilon, \omega) \in I\!N$ such that

$$\sum_{i=1}^{n} \left| P_i^{(N)}(\omega) - P(\Gamma_i) \right| \le \delta(\epsilon)$$

holds for all $N \ge N_0$.

Applying Lemma 8.25(iv) and Theorem 8.10, we obtain for all $N \ge N_0$ and $\alpha \in (0,1)$:

$$d_H\left[\left[\Phi_\alpha\left[P_1^{(N)}(\omega), \ldots, P_n^{(N)}(\omega)\right], \Psi_\alpha\left[P_1^{(N)}(\omega), \ldots, P_n^{(N)}(\omega)\right]\right], \tilde{B}_\alpha\right]$$
$$= \max \left\{ \begin{array}{l} \left|\Phi_\alpha\left[P_1^{(N)}(\omega), \ldots, P_n^{(N)}(\omega)\right] - \inf \tilde{B}_\alpha\right|, \\ \left|\Psi_\alpha\left[P_1^{(N)}(\omega), \ldots, P_n^{(N)}(\omega)\right] - \sup \tilde{B}_\alpha\right| \end{array} \right\} \le \epsilon,$$

as $\tilde{B}_\alpha$ is convex.

Obviously

$$\left\{\left[\Phi_\alpha\left[P_1^{(N)}(\omega), \ldots, P_n^{(N)}(\omega)\right], \Psi_\alpha\left[P_1^{(N)}(\omega), \ldots, P_n^{(N)}(\omega)\right]\right] \mid \alpha \in (0,1)\right\}.$$

is a set representation of $B_\omega^{(N)}$ for all $N \in \mathbb{N}$, and by Lemma 8.7 we know that

$$\left\{ \left[\inf \tilde{B}_\alpha, \sup \tilde{B}_\alpha \right] \mid \alpha \in (0,1) \right\}$$

is a set representation of $\widetilde{\text{Var}}\, X$. We can apply Lemma 6.8 and conclude the assertion. •

Theorem 10.15 *Let $\Omega = \{\omega_1, \ldots, \omega_n\}$ and $\mathfrak{A} = \mathfrak{P}(\Omega)$. Let $\{X_j\}_{j \in \mathbb{N}}$ be a p.i.i.s.-sequence of finite discrete f.r.v.'s on the probability space $(\Omega, \mathfrak{A}, P)$. Let $\{\mu_1, \ldots, \mu_n\} \subseteq N(\mathbb{R})$ contain the (common) codomain of $\{X_j\}_{j \in \mathbb{N}}$. Define for $\omega \in \Omega$, $N \in \mathbb{N}$, $i \in \{1, \ldots, n\}$ $P_i^{(N)}(\omega)$ as in Theorem 10.14 and $C_\omega^{(N)} \in U(\mathbb{R})$ by*

$$C_\omega^{(N)}(t) \stackrel{\mathrm{d}}{=}$$

$$\sup \left\{ \alpha \mathrm{I}_{\left[\Phi_\alpha \left[P_1^{(N)}(\omega), \ldots, P_n^{(N)}(\omega) \right], \Sigma_\alpha \left[P_1^{(N)}(\omega), \ldots, P_n^{(N)}(\omega) \right] \right]}(t) \mid \alpha \in (0,1) \right\}$$

for $t \in \mathbb{R}$ where Φ_α is defined as in Definition 8.27(iv) and Σ_α as in Theorem 8.18. Then there exists a zero set M such that for all $\omega \in \Omega \backslash M$

$$\left\{ C_\omega^{(N)} \right\}_{N \in \mathbb{N}} \xrightarrow{d_\infty} \text{Var co } X_1$$

is valid.

10.3 GLIWENKO-CANTELLI THEOREM

The Gliwenko-Cantelli Theorem, often called "Central Theorem of Statistics", states, that the sequence of the maximal deviation of the empirical distribution function and the "real" distribution functions converges almost sure against zero. We are going to prove an analogue theorem for vague data. A step in this direction is the following generalization of a Lemma of Gliwenko-Cantelli:

Theorem 10.16 *Let $(\Omega, \mathfrak{A}, P)$ be a probability space and $I \subseteq \mathbb{R}$ a convex set. For $x \in I$ and $i \in \mathbb{N}$ let $A_i(x) \in \mathfrak{A}$ be such that*

(i) $\forall\, i \in \mathbb{N}$, $x \in I$, $y \in I$: $x \leq y$ implies $A_i(x) \subseteq A_i(y)$

(ii) $\forall x \in I, \forall\, i \in \mathbb{N}, j \in \mathbb{N}$: $i \neq j$ implies that $A_i(x)$ and $A_j(x)$ are independent, and

(iii) $\forall x \in I, \forall i \in \mathbb{N}$: $P[A_i(x)] = P[A_1(x)]$.

Then there exists an $M \in \mathfrak{A}$ with $P(M) = 0$ such that

$$\lim_{n\to\infty} \sup_{x\in I} \left| \frac{1}{n} \sum_{i=1}^{n} \mathrm{I}_{A_i(x)}(\omega) - P[A_1(x)] \right| = 0$$

holds for all $\omega \in \Omega \setminus M$.

Proof

(i) As the probability measure P is continuous, we have for $i \in I\!N$, $x \in I$:

$$P\left[\bigcup_{y<x} A_i(y)\right] = \lim_{\substack{h\to 0\\ h>0}} P[A_i(x-h)] = \lim_{\substack{h\to 0\\ h>0}} P[A_1(x-h)]$$

and

$$P\left[\bigcap_{y>x} A_i(y)\right] = \lim_{\substack{h\to 0\\ h>0}} P[A_i(x+h)] = \lim_{\substack{h\to 0\\ h>0}} P[A_1(x+h)].$$

Let (in I) be the set of all inner points of I. Then we have:

$$P\left[\bigcup_{y\in \text{in } I} A_i(y)\right] = \lim_{\substack{y\to \sup I\\ y<\sup I}} P[A_i(y)] = \lim_{\substack{y\to \sup I\\ y<\sup I}} P[A_1(y)]$$

and

$$P\left[\bigcap_{y\in \text{in } I} A_i(y)\right] = \lim_{\substack{y\to \inf I\\ y>\inf I}} P[A_i(y)] = \lim_{\substack{y\to \inf I\\ y>\inf I}} P[A_1(y)].$$

All these sets belong to $\mathfrak{A}$.
Let $(i,j) \in I\!N^2$ such that $i \neq j$ holds. Because of assumption (i), we can follow

$$\left\{\bigcup_{y<x} A_i(y)\right\} \cap \left\{\bigcup_{y<x} A_j(y)\right\} = \bigcup_{y<x} \{A_i(y) \cap A_j(y)\},$$

and we can conclude

$$P\left[\left\{\bigcup_{y<x} A_i(y)\right\} \cap \left\{\bigcup_{y<x} A_j(y)\right\}\right] = P\left[\bigcup_{y<x} \{A_i(y) \cap A_j(y)\}\right]$$

$$= \lim_{\substack{h\to 0\\ h>0}} P[A_i(x-h)\cap A_j(x-h)]$$
$$= \lim_{\substack{h\to 0\\ h>0}} \{P[A_i(x-h)]\cdot P[A_j(x-h)]\}$$
$$= P\left(\bigcup_{y<x} A_i(y)\right)\cdot P\left(\bigcup_{y<x} A_j(y)\right).$$

We can conclude:

$$\left\{ \mathrm{I}_{\bigcup_{y<x} A_i(y)} \right\}_{i\in\mathbb{N}}$$

is a p.i.i.d.- sequence for all $x \in I$. In the same way we can show:

$$\left\{ \mathrm{I}_{\bigcup_{y\in \text{in } I} A_i(y)} \right\}_{i\in\mathbb{N}} \quad \text{and} \quad \left\{ \mathrm{I}_{\bigcap_{y\in \text{in } I} A_i(y)} \right\}_{i\in\mathbb{N}}$$

are p.i.i.d.-sequences, and

$$\left\{ \mathrm{I}_{\bigcap_{y>x} A_i(y)} \right\}_{i\in\mathbb{N}}$$

is a p.i.i.d.-sequence for all $x \in I$.
We can apply a strong law of large numbers from N. Etemadi [43], only asserting pairwise independent random variables, and obtain:
For all $x \in I$ there exists a zero set $M(x)$ such that

$$\lim_{n\to\infty} \frac{1}{n}\sum_{i=1}^{n} \mathrm{I}_{A_i(x)}(\omega) = P\left[A_1(x)\right] ,$$

$$\lim_{n\to\infty} \frac{1}{n}\sum_{i=1}^{n} \mathrm{I}_{\bigcup_{y<x} A_i(y)}(\omega) = \lim_{\substack{h\to 0\\ h>0}} P\left[A_1(x-h)\right] , \text{ and}$$

$$\lim_{n\to\infty} \frac{1}{n}\sum_{i=1}^{n} \mathrm{I}_{\bigcap_{y>x} A_i(y)}(\omega) = \lim_{\substack{h\to 0\\ h>0}} P\left[A_1(x+h)\right] .$$

is valid for $\omega \in \Omega \backslash N(x)$. And there is another zero set M_1 such that

$$\lim_{n\to\infty} \frac{1}{n} \sum_{i=1}^{n} \mathrm{I}_{\bigcup_{y \in \text{in } I} A_i(y)}(\omega) = P\left(\bigcup_{y \in \text{in } I} A_i(y)\right) \quad \text{and}$$

$$\lim_{n\to\infty} \frac{1}{n} \sum_{i=1}^{n} \mathrm{I}_{\bigcap_{y \in \text{in } I} A_i(y)}(\omega) = P\left(\bigcap_{y \in \text{in } I} A_i(y)\right)$$

holds for $\omega \in \Omega \backslash M_1$.

(ii) For abbreviation, define

$$A \stackrel{\mathrm{d}}{=} \inf\{P[A_1(x)] \mid x \in I\} \geq 0 \quad \text{and} \quad B \stackrel{\mathrm{d}}{=} \sup\{P[A_1(x)] \mid x \in I\} \leq 1 .$$

Let $R \in I\!N$ be arbitrary. Define

$$y_j^{(R)} \stackrel{\mathrm{d}}{=} A + \frac{j}{R}(B - A)$$

for $j \in \{0, 1, \ldots, R\}$. The set

$$\left\{ x \in I \,\middle|\, P\left[\bigcup_{y<x} A_1(y)\right] \leq y_j^{(R)} \leq P\left[\bigcap_{y>x} A_1(y)\right] \right\}$$

is for all $R \in I\!N$ and $j \in \{1, \ldots, R-1\}$ not empty. Choose $x_j^{(R)}$ arbitrary from this set for $j \in \{1, \ldots, R-1\}$ such that

$$x_1^{(R)} \leq x_2^{(R)} \leq \ldots \leq x_{R-1}^{(R)}$$

is valid. Define for $\omega \in \Omega$, $n \in I\!N$

$$T_n^{(R)}(\omega) \stackrel{\mathrm{d}}{=}$$

$$
\max \left\{ \begin{array}{l}
\left| \frac{1}{n} \sum_{i=1}^{n} \mathrm{I}_{\bigcap_{y \in \operatorname{in} I} A_y}(\omega) - P\left(\bigcap_{y \in \operatorname{in} I} A_1(y) \right) \right|, \\
\left| \frac{1}{n} \sum_{i=1}^{n} \mathrm{I}_{\bigcup_{y \in \operatorname{in} I} A_y}(\omega) - P\left(\bigcup_{y \in \operatorname{in} I} A_1(y) \right) \right|, \\
\max\limits_{1 \le j \le R-1} \max \left[\begin{array}{l}
\left| \frac{1}{n} \sum_{i=1}^{n} \mathrm{I}_{A_i\left(x_j^{(R)}\right)}(\omega) - P\left(A_1\left(x_j^{(R)}\right) \right) \right|, \\
\left| \frac{1}{n} \sum_{i=1}^{n} \mathrm{I}_{\bigcup_{y < x_j^{(R)}} A_y}(\omega) - P\left(\bigcup_{y < x_j^{(R)}} A_1(y) \right) \right|, \\
\left| \frac{1}{n} \sum_{i=1}^{n} \mathrm{I}_{\bigcap_{y > x_j^{(R)}} A_y}(\omega) - P\left(\bigcap_{y > x_j^{(R)}} A_1(y) \right) \right|,
\end{array} \right]
\end{array} \right\}
$$

and

$$
S_n^{(R)}(\omega) \stackrel{\mathrm{d}}{=}
$$

$$
\begin{cases}
T_n^{(R)}(\omega) & \text{, if } I = \operatorname{in} I \\
\max \left\{ T_n^{(R)}(\omega), \left| \frac{1}{n} \sum\limits_{i=1}^{n} \mathrm{I}_{A_i(\inf I)}(\omega) - P\left[A_1(\inf I)\right] \right| \right\} & \text{, if } \inf I \in I \text{ and } \sup I \notin I \\
\max \left\{ T_n^{(R)}(\omega), \left| \frac{1}{n} \sum\limits_{i=1}^{n} \mathrm{I}_{A_i(\sup I)}(\omega) - P\left[A_1(\sup I)\right] \right| \right\} & \text{, if } \inf I \notin I \text{ and } \sup I \in I \\
\max \left\{ \begin{array}{l} T_n^{(R)}(\omega), \\ \left| \frac{1}{n} \sum\limits_{i=1}^{n} \mathrm{I}_{A_i(\inf I)}(\omega) - P\left[A_1(\inf I)\right] \right|, \\ \left| \frac{1}{n} \sum\limits_{i=1}^{n} \mathrm{I}_{A_i(\sup I)}(\omega) - P\left[A_1(\sup I)\right] \right| \end{array} \right\} & \text{, if } \inf I \in I \text{ and } \sup I \in I
\end{cases}
$$

(iii) Let $R \in \mathbb{N}$ be arbitrary. We want to show that

$$\left| \frac{1}{n} \sum_{i=1}^{n} \mathrm{I}_{A_i(x)}(\omega) - P[A_1(x)] \right| \leq S_n^{(R)}(\omega) + \frac{1}{R}$$

is valid for $\omega \in \Omega$, $n \in \mathbb{N}$, $x \in I$. Let $\omega \in \Omega$, $n \in \mathbb{N}$, $x \in I$ be arbitrary.

If there is a $j \in \{1, \ldots, R-1\}$ with $x = x_j^{(R)}$, or if $x = \inf I$ or $x = \sup I$ holds, the assertion is obvious.

Let $x \in \left(\inf I, x_1^{(R)}\right)$. Then

$$\frac{1}{n} \sum_{i=1}^{n} \mathrm{I}_{A_i(x)}(\omega) - P[A_1(x)] \leq \frac{1}{n} \sum_{i=1}^{n} \mathrm{I}_{\bigcup\limits_{y < x_1^{(R)}} A_i(y)}(\omega) - A$$

$$= \frac{1}{n} \sum_{i=1}^{n} \mathrm{I}_{\bigcup\limits_{y < x_1^{(R)}} A_i(y)}(\omega)$$

$$- P\left[\bigcup_{y < x_1^{(R)}} A_1(y) \right] + P\left[\bigcup_{y < x_1^{(R)}} A_1(y) \right] - A$$

$$\leq S_n^{(R)}(\omega) + y_1^{(R)} - A \leq S_n^{(R)}(\omega) + \frac{1}{R}$$

as well as

$$\frac{1}{n} \sum_{i=1}^{n} \mathrm{I}_{A_i(x)}(\omega) - P[A_1(x)]$$

$$\geq \frac{1}{n} \sum_{i=1}^{n} \mathrm{I}_{\bigcap\limits_{x \in \text{in } I} A_i(y)}(\omega) \; - \; P\left[\bigcup_{y < x_1^{(R)}} A_1(y) \right]$$

$$= \frac{1}{n} \sum_{i=1}^{n} \mathrm{I}_{\bigcap\limits_{x \in \text{in } I} A_i(y)}(\omega) \; - \; P\left[\bigcap_{y \in \text{in } I} A_1(y) \right]$$

$$+ P\left[\bigcap_{y \subset \text{in } I} A_1(y) \right] - P\left[\bigcup_{y < x_1^{(R)}} A_1(y) \right]$$

$$\geq - S_n^{(R)}(\omega) + A - y_1^{(R)} \geq -S_n^{(R)}(\omega) - \frac{1}{R}$$

is valid.
If there is a $j \in \{2, \ldots, R-1\}$ with $x \in \left(x_{j-1}^{(R)}, x_j^{(R)}\right)$, then

$$\begin{aligned}
&\frac{1}{n}\sum_{i=1}^{n} \mathrm{I}_{A_i(x)}(\omega) - P[A_1(x)] \\
&\leq \frac{1}{n}\sum_{i=1}^{n} \mathrm{I}_{\bigcup_{y<x_j^{(R)}} A_y}(\omega) - P\left[\bigcup_{y<x_j^{(R)}} A_1(y)\right] \\
&\quad + P\left[\bigcup_{y<x_j^{(R)}} A_1(y)\right] - P\left[\bigcap_{y>x_{j-1}^{(R)}} A_1(y)\right] \\
&\leq S_n^{(R)}(\omega) + y_j^{(R)} - y_{j-1}^{(R)} \\
&\leq S_n^{(R)}(\omega) + \frac{1}{R}
\end{aligned}$$

as well as

$$\begin{aligned}
&\frac{1}{n}\sum_{i=1}^{n} \mathrm{I}_{A_i(x)}(\omega) - P[A_1(x)] \\
&\geq \frac{1}{n}\sum_{i=1}^{n} \mathrm{I}_{\bigcap_{y>x_{j-1}^{(R)}} A_y}(\omega) - P\left[\bigcap_{y>x_{j-1}^{(R)}} A_1(y)\right] \\
&\quad + P\left[\bigcap_{y>x_{j-1}^{(R)}} A_1(y)\right] - P\left[\bigcup_{y<x_j^{(R)}} A_1(y)\right] \\
&\geq -S_n^{(R)}(\omega) + y_{j-1}^{(R)} - y_j^{(R)} \\
&\geq -S_n^{(R)}(\omega) - \frac{1}{R}\,.
\end{aligned}$$

The case $y \in \left(y_{R-1}^{(R)}, \sup I\right)$ can be treated in a similar way.

(iv) Define

$$M_2 \stackrel{\mathrm{d}}{=} \begin{cases} N[\inf I] & \text{, if } \inf I \in I \\ \emptyset & \text{, otherwise} \end{cases},$$

$$M_3 \stackrel{\mathrm{d}}{=} \begin{cases} N[\sup I] & \text{, if } \inf I \in I \\ \emptyset & \text{, otherwise} \end{cases},$$

and

$$M \stackrel{\mathrm{d}}{=} \bigcup_{R=1}^{\infty} \bigcup_{j=1}^{R-1} M\left[x_j^{(R)}\right] \cup M_1 \cup M_2 \cup M_3;$$

M is a zero set.
It is easy to show that $\left\{a_i^{(j)}\right\}_{i \in I\!N} \to a_j$ for all $j \in \{1, \ldots, s\}$ implies

$$\left\{\max_{1 \leq j \leq s} \left|a_i^{(j)} - a_j\right|\right\}_{i \in I\!N} \to 0$$

So we can follow from part (i):

$$\lim_{n \to \infty} S_n^{(R)}(\omega) = 0.$$

As we have shown in part (iii),

$$\sup_{x \in I} \left|\frac{1}{n} \sum_{i=1}^{n} \mathrm{I}_{A_i(x)}(\omega) - P[A_1(x)]\right| \leq S_n^{(R)}(\omega) + \frac{1}{R} \quad \text{holds.}$$

Let $\omega \in \Omega \backslash M$ and $\epsilon > 0$ be arbitrary. Define

$$R = R(\epsilon) \stackrel{\mathrm{d}}{=} \left[\frac{2}{\epsilon}\right] + 1 \in I\!N .$$

There exists an $N_0 = N_0(\omega, \epsilon)$ such that

$$S_n^{(R)}(\omega) \leq \frac{\epsilon}{2}$$

holds for all $n \geq N_0$. We can follow:

$$\sup_{x \in I} \left|\frac{1}{n} \sum_{i=1}^{n} \mathrm{I}_{A_i(x)}(\omega) - P[A_1(x)]\right| \leq \frac{\epsilon}{2} + \frac{1}{R(\epsilon)} \leq \epsilon. \quad \bullet$$

Theorem 10.15 can be used for proving the Gliwenko-Cantelli Theorem for vague data.

Remark J. Elker, D. Pollard and W. Stute [42] and P. Gaenssler and W. Stute [59] proved Gliwenko-Cantelli Theorems for classes of convex sets. It is, however, not possible to apply their results here directly, as their results require measures on a measure space $(\Omega, \mathfrak{A})$. If the conditions of Theorem 10.15 are fulfilled, we can define

$$\mu[(-\infty, x]] \stackrel{\mathrm{d}}{=} P[A_1(x)] \text{ for } x \in I\!R \,,$$

but μ is not a measure on the set of all half infinite intervals. So it cannot be extended uniquely to the generated ring.

Theorem 10.17 *Let $n \in I\!N$ and $\{\mu_1, \ldots, \mu_n\} \in [F(I\!R)]^n$.*
Define for $x \in I\!R$ $S_n[\mu_1, \ldots, \mu_n](x)$ as in Definition 8.36 and $T_n[\mu_1, \ldots, \mu_n](x) \in U_C(I\!R)$ by

$$\{T_n[\mu_1, \ldots, \mu_n](x)\}(t) \stackrel{\mathrm{d}}{=}$$
$$\sup\left\{\alpha \mathrm{I}_{[S_n[\sup(\mu_1)_\alpha, \ldots, \sup(\mu_n)_\alpha](x), S_n[\inf(\mu_1)_\alpha, \ldots, \inf(\mu_n)_\alpha](x)]}(t) \mid \alpha \in (0,1)\right\}$$

for $t \in I\!R$.

Let $(\Omega, \mathfrak{A}, P)$ be a probability space and $\{X_i\}_{i \in I\!N}$ be a p.i.i.d. sequence of fuzzy random variables on $(\Omega, \mathfrak{A}, P)$ with the common distribution function F_X. Then

(i) For all $x \in I\!R$ there exists an zero set $M(x) \in \mathfrak{A}$ such that

$$\{S_n[(X_1)_\omega, \ldots (X_n)_\omega](x)\}_{n \in I\!N} \stackrel{d_\infty}{\rightarrow} F_X(x)$$

and

$$\{T_n[(X_1)_\omega, \ldots (X_n)_\omega](x)\}_{n \in I\!N} \stackrel{d_\infty}{\rightarrow} F_X(x)$$

is valid for all $\omega \in \Omega \backslash M(x)$.

(ii) There exists a zero set $M \in \mathfrak{A}$ such that

$$\{S_n[(X_1)_\omega, \ldots (X_n)_\omega](.)\}_{n \in I\!N} \stackrel{d_{H,Q}}{\rightarrow} F_X(.)$$

and

$$\{T_n[(X_1)_\omega, \ldots (X_n)_\omega](.)\}_{n \in I\!N} \stackrel{d_{H,Q}}{\rightarrow} F_X(.)$$

is valid for $\omega \in \Omega \backslash M$ (in the sense of Definition 6.15 on page 60).

Proof Define

$$\left(\underline{Y_i}\right)_\alpha(\omega) \stackrel{\text{d}}{=} \inf\left(\left(X_i\right)_\omega\right)_\alpha \quad \text{and} \quad \left(\overline{Y_i}\right)_\alpha(\omega) \stackrel{\text{d}}{=} \sup\left(\left(X_i\right)_\omega\right)_\alpha$$

for $i \in I\!N$, $\omega \in \Omega$, and $\alpha \in [0,1)$.

(i) Let $x \in I\!R$ be arbitrary. $\left\{\left(\underline{Y_i}\right)_\alpha\right\}_{i \in I\!N}$ and $\left\{\left(\overline{Y_i}\right)_\alpha\right\}_{i \in I\!N}$ are p.i.i.d. sequences for all $\alpha \in [0,1)$. Define $I \stackrel{\text{d}}{=} [0,1)$ and

$$A_i(\alpha) \stackrel{\text{d}}{=} \left\{\omega \in \Omega \mid \left(\overline{Y_i}\right)_\alpha(\omega) \le x\right\}$$

for $\alpha \in I$, $i \in I\!N$. The assumptions of Theorem 10.16 are satisfied. We conclude: There exists a zero set $M_1(x)$ such that

$$\lim_{n\to\infty} \sup_{\alpha\in[0,1)} \left| \frac{1}{n} \sum_{i=1}^{n} \mathrm{I}_{A_i(\alpha)}(\omega) - P[A_1(\alpha)] \right| = 0$$

is valid for all $\omega \in \Omega \backslash M_1(x)$. We conclude:

$$\lim_{n\to\infty} \sup_{\alpha\in[0,1)} \left| S_n\left[\left(\overline{Y}_1\right)_\alpha(\omega), \ldots, \left(\overline{Y}_n\right)_\alpha(\omega)\right](x) - P\left[\left(\overline{Y}_1\right)_\alpha \le x\right] \right| = 0$$

for $\omega \in \Omega \backslash M_1(x)$.

The Theorem 10.16 is applied a second time. Define $I \stackrel{\text{d}}{=} [0,1)$ and

$$A_i(\alpha) \stackrel{\text{d}}{=} \Omega \backslash \left\{\omega \in \Omega \mid \left(\underline{Y_i}\right)_\alpha(\omega) \le x\right\}$$

for $\alpha \in I$, $i \in I\!N$.

For two independent sets $A \in \mathfrak{A}$, $B \in \mathfrak{A}$

$$P[(\Omega\backslash A) \cap (\Omega\backslash B)] = P[\Omega\backslash(A \cup B)] = 1 - P(A \cup B)$$

$$= 1 - P(A) - P(B) + P(A) \cdot P(B) = P(\Omega\backslash A) \cdot P(\Omega\backslash B)$$

is valid. So the three assertions of the theorem are fulfilled, and we follow:

$$\lim_{n\to\infty} \sup_{\alpha\in[0,1)} \left| \frac{1}{n} \sum_{i=1}^{n} \mathrm{I}_{A_i(\alpha)}(\omega) - P[A_1(\alpha)] \right| = 0.$$

We conclude:

$$\lim_{n\to\infty} \sup_{\alpha\in[0,1)} \left| \frac{1}{n} \sum_{i=1}^{n} \left[1 - \mathrm{I}_{\left[\left(\underline{Y_i}\right)_\alpha \le x\right]}(\omega)\right] - \left[1 - P\left(\left(\underline{Y_1}\right)_\alpha \le x\right)\right] \right| = 0$$

and

$$\lim_{n\to\infty} \sup_{\alpha\in[0,1)} |S_n[(\underline{Y}_1)_\alpha(\omega),\dots,(\underline{Y}_n)_\alpha(\omega)](x) - P((\underline{Y}_1)_\alpha \le x)| = 0$$

for $\omega \in \Omega\backslash M_2(x)$ where $M_2(x)$ is a zero set.

Defining $M(x) \stackrel{\mathrm{d}}{=} M_1(x) \cup M_2(x)$ and applying Lemma 6.2 we obtain:

$$\lim_{n\to\infty} \sup_{\alpha\in[0,1)} d_H \left[\left\{ \begin{array}{c} [S_n[(\overline{Y}_1)_\alpha(\omega),\dots,(\overline{Y}_n)_\alpha(\omega)](x), \\ S_n[(\underline{Y}_1)_\alpha(\omega),\dots,(\underline{Y}_n)_\alpha(\omega)](x)], \\ [P\{(\overline{Y}_1)_\alpha \le x\}, P\{(\underline{Y}_1)_\alpha \le x\}] \end{array} \right\} \right] = 0$$

holds for $\omega \in \Omega\backslash M(x)$. Obviously

$$\left\{ \begin{array}{l} [S_n[(\overline{Y}_1)_\alpha(\omega),\dots,(\overline{Y}_n)_\alpha(\omega)](x), \\ S_n[(\underline{Y}_1)_\alpha(\omega),\dots,(\underline{Y}_n)_\alpha(\omega)](x)] \end{array} \;\middle|\; \alpha \in (0,1) \right\}$$

is a set representation for $T_n[X_1,\dots,X_n](x)$ for all $n \in I\!N$, $\omega \in \Omega$. Theorem 9.3(ii) shows that

$$\{[P\{(\overline{Y}_1)_\alpha \le x\}, P\{(\underline{Y}_1)_\alpha \le x\}] \mid \alpha \in (0,1)\}$$

is a set representation for $F_X(x)$. So the first assertion follows from Lemma 6.8, while Lemma 8.39 yields the second one.

(ii.i) Let $\alpha \in [0,1)$ be fixed. We can apply Theorem 10.16 two more times, first with $I \stackrel{\mathrm{d}}{=} I\!R$ and

$$A_i(x) \stackrel{\mathrm{d}}{=} \{\omega \in \Omega \mid (\underline{Y}_i)_\alpha(\omega) \le x\}$$

for $x \in I$, $i \in I\!N$, and second

$$A_i(x) \stackrel{\mathrm{d}}{=} \bigcup_{r=1}^{\infty} \left\{\omega \in \Omega \;\middle|\; (\underline{Y}_i)_{\alpha+(1-\alpha)/(2r)}(\omega) \le x\right\}.$$

Therefore there exists a zero set $N(\alpha)$ such that

$$\lim_{n\to\infty} \sup_{x\in I\!R} |S_n[(\overline{Y}_1)_\alpha(\omega),\dots,(\overline{Y}_n)_\alpha(\omega)](x) - P\{(\overline{Y}_1)_\alpha \le x\}| = 0$$

and

$$\lim_{n\to\infty} \sup_{x\in I\!R} \left| \lim_{r\to\infty} S_n[(\underline{Y}_1)_{\alpha_r}(\omega),\dots,(\underline{Y}_n)_{\alpha_r}(\omega)](x) - \right.$$

$$\lim_{r\to\infty} P\{(\underline{Y}_1)_{\alpha_r} \leq x\}\Big| = 0$$

holds for $\omega \in \Omega\backslash M(\alpha)$ where $\alpha_r \stackrel{\mathrm{d}}{=} \alpha + (1-\alpha)/(2r)$ for $r \in \mathbb{N}$ With the help of Lemma 8.39 and Theorem 9.4 we conclude that

$$\lim_{n\to\infty} \sup_{x\in\mathbb{R}} d_H\left[(T_n[(X_1)_\omega, \ldots, (X_n)_\omega](x))_\alpha, (F_X(x))_\alpha\right] = 0$$

is valid for $\omega \in \Omega\backslash M(\alpha)$.

(ii.ii) Defining

$$M \stackrel{\mathrm{d}}{=} \bigcup_{\alpha\in[0,1)\cap\mathbb{Q}} N(\alpha)$$

we obtain the two assertions with the help of Definition 6.14 and Lemma 8.42.

10.4 RELATED RESULTS

A slightly different approach to the theory of fuzzy random variables is presented by E.P. Klement, M.L. Puri, and D.A. Ralescu [108,109,174,175]. Their approach relies heavily on probability techniques in Banach spaces, and it generalizes the work on random sets of Z. Artstein and R.A. Vitale [2], W. Weil [267], and E. Gine, M.G. Hahn and J. Zinn [63].

The key tools being embedding theorems, the result cannot be directly implemented in a software tool as it is possible with our results.

Random sets were first considered by H.E. Robbins [182] and later by D.G. Kendall [104,105] and G. Matheron [135]. We recall that a random set is a Borel measurable function X from a probability space $(\Omega, \mathfrak{A}, P)$ to $(\mathfrak{C}(\mathbb{R}), d_H)$, where $\mathfrak{C}(\mathbb{R})$ is the set of all nonempty compact subsets of $\mathbb{R}$, and d_H is the Hausdorff metric (compare Definition 6.1). More generally, it is a mapping X from $(\Omega, \mathfrak{A}, P)$ to $(\mathfrak{C}(\Sigma), d_H)$, where $(\Sigma, \| \ \|)$ is a normed space, and $\mathfrak{C}(\Omega)$ is the class of all nonempty compact subsets of Σ.

A selection f of a random set X is a measurable mapping from $(\Omega, \mathfrak{A}, P)$ to $(\Sigma, \| \ \|)$ such that $f(\omega) \in X(\omega)$ holds almost everywhere. If $\Sigma = \mathbb{R}^d$ with $d \in \mathbb{N}$, then a selector exists as shown by R.J. Aumann [3] as well as by K. Kuratowski and C. Ryll-Nardzewsky [126]. R.J. Aumann defined the expected value $E\,X$ of a random set X by

$$E\,X \stackrel{\mathrm{d}}{=} \{E\,U \,|\, U \text{ is a selection of } X \text{ and } E\,U \text{ exists } \}.$$

It is possible to define the notion "i.i.d. sequence of random sets" (compare P. Billingsley [12]).

The strong law of large numbers for random sets was derived by Z. Artstein and R.A. Vitale [2] for the case $\Sigma = \mathbb{R}^d$. This result was extended to Banach spaces fulfilling a special condition by M. Puri and D.A. Ralescu [179]. Both proofs were given in three steps. In the first step, the collection of all nonempty convex compact subsets of Σ was isometrically embedded into a Banach space Y. In the second step, a limit theorem for this Banach space was applied, and then the constraint "convexity" was omitted. The Banach space Y was the set of all bounded continuous functions from S to $\mathbb{R}$ where $S \stackrel{\mathrm{d}}{=} \{x \in \Sigma \,|\, \|x\| = 1\}$. The limit theorems of E.P. Klement, M.L. Puri and D.A. Ralescu [100,109] are also proved with the help of these three steps.

If X is a random set with values in $(\mathfrak{C}(\Sigma), d_H)$, define $\|X\|$ by

$$\|X\| \stackrel{\mathrm{d}}{=} \sup\{\|X_\omega\| \,|\, \omega \in \Omega\}$$

where

$$\|A\| \stackrel{\mathrm{d}}{=} d_H[A,\{0\}] = \sup\{\|a\| \mid a \in A\}$$

for $A \in \mathfrak{S}(\Sigma)$.

In this section a fuzzy random variable in the sense of Klement, Puri and Ralescu (f.r.v.*) is a Borel measurable function

$$X : (\Omega, \mathfrak{A}, P) \to (F_C(\mathbb{R}), d_\infty).$$

We define the mapping supp X by

$$(\text{supp } X)_\omega \stackrel{\mathrm{d}}{=} \text{cl } (X_\omega)_0$$

for $\omega \in \Omega$ which is a (compact set valued) random set.

If X is f.r.v.* such that $E\,\|\text{supp } X\| < \infty$ ($\|\text{supp } X\|$ is a usual random variable), then the expected value $E\,X \in F_C(\mathbb{R})$ is uniquely defined by

$$(E\,X)_{\overline{\alpha}} \stackrel{\mathrm{d}}{=} E\,(X_{\overline{\alpha}})$$

for $\alpha \in (0,1]$. (We define $X_{\overline{\alpha}}(\omega) \stackrel{\mathrm{d}}{=} (X_\omega)_{\overline{\alpha}}$ for $\omega \in \Omega$ and $\alpha \in (0,1]$; it is a random set).

It makes sense to talk about independent and identically distributed f.r.v.*'s, since $(F_C(\mathbb{R}), d_\infty)$ is a metric space (see P. Billingsley [12]). The strong law of large numbers was formulated by E.P. Klement, M.L. Puri and D.A. Ralescu [108] in the following way:

Theorem 10.18 *Let $\{X_j\}_{j\in\mathbb{N}}$ be an i.i.d. sequence of f.r.v.* 's such that $E\,\|\text{supp } X_1\| < \infty$. Then*

$$\left\{\frac{1}{n}\sum_{i=1}^{n} X_i\right\}_{n\in\mathbb{N}} \xrightarrow{d_1} E\,(\text{co } X_1) \text{ a.e.}$$

Proof It is done in the three steps mentioned above.

(i) $\{\text{co } X_j\}_{j\in\mathbb{N}}$ is an i.i.d. sequence of f.r.v.* 's with values in $U_C(\mathbb{R})$. As it was shown in Theorem 6.10, there exists a linear isometry Φ between $(U_C(\mathbb{R}), d_1)$ and $(Z, \|\ \|)$ where $(Z, \|\ \|)$ is a separable Banach space.

(ii) $\{\Phi(\text{co } X_i)\}_{i\in\mathbb{N}}$ is an i.i.d. sequence of Z-valued random elements. By a strong law of large numbers in separable Banach spaces from M.L. Puri and D.A. Ralescu [172] follows that

$$\left\{\frac{1}{n}\sum_{i=1}^{n} (\Phi(\text{co } X_i))\right\}_{n\in\mathbb{N}} \xrightarrow{\|\ \|} E\,(\Phi(\text{co } X_1)) \text{ a.e.}$$

After some calculations we obtain the equality

$$E\,(\Phi\,(\mathrm{co}\ X_1)) = \Phi\,(E\,(\mathrm{co}\ X_1)),$$

and therefore

$$\left\{\frac{1}{n}\sum_{i=1}^{n} \mathrm{co}\ X_i\right\}_{n\in\mathbb{N}} \xrightarrow{d_1} E\ \mathrm{co}\ X_1 \ \text{ holds a.e.}$$

(iii) The general case can be solved by applying the Shapley-Folkmann-Starr-theorem (see K.T. Arrow and F.H. Hahn [13]). We have

$$d_H\left[\left(\frac{1}{n}\sum_{i=1}^{n} X_i\right)_{\overline{\alpha}}, \left(\frac{1}{n}\sum_{i=1}^{n} \mathrm{co}\ X_i\right)_{\overline{\alpha}}\right] \leq \frac{1}{n}\max_{1\leq i\leq n} \|(X_i)_{\overline{\alpha}}\|$$

for every $\alpha \in (0,1]$. This implies immediately:

$$d_\infty\left(\frac{1}{n}\sum_{i=1}^{n} X_i, \frac{1}{n}\sum_{i=1}^{n} \mathrm{co}\ X_i\right) \leq \frac{1}{n}\max_{1\leq i\leq n} \|\mathrm{supp}\ X_i\|.$$

It follows

$$d_1\left(\frac{1}{n}\sum_{i=1}^{n} X_i, \frac{1}{n}\sum_{i=1}^{n} \mathrm{co}\ X_i\right) \longrightarrow 0 \ \text{ a.e.}$$

(see Y.S. Chow and H. Teicher [19], e.g.), and this yields the assertion •

It has become clear from the preceding considerations that we have to restrict the range of fuzzy random variables in order to obtain meaningful results. For the central limit theorem, we consider the space $F_{\mathrm{Lip}}\,(\mathbb{R})$ of fuzzy sets $\mu \in F_C\,(\mathbb{R})$ such that the map $\alpha \mapsto \mu_{\overline{\alpha}}$ satisfies a Lipschitz condition, i.e. there exists a constant $c > 0$ such that $d_H\left[\mu_{\overline{\alpha}}, \mu_{\overline{\beta}}\right] \leq c\cdot|\alpha-\beta|$ for every $(\alpha,\beta) \in (0,1]^2$. Define the space $U_{CL}\,(\mathbb{R}) \stackrel{\mathrm{d}}{=} U_C\,(\mathbb{R}) \cap F_{\mathrm{Lip}}\,(\mathbb{R})$.

We can embed this space into the space $Z = \mathfrak{C}\,([0,1]\times\{-1,+1\})$, the Banach space of all continuous functions on $[0,1]\times\{-1,+1\}$, with the Tschebychev norm

$$\|f\|_\infty \stackrel{\mathrm{d}}{=} \sup\,\{|f(t,u)|\,|\,t\in[0,1], u\in\{-1,+1\}\}$$

for $f \in Z$. The embedding function Φ is a linear isometry from $(U_{CL}(\mathbb{R}), d_\infty)$ to $(Z, \|\ \|_\infty)$ and is given by

$$\Phi(\mu)(\alpha,u) \stackrel{\mathrm{d}}{=} \begin{cases} \sup\limits_{x\in\mu_{\overline{\alpha}}} x\cdot u & \text{, if } \alpha\in(0,1] \\ \sup\limits_{x\in\mathrm{cl}\ \mu_0} x\cdot u & \text{, if } \alpha=0 \end{cases}$$

for $\alpha \in [0,1]$ and $u \in \{-1,+1\}$.

Theorem 10.19 *Let $\{X_j\}_{j\in\mathbb{N}}$ be an i.i.d. sequence of f.r.v.* 's with values in $F_{Lip}(\mathbb{R})$ such that*
(i) $E\,\|\text{supp}\ X_1\|^2 < \infty$, and
(ii) $E\left[\sup\limits_{\alpha\neq\beta} \frac{d_H\left[(X_1)_{\overline{\alpha}},(X_1)_{\overline{\beta}}\right]}{|\alpha-\beta|}\right]^2 < +\infty$ hold.

Then there exists a Gaussian random element Y in $\mathfrak{C}([0,1]\times\{-1,+1\})$ such that

$$\sqrt{n}\cdot d_\infty\left(\frac{1}{n}\sum_{i=1}^{n} X_i, E\ (\text{co}\ X_1)\right) \longrightarrow \|Y\|_\infty$$

in distribution.

Proof We consider first the convex case, i.e. $X_k : \Omega \to U_{CL}(\mathbb{R})$. $Y_k \stackrel{\mathrm{d}}{=} \Phi(X_k)$ can be considered as random element in $\mathfrak{C}([0,1]\times\{-1,1\})$. A short calculation shows that the central limit theorem of N.C. Jain and M.B. Marcus [100] can be applied. Thus

$$\sqrt{n}\left[\frac{1}{n}\sum_{i=1}^{n}\Phi(X_i) - E\ \Phi(X_1)\right] \to Y \text{ weakly,}$$

where Z is Gaussian in $\mathfrak{C}([0,1]\times\{-1,+1\})$. It is not difficult to show that $E\ \Phi(X_1) = \Phi(E\ X_1)$ holds. This implies

$$\sqrt{n}\ d_\infty\left(\frac{1}{n}\sum_{i=1}^{n} X_i, E\ X_1\right) \to \|Y\|_\infty \text{ weakly.}$$

Assume now that $X_k : \Omega \to F_L(\mathbb{R})$, and consider the f.r.v.* 's co X_k. The Shapley-Folkman-Starr-theorem gives

$$\sqrt{n}d_\infty\left(\frac{1}{n}\sum_{i=1}^{n} X_i, E\ (\text{co}\ X_1)\right) \leq$$

$$\frac{1}{\sqrt{n}} \max_{1 \leq i \leq n} \|\text{supp } X_i\| + \sqrt{n}\, d_\infty \left(\frac{1}{n} \sum_{i=1}^{n} \text{co } X_i \,,\; E\,(\text{co } X_i) \right).$$

$E\,\|\text{supp } X_1\|^2 < \infty$ implies that $\frac{1}{\sqrt{n}} \max_{1 \leq i \leq n} \|\text{supp } X_1\| \to 0$ weakly. With the notion

$$\|\mu\|_L = \sup_{\alpha \neq \beta} \frac{d_H\left[\mu_{\overline{\alpha}}, \mu_{\overline{\beta}}\right]}{|\alpha - \beta|} + \sup_{\alpha > 0} \|\mu_{\overline{\alpha}}\|,$$

we obtain the inequalities

$$\|\text{co } \mu\|_L \leq \|\mu\|_L \text{ and } E\,\|\text{co } \|X_1|_L^2 < \infty.$$

Therefore the second term of the above inequality converges weakly to $\|Y\|_\infty$ by part 1 of the proof. •

11 Some aspects of statistical inference

11.1 PARAMETRIC POINT ESTIMATION

We now begin the study of problems of statistical inference. In this section we shall study the problems associated with parametric estimation when the sample size is fixed. Suppose for example that a random variable U is known to have a normal distribution $N(\mu, \sigma^2)$, but one of the parameters is not known, say μ. Suppose further that a sample $U_1, \ldots, U_n$ is taken on U. The problem of point estimation is to pick a (one dimensional) statistic $T_n(U_1, \ldots, U_n)$ that estimates best the parameter μ. The numerical value of T_n, when the realization is $x_1, \ldots, x_n$, is called an estimate of μ, while the statistic $T_n(x_1, \ldots, x_n)$ is called estimator of μ.

Let Y denote in the following a class of distribution functions depending on r parameters with $r \in I\!N$. We consider a random experiment which can be described by a random variable U, where we assume that the distribution function of U, say D_U, belongs to Y. The problem of point estimation is to give a good guess for one of the r parameters of D_U, say $\Gamma_Y(D_U)$. Γ_Y can be viewed as a mapping that assigns to each distribution function of Y its parameter.

If, for example, Y is the class of all normal distributions $N(\mu, \sigma^2)$, one interesting parameter would be the expected value μ, i.e. Γ_Y assigns to each $N(\mu, \sigma^2)$- distribution its parameter μ. Taking a simple sample of size n means to consider a realization of an i.i.d. random vector $(U_1, \ldots, U_n)$ where U_i has the distribution function D_U for $i \in \{1, \ldots, n\}$.

We want to consider the situation that the data are not precise but vague. Is it possible to decide, on the basis of these vague observations, which parameter Γ_Y may represent f.e. the expected value ? Our proposed solution is to describe the random experiment with vague outcomes by a mapping

$$(X_1, \ldots, X_n) : \Omega \to [F(I\!R)]^n,$$

which we call a vague random sample. We then define the notion of a vague parameter Γ_Y of this fuzzy random variable which generalizes the above described notion. This vague parameter may be viewed as a perception of the unknown parameter. Point estimation then deals with the problem of giving good guesses for the vague parameter.

In order to work out a reasonable definition for this value parameter Γ_Y, we extend the concepts of Chapter 7. Let $(\Omega, \mathfrak{A}, P)$ (the sample space) and $(\Omega', \mathfrak{A}', P')$ (measure errors or influences of the environment) are two arbitrary probability spaces. A mapping $X : \Omega \to F(I\!R)$ is considered as a perception of a (usual) random variable

$$U : \Omega \times \Omega' \to I\!R.$$

If Ω' has only one element, we can ignore Ω', it has no influence, and $U \in \chi$ holds. If for all $\lambda \in [0, 1]$ there is an $A \in \mathfrak{A}'$ with $P'(A) = \lambda$, then $U \in \tilde{\chi}$ is valid.

The observed realizations of a fuzzy random vector

$$(X_1, \ldots, X_n) : \Omega \to [F(I\!R)]^n$$

can be considered as the fuzzy perception of an i.i.d. random vector

$$(U_1, \ldots, U_n) : \Omega \times \Omega' \to I\!R^n.$$

We assume that the distribution function of U_1 belongs to Y. So the set of all possible originals is χ_Y^n. χ_Y^n is the set of all i.i.d. random vectors $(V_1, \ldots, V_n)$ of size n on $\Omega \times \Omega'$ which are $\mathfrak{A} \otimes \mathfrak{A}' \to \mathfrak{B}_n$-measurable such that $D_{V_1} \in Y$ holds. $\mathfrak{B}_n$ denotes the σ-algebra on $I\!R^n$.

The acceptability that a given i.i.d.-random vector $(V_1, \ldots, V_n)$ on $\Omega \times \Omega'$ is original of $(X_1, \ldots, X_n)$ is (Definition 7.3)

$$\tilde{\mu}_{(X_1, \ldots, X_n)}(V_1, \ldots, V_n) \stackrel{\mathrm{d}}{=} \min_{i=1,\ldots,n} \inf \{(X_i)_\omega [V_i(\omega, \omega')] \,|\, \omega \in \Omega, \omega' \in \Omega'\}.$$

With the help of L.A. Zadeh's extension principle [214], we can give a canonical definition of a fuzzy parameter.

Definition 11.1 *Let $n \in I\!N$. The vague parameter $\Gamma_Y[X_1, \ldots, X_n]$ of the vague random sample (i.e. fuzzy random vector)*

$$(X_1,\ldots,X_n) : \Omega \to [F(\mathbb{R})]^n$$

with respect to $\Gamma_Y : Y \to \mathbb{R}$ *is a fuzzy set defined by*

$$\Gamma_Y[X_1,\ldots,X_n](t) \overset{\text{d}}{=} \sup\left\{ \tilde{\mu}_{(X_1,\ldots,X_n)}(V_1,\ldots,V_n) \,\middle|\, \begin{array}{r} (V_1,\ldots,V_n) \in \chi_Y^n, \\ \Gamma_Y(D_{V_1}) = t \end{array} \right\}$$

for $t \in \mathbb{R}$ *.*

If there exists an $(U_1,\ldots,U_n) \in \chi_Y^n$ with $(X_i)_\omega = \mathrm{I}_{\{U_i(\omega)\}}$ for $i \in \{1,\ldots,n\}$ and $\omega \in \Omega$, then

$$\Gamma_Y[X_1,\ldots,X_n] = \mathrm{I}_{\Gamma_Y(D_{U_1})} \text{ holds.}$$

It is easy to show that under the assumptions of Definition 11.1 the following equality holds for all $\alpha \in [0,1)$:

$$(\Gamma_Y[X_1,\ldots,X_n])_\alpha =$$

$$\left\{ t \in \mathbb{R} \,\middle|\, \begin{array}{l} \exists (V_1,\ldots,V_n) \in \chi_Y^n \text{ with } \Gamma_Y(D_{V_1}) = t \text{ such that} \\ V_i(\omega,\omega') \in ((X_i)_\omega)_\alpha \text{ holds for } \omega \in \Omega, \omega' \in \Omega', i \in \{1,\ldots,n\} \end{array} \right\}$$

Unfortunately, if two fuzzy random variables are identically distributed, then their parameters generally do not coincide. We have to require some assumptions to assert this desired property.

Theorem 11.2 *Let* Y *be a class of distribution functions and* Γ_Y *be a non-increasing mapping from* Y *to* $\mathbb{R}$ *, i.e.* $D_1 \in Y$ *and* $D_2 \in Y$ *and* $D_1(t) \geq D_2(t)$ *for* $t \in \mathbb{R}$ *implies* $\Gamma_Y(D_1) \leq \Gamma_Y(D_2)$*.*

Let $n \in \mathbb{N}$ *and* $(X_1,\ldots,X_n)$ *be a completely independent and identically distributed fuzzy random vector with values in* $[F_C(\mathbb{R})]^n$ *such that* $D_{(\underline{Y_1})_\alpha} \in Y$ *and* $D_{(\overline{Y_1})_\alpha} \in Y$ *holds for all* $\alpha \in [0,1)$ *.*
Then the following properties hold:

(i) $\left\{ \left[\Gamma_Y\left(D_{(\underline{Y_1})_\alpha}\right), \Gamma_Y\left(D_{(\overline{Y_1})_\alpha}\right) \right] \mid \alpha \in (0,1) \right\}$ *is a set representation of* co $\Gamma_Y[X_1,\ldots,X_n]$ *as well as of* co $\Gamma_Y[X_i]$ *for* $i \in \{1,\ldots,n\}$

(ii) co $\Gamma_Y[X_1,\ldots,X_n] =$ co $\Gamma_Y[X_i]$ *holds for* $i \in \{1,\ldots,n\}$

(iii) There exists a $t \in \mathbb{R}$ *with* co $\Gamma_Y[X_1](t) = 1$.

Proof We can apply Lemma 3.17(ii) and follow for $\omega \in \Omega$, $\alpha \in (0,1)$, $i \in \{1,\ldots,n\}$:

$$((X_i)_\omega)_\alpha \subseteq \text{cl}\ ((X_i)_\omega)_\alpha \subseteq ((X_i)_\omega)_{\overline{\alpha}}.$$

For all $\alpha \in (0,1)$ it follows:

$$\begin{aligned}
&(\Gamma_Y[X_1,\ldots,X_n])_\alpha \\
&= \left\{ t \in I\!R \;\middle|\; \begin{array}{l} \exists (V_1,\ldots,V_n) \in \chi_Y^n \text{ with } \Gamma_Y(D_{V_1}) = t \text{ such that} \\ V_i(\omega,\omega') \in ((X_i)_\omega)_\alpha \text{ for all } \omega \in \Omega,\ \omega' \in \Omega',\ i \in \{1,\ldots,n\} \end{array} \right\} \\
&\subseteq A_\alpha \overset{\text{d}}{=} \left\{ t \in I\!R \;\middle|\; \begin{array}{l} \exists (V_1,\ldots,V_n) \in \chi_Y^n \text{ with } \Gamma_Y(D_{V_1}) = t \text{ such that} \\ V_i(\omega,\omega') \in \text{cl}\ ((X_i)_\omega)_\alpha \\ \quad \text{for all } \omega \in \Omega,\ \omega' \in \Omega',\ i \in \{1,\ldots,n\} \end{array} \right\} \\
&\subseteq \left\{ t \in I\!R \;\middle|\; \begin{array}{l} \exists (V_1,\ldots,V_n) \in \chi_Y^n \text{ with } \Gamma_Y(D_{V_1}) = t \text{ such that} \\ V_i(\omega,\omega') \in ((X_i)_\omega)_{\overline{\alpha}} \text{ for all } \omega \in \Omega,\ i \in \{1,\ldots,n\} \end{array} \right\} \\
&\subseteq (\Gamma_Y[X_1,\ldots,X_n])_{\overline{\alpha}}.
\end{aligned}$$

So $\{A_\alpha \mid \alpha \in (0,1)\}$ is a set representation of $\Gamma_Y[X_1,\ldots,X_n]$ because of Theorem 3.5.

On the one hand, $(\underline{Y_i})_\alpha(\omega) \in \text{cl}\ ((X_i)_\omega)_\alpha$ and $(\overline{Y}_i)_\alpha(\omega) \in \text{cl}\ ((X_i)_\omega)_\alpha$ is valid for all $\omega \in \Omega$, $i \in \{1,\ldots,n\}$, $\alpha \in (0,1)$. As $((\underline{Y}_1)_\alpha,\ldots,(\underline{Y}_n)_\alpha) \in \chi_Y^n$ and $((\overline{Y}_1)_\alpha,\ldots,(\overline{Y}_n)_\alpha) \in \chi_Y^n$ holds for all $\alpha \in [0,1)$ (compare Theorem 9.13(i)), we can conclude:

$$\Gamma_Y\left(D_{(\underline{Y}_1)_\alpha}\right) \in A_\alpha$$

and

$$\Gamma_Y\left(D_{(\overline{Y}_1)_\alpha}\right) \in A_\alpha$$

is valid for $\alpha \in (0,1)$.

On the other hand, let $\alpha \in (0,1)$ and $(V_1,\ldots,V_n) \in \chi_Y^n$ such that $V_i(\omega,\omega') \in \text{cl}\ ((X_i)_\omega)_\alpha$ holds for $(\omega,\omega') \in \Omega \times \Omega'$, $i \in \{1,\ldots,n\}$. We conclude:

$$(\underline{Y}_i)_\alpha(\omega) \leq V_i(\omega,\omega') \leq (\overline{Y}_i)_\alpha(\omega)$$

for $\omega \in \Omega$, $i \in \{1,\ldots,n\}$. Therefore

$$D_{(\underline{Y}_1)_\alpha}(t) \geq D_{V_1}(t) \geq D_{(\overline{Y}_1)_\alpha}(t)$$

holds for $t \in \mathbb{R}$. This implies:

$$\Gamma_Y\left(D_{(\underline{Y}_i)_\alpha}\right) \leq \Gamma_Y\left(D_{V_1}\right) \leq \Gamma_Y\left(D_{(\overline{Y}_1)_\alpha}\right).$$

So

$$A_\alpha \subseteq \left[\Gamma_Y\left(D_{(\underline{Y}_1)_\alpha}\right), \Gamma_Y\left(D_{(\overline{Y}_1)_\alpha}\right)\right]$$

holds by definition.
Both steps combined show:

$$\text{co } A_\alpha = \left[\Gamma_Y\left(D_{(\underline{Y}_1)_\alpha}\right), \Gamma_Y\left(D_{(\overline{Y}_1)_\alpha}\right)\right].$$

All other assertions can be shown in a similar way. •

Definition 11.3 *Let $n \in \mathbb{N}$, Y be a class of distribution functions and $\Gamma_Y : Y \to \mathbb{R}$. Let $(X_1, \ldots, X_n)$ be a vague random sample from Y. Each mapping*

$$\begin{aligned} T_n : \ \Omega \ &\rightarrow \ E(\mathbb{R}) \\ \omega \ &\mapsto \ T_n[(X_1)_\omega, \ldots, (X_n)_\omega] \end{aligned}$$

is called point estimate for the unknown parameter Γ_Y.

The problem of parametric point estimation is to find an estimate T_n for the unknown parameter Γ_Y that has some useful properties.

Example Let Y denote the set of all normal distributions $N(\mu, 1)$ with $\mu \in \mathbb{R}$. Let $(X_1, \ldots, X_n)$ be a vague random sample from Y. Then

$$T_n = \overline{X}_n \stackrel{\mathrm{d}}{=} \frac{1}{n} \sum_{i=1}^{n} X_i$$

is an estimate of Γ_Y. Some other estimates are $T_n = \mathrm{I}_{\{0.5\}}$ and $T_n = X_1$ and $T_n = \frac{1}{2}(X_1 + X_n)$

These examples show that we need some criteria to choose among possible estimates. In the following we will consider some properties that we may require our estimates to possess, and we will discuss some common methods of estimation. Let us first introduce the concept of consistency.

Definition 11.4 *Let $T_n : \Omega \to F(\mathbb{R})$ be a f.r.v. for all $n \in \mathbb{N}$, and let $\mu \in F(\mathbb{R})$.*

(a) $\{T_n\}_{n \in I\!N}$ *converges in probability against* μ — *in signs:* $\{T_n\}_{n \in I\!N} \xrightarrow{P} \mu$ — *if and only if*

$$\lim_{n \to \infty} P(\{\omega \in \Omega \mid d_\infty(T_n(\omega), \mu) > \epsilon\}) = 0$$

holds for all $\epsilon > 0$.

(b) $\{T_n\}_{n \in I\!N}$ *is called unbiased if* $E(T_n) = \mu$ *is valid for* $n \in I\!N$.

If we want to estimate an unknown parameter by a sequence of fuzzy random variables, some assumptions must be fulfilled, so we consider classes of such sequences.

Definition 11.5 *Let* Y *be a class of distribution functions, and* $\Gamma_Y : Y \to I\!R$ *be a mapping. Let* $T_n : [F(I\!R)]^n \to F(I\!R)$ *be a mapping for all* $n \in I\!N$. *Let* $\mathfrak{S}$ *be a class of sequences* $\{X_j\}_{j \in I\!N}$ *of f.r.v.'s such that*

$$\text{co } (\Gamma_Y[X_i]) = \text{co } (\Gamma_Y[X_1, \ldots, X_n])$$

is valid for all $\{X_j\}_{j \in I\!N} \in \mathfrak{S}$, $i \in I\!N$, $n \in I\!N$.

(a) $\{T_n\}_{n \in I\!N}$ *is called a consistent estimator for* Y *and* Γ_Y *with respect to* $\mathfrak{S}$ *if for all sequences* $\{X_j\}_{j \in I\!N}$ *in* $\mathfrak{S}$

$$\{T_n[X_1, \ldots, X_n]\}_{n \in I\!N} \xrightarrow{P} \text{co } \Gamma_Y[X_1]$$

holds.

(b) $\{T_n\}_{n \in I\!N}$ *is called a strong consistent estimator for* Y *and* Γ_Y *with respect to* $\mathfrak{S}$ *if for all sequences* $\{X_j\}_{j \in I\!N}$ *in* $\mathfrak{S}$

$$\{T_n[X_1, \ldots, X_n]\}_{n \in I\!N} \xrightarrow{d_\infty} \text{co } \Gamma_Y[X_1]$$

holds almost sure.

(c) $\{T_n\}_{n \in I\!N}$ *is called unbiased with respect to* $\mathfrak{S}$ *if for all sequences* $\{X_j\}_{j \in I\!N}$ *in* $\mathfrak{S}$ *and all* $n \in I\!N$

$$\tilde{E}\{T_n[X_1, \ldots, X_n]\} = \text{co } \Gamma_Y[X_1]$$

holds.

Suppose that the age of a population in a given region follows a Poisson distribution with the unknown parameter $\lambda > 0$. Then the arithmetic mean is a good, i.e. unbiased and consistent, estimate of λ. This is the classical situation. Now consider the case of vague data. How can we construct a reasonable estimate in this case ?

If the unknown parameter is the expected value, the mean value $\overline{X}_n$ is a good estimator as the strong law of large numbers for f.r.v.'s shows. In a generalization, the following theorem shows that if a good estimate t_n in the classical case is known, then we can construct a good estimate T_n in the presence of vague data.

Theorem 11.6 *Let Y be a class of distribution functions and $\Gamma_Y : Y \to \mathbb{R}$ be non-increasing and continuous, i.e. if $\{Z_n\}_{n\in\mathbb{N}}$ is a sequence of usual random variables such that $D_{Z_n} \in Y$ holds for $n \in \mathbb{N}$, and if Z is a usual random variable with $D_Z \in Y$, then $\{Z_n(\omega)\}_{n\in\mathbb{N}} \to Z(\omega)$ for $\omega \in \Omega$ implies*

$$\{\Gamma_Y(D_{Z_n})\}_{n\in\mathbb{N}} \to \Gamma_Y(D_Z).$$

Let $\{t_n\}_{n\in\mathbb{N}}$ be a consistent estimator in the classical sense for Y and Γ_Y, i.e.

$$\{t_n(V_1,\ldots,V_n)\}_{n\in\mathbb{N}} \xrightarrow{P} \Gamma_Y(D_{V_1})$$

if $\{V_i\}_{i\in\mathbb{N}}$ i.i.d. and $D_{V_1} \in Y$, such that $t_n : \mathbb{R}^n \to \mathbb{R}$ is non-decreasing in each component and continuous for all $n \in \mathbb{N}$. Define

$$\begin{array}{llcl} T_n : & [F(\mathbb{R})]^n & \to & F(\mathbb{R}) \\ & (\mu_1,\ldots,\mu_n) & \mapsto & \text{co } t_n[\mu_1,\ldots,\mu_n] \end{array}$$

for $n \in \mathbb{N}$ (compare Definition 4.4 and Definition 3.10).

Then $\{T_n\}_{n\in\mathbb{N}}$ is a consistent estimator for Y and Γ_Y with respect to $\mathfrak{S}$ where $\{X_j\}_{j\in\mathbb{N}} \in \mathfrak{S}$ holds if and only if $\{X_j\}_{j\in\mathbb{N}}$ is an i.i.d. sequence of f.r.v.'s with values in $F_C(\mathbb{R})$ such that $D_{(\underline{Y}_1)_\alpha} \in Y$, $D_{(\overline{Y}_1)_\alpha} \in Y$, $D_{(\underline{X}_1)_\beta} \in Y$, and $D_{(\overline{X}_1)_\beta} \in Y$ holds for all $\alpha \in [0,1)$ and $\beta \in (0,1]$. We define:

$$(\underline{Y}_1)_\alpha(\omega) \stackrel{\text{d}}{=} \inf((X_1)_\omega)_\alpha \quad \text{and} \quad (\overline{Y}_1)_\alpha(\omega) \stackrel{\text{d}}{=} \sup((X_1)_\omega)_\alpha$$

$$(\underline{X}_1)_\alpha(\omega) \stackrel{\text{d}}{=} \inf((X_1)_\omega)_{\overline{\beta}} \quad \text{and} \quad (\overline{X}_1)_\alpha(\omega) \stackrel{\text{d}}{=} \sup((X_1)_\omega)_{\overline{\beta}}.$$

Proof

(a) Let $n \in \mathbb{N}$ and $(\mu_1,\ldots,\mu_n) \in [F_C(\mathbb{R})]^n$. Lemma 3.5(i) and Lemma 4.5(i) show that

$$\{t_n[\text{cl }(\mu_1)_\alpha,\ldots,\text{cl }(\mu_n)_\alpha] \mid \alpha \in (0,1)\}$$

is a set representation of $t_n[\mu_1,\ldots,\mu_n]$.
Let $\alpha \in (0,1)$.
On the one hand,

$$T_n[\inf(\mu_1)_\alpha,\ldots,\inf(\mu_n)_\alpha] \in t_n[\text{cl}\ (\mu_1)_\alpha,\ldots,\text{cl}\ (\mu_n)_\alpha] \text{ and}$$
$$T_n[\sup(\mu_1)_\alpha,\ldots,\sup(\mu_n)_\alpha] \in t_n[\text{cl}\ (\mu_1)_\alpha,\ldots,\text{cl}\ (\mu_n)_\alpha]$$

holds. On the other hand, the monotony of t_n shows

$$t_n[\text{cl}\ (\mu_1)_\alpha,\ldots,\text{cl}\ (\mu_n)_\alpha]$$
$$\subseteq [t_n[\inf(\mu_1)_\alpha,\ldots,\inf(\mu_n)_\alpha], t_n[\sup(\mu_1)_\alpha,\ldots,\sup(\mu_n)_\alpha]].$$

Therefore

$$\{[t_n[\inf(\mu_1)_\alpha,\ldots,\inf(\mu_n)_\alpha], t_n[\sup(\mu_1)_\alpha,\ldots,\sup(\mu_n)_\alpha]] \mid \alpha \in (0,1)\}$$

is a set representation of $T_n[\mu_1,\ldots,\mu_n]$.
It follows for $\alpha \in [0,1)$:

$$\inf(T_n[\mu_1,\ldots,\mu_n])_\alpha = t_n[\inf(\mu_1)_\alpha,\ldots,\inf(\mu_n)_\alpha] \text{ and}$$
$$\sup(T_n[\mu_1,\ldots,\mu_n])_\alpha = t_n[\sup(\mu_1)_\alpha,\ldots,\sup(\mu_n)_\alpha]$$

and for $\alpha \in (0,1]$ we have with Lemma 3.15(ii) and as t_n is continuous

$$(T_n[\mu_1,\ldots,\mu_n])_{\overline{\alpha}}$$
$$= [t_n[\inf(\mu_1)_\alpha,\ldots,\inf(\mu_n)_\alpha], t_n[\sup(\mu_1)_\alpha,\ldots,\sup(\mu_n)_\alpha]].$$

(b) Let $\{X_j\}_{j\in\mathbb{N}}$ be an i.i.d. sequence of f.r.v.'s with values in $F_C(\mathbb{R})$ such that $D_{(\underline{Y}_1)_\alpha} \in Y$ and $D_{(\overline{Y}_1)_\alpha} \in Y$ for $\alpha \in [0,1)$ and $D_{(\underline{X}_1)_\alpha} \in Y$ and $D_{(\overline{X}_1)_\alpha} \in Y$ for $\alpha \in (0,1]$ holds. We know that

$$\{t_n[(\underline{Y}_1)_\alpha,\ldots,(\underline{Y}_n)_\alpha]\}_{n\in\mathbb{N}} \xrightarrow{P} \Gamma_Y\left(D_{(\underline{Y}_1)_\alpha}\right)$$

and

$$\{t_n[(\overline{Y}_1)_\alpha,\ldots,(\overline{Y}_n)_\alpha]\}_{n\in\mathbb{N}} \xrightarrow{P} \Gamma_Y\left(D_{(\overline{Y}_1)_\alpha}\right)$$

is valid.
With Lemma 6.2 and Theorem 11.2 we can follow for $\alpha \in [0,1)$:

$$\lim_{n\to\infty} P\left(\{\omega \in \Omega \mid d_H\left[(T_n[(X_1)_\omega,\ldots,(X_n)_\omega])_\alpha, \text{co}\ (\Gamma_Y(X_1))_\alpha\right] > \epsilon\}\right)$$

$$= 0$$

holds for all $\epsilon > 0$.
For all $\alpha \in (0,1]$ $\{(\underline{X}_i)_\alpha\}_{i\in\mathbb{N}}$ and $\{(\overline{X}_i)_\alpha\}_{i\in\mathbb{N}}$ are i.i.d. sequences. It is possible to show that

$$\inf(\Gamma_Y[X_1])_{\overline{\alpha}} = \Gamma_Y\left(D_{(\underline{X}_1)_\alpha}\right) \quad \text{and}$$

$$\sup(\Gamma_Y[X_1])_{\overline{\alpha}} = \Gamma_Y\left(D_{(\overline{X}_1)_\alpha}\right)$$

holds for $\alpha \in (0,1]$. We conclude for $\alpha \in (0,1]$:

$$\lim_{n\to\infty} P\left[\left\{\omega \in \Omega \mid d_H\left[(T_n[(X_1)_\omega,\ldots,(X_n)_\omega])_{\overline{\alpha}}, (\text{co } \Gamma_Y[X_1])_{\overline{\alpha}}\right] > \epsilon\right\}\right] = 0$$

for all $\epsilon > 0$.
We can apply a technique similar to the proof of Theorem 10.5 and obtain the assertion.

Theorem 11.7 *Let Y be a class of distribution functions and $\Gamma_Y : Y \to \mathbb{R}$ be a non-increasing mapping. Let $\{t_n\}_{n\in\mathbb{N}}$ be an unbiased estimator in the classical sense for Y and Γ_Y. Define T_n as in Theorem 11.6 for $n \in \mathbb{N}$. Then $\{T_n\}_{n\in\mathbb{N}}$ is an unbiased estimator for Y and Γ_Y with respect to $\mathfrak{S}$ where $\mathfrak{S}$ is the class of all i.i.d. sequences $\{X_j\}_{j\in\mathbb{N}}$ of f.r.v.'s with values in $F_C(\mathbb{R})$ such that $D_{(\underline{Y}_1)_\alpha} \in Y$ and $D_{(\overline{Y}_1)_\alpha} \in Y$ is valid for all $\alpha \in [0,1)$.*

Proof Let $\{X_j\}_{j\in\mathbb{N}}$ be an i.i.d. sequence of f.r.v.'s with values in $F_C(\mathbb{R})$ such that $D_{(\underline{Y}_1)_\alpha} \in Y$ and $D_{(\overline{Y}_1)_\alpha} \in Y$ holds for $\alpha \in [0,1)$. It follows for $\omega \in \Omega$, $\alpha \in [0,1)$, $n \in \mathbb{N}$:

$$\inf(T_n[(X_1)_\omega,\ldots,(X_n)_\omega])_\alpha = t_n[(\underline{Y}_1)_\alpha(\omega),\ldots,(\underline{Y}_n)_\alpha(\omega)]$$

and

$$\sup(T_n[(X_1)_\omega,\ldots,(X_n)_\omega])_\alpha = t_n\left[(\overline{Y}_1)_\alpha(\omega),\ldots,(\overline{Y}_n)_\alpha(\omega)\right].$$

As t_n is unbiased,

$$E\{t_n[(\underline{Y}_1)_\alpha,\ldots,(\underline{Y}_n)_\alpha]\} = \Gamma_Y\left(D_{(\underline{Y}_1)_\alpha}\right) \quad \text{and}$$

$$E\left\{t_n\left[(\overline{Y}_1)_\alpha,\ldots,(\overline{Y}_n)_\alpha\right]\right\} = \Gamma_Y\left(D_{(\overline{Y}_1)_\alpha}\right)$$

is valid for $\alpha \in [0,1)$, $n \in \mathbb{N}$. We can combine theorem 8.3 and Theorem 11.2(i), and the assertion follows. •

Often the parameter which we are looking for, is the expected value of the characteristic of the population. The strong law of large numbers allows us to derive a strong consistent estimator. Let E assign to each distribution function D its expected value (if it exists), i.e. $E\,(D) = E\,U$, if $D_U = D$ and $E\,|U| < \infty$ holds.

Theorem 11.8 *Let Y be a class of distribution functions. Define for $n \in I\!N$*

$$\begin{array}{cccc} T_n : & [F\,(I\!R)]^n & \to & F\,(I\!R) \\ & (\mu_1, \ldots, \mu_n) & \mapsto & \text{co } \frac{1}{n} \sum_{i=1}^{n} \mu_i \end{array}.$$

Then $\{T_n\}_{n \in I\!N}$ is a strong consistent and unbiased estimator for Y and E with respect to $\mathfrak{S}$ where $\mathfrak{S}$ is defined as in Theorem 11.7.

Proof Let $\{X_j\}_{j \in I\!N} \in \mathfrak{S}$ be arbitrary. As $\Gamma_Y = E$ holds, Theorem 11.2(i) shows that

$$\left\{ \left[E\,(\underline{Y}_1)_\alpha, E\,(\overline{Y}_1)_\alpha \right] \mid \alpha \in (0,1) \right\}$$

is a set representation of co $\Gamma_Y[X_1]$. By Theorem 8.3(iv) we follow:

$$\text{co }(\Gamma_Y[X_1]) = \text{co } E\,X_1.$$

Theorem 10.5 shows that

$$\left\{ \frac{1}{n} \sum_{i=1}^{n} X_i \right\}_{n \in I\!N} \xrightarrow{d_\infty} E \text{ co } X_1 \quad \text{almost sure},$$

it follows:

$$\left\{ \text{co } \frac{1}{n} \sum_{i=1}^{n} X_i \right\}_{n \in I\!N} \xrightarrow{d_\infty} \text{co } E \text{ co } X_1 = \text{co } E\,X_1 \quad \text{almost sure.}$$

With the help of Theorem 11.7, we can show that T_n is unbiased. •

Applying Theorem 10.14, we can also obtain a strong consistent estimator $\{T_n\}_{n \in I\!N}$ for Y and Var with respect to a suitable class $\mathfrak{S}$. From Theorem 11.8 follows that

$$T_n(X_1, \ldots, X_n) = \frac{1}{n} \sum_{i=1}^{n} X_i$$

is a good estimate in the case of an underlying Poisson distribution. Even if the assumptions of the theorem are not satisfied, the methods presented here can be used to construct reasonable estimates, for example maximum likelihood estimates.

11.2 CONFIDENCE ESTIMATION

In many problems of statistical inference, the experimenter is interested in constructing a family of sets that contains the true (unknown) parameter with a specified (high) probability. If Y, f.e., represents the length of life of a piece of equipment, the experimenter is interested in a lower bound $\underline{\theta}$ for the mean θ of U. Since $\underline{\theta}$ will be a function of the observations, one cannot ensure with probability 1 that $\underline{\theta} \leq \theta$. We can only choose a number $1-\delta$ being close to 1 such that

$$P_\theta[\underline{\theta}(X_1, \ldots, X_n) \leq \theta] \geq 1-\delta$$

for all θ.

Problems of this type are problems of confidence estimation. In the following we will derive vague confidence intervals for our vague parameter. We repeat the following notions of classical statistics (compare [20,52,83,183,187]).

Let $\delta \in (0,1)$ and $n \in I\!N$. A(usual) one-sided $100 * (1-\delta)\%$confidence interval $[T_n, +\infty)$ for Y and Γ_Y is a $\mathfrak{B}_n \rightarrow \mathfrak{B}_1$-measurable mapping $T_n : I\!R^n \rightarrow I\!R$ such that $(V_1, \ldots, V_n) \in \chi_Y^n$ implies

$$(P \otimes P')(\{(\omega, \omega') \in \Omega \times \Omega' \mid T_n[V_1, (\omega, \omega'), \ldots, V_n(\omega, \omega')] \leq \Gamma_Y(D_{V_1})\})$$
$$\geq 1-\delta.$$

In a similar way a confidence interval $(-\infty, U_n]$ for Y and Γ_Y can be defined.

We repeat that the fuzzy parameter $\Gamma_Y[X_1, \ldots, X_n]$ can be considered as a fuzzy perception of our unknown parameter $\Gamma_Y(D_{U_1})$, where $(U_1, \ldots, U_n) \in \chi_Y^n$ is the unknown original of our fuzzy random variable $(X_1, \ldots, X_n)$.

A usual confidence interval assigns to each $(x_1, \ldots, x_n) \in I\!R^n$ an interval of the real line. So a fuzzy confidence interval assigns to each $(\mu_1, \ldots, \mu_n) \in [F(I\!R)]^n$ a convex fuzzy set $K_n[\mu_1, \ldots, \mu_n]$. We want to ensure that

$$\tilde{\mu}_{(X_1,\ldots,X_n)}(U_1, \ldots, U_n) > \alpha,$$

where $(U_1, \ldots, U_n) \in \chi_Y^n$, implies

$$\Gamma_Y(D_{U_1}) \in (K_n[X_1,\ldots,X_n])_\alpha$$

with a given probability. So we define the following:

Definition 11.9 *Let Y be a class of distribution functions and $\Gamma_Y : Y \to \mathbb{R}$ be a mapping. Let $n \in \mathbb{N}$ and $\delta \in (0,1)$.*
A mapping $K_n : [F(\mathbb{R})]^n \to U(\mathbb{R})$ is called a $100 * (1-\delta)\%$ fuzzy confidence interval for Y and Γ_Y if for all fuzzy random vectors

$$(X_1,\ldots,X_n) : \Omega \to [Q(\mathbb{R})]^n$$

and for all $\alpha \in [0,1)$

$$P\left(\{\omega \in \Omega \,|\, (\Gamma_Y[X_1,\ldots,X_n])_\alpha \subseteq \text{cl}\ (K_n[(X_1)_\omega,\ldots,(X_n)_\omega])_\alpha\}\right) \geq 1-\delta$$

is valid.

It may occur that we have to deal with subsets A of Ω not belonging to the σ-algebra $\mathfrak{A}$ of all measurable sets. The notion $P(A) \geq 1-\delta$ for an $A \subseteq \Omega$ means:

$$\exists\, B \in \mathfrak{A} \otimes \mathfrak{A}' \ \text{ with } \ (P \otimes P')(B) \geq 1-\delta \ \text{ such that } \ B \subseteq A \times \Omega'.$$

Theorem 11.10 *Let Y be a class of distribution functions and $\Gamma_Y : Y \to \mathbb{R}$ be a mapping. Let $n \in \mathbb{N}$ and $\delta \in (0,1)$. Let $[T_n, +\infty)$ and $(-\infty, U_n]$ be two (usual) one-sided confidence intervals for Y and Γ_Y such that $\delta_1 + \delta_2 = \delta$ and $T_n \leq U_n$ (pointwise) is valid.*
Define for $(\mu_1,\ldots,\mu_n) \in [F(\mathbb{R})]^n$, $\alpha \in [0,1)$, and $t \in \mathbb{R}$:

$$A_\alpha[\mu_1,\ldots,\mu_n] \stackrel{\mathrm{d}}{=} \inf\left\{u \in \mathbb{R} \;\middle|\; \begin{array}{l} \exists(x_1,\ldots,x_n) \in (\mu_1)_\alpha \times \ldots \times (\mu_n)_\alpha \\ \text{such that } T_n(x_1,\ldots,x_n) \le u \end{array}\right\}$$

$$B_\alpha[\mu_1,\ldots,\mu_n] \stackrel{\mathrm{d}}{=} \sup\left\{u \in \mathbb{R} \;\middle|\; \begin{array}{l} \exists(x_1,\ldots,x_n) \in (\mu_1)_\alpha \times \ldots \times (\mu_n)_\alpha \\ \text{such that } U_n(x_1,\ldots,x_n) \ge u \end{array}\right\}$$

$$E_{\alpha}[\mu_1,\ldots,\mu_n] \stackrel{\mathrm{d}}{=}$$

$$\begin{cases} [A_\alpha[\mu_1,\ldots,\mu_n], B_\alpha[\mu_1,\ldots,\mu_n]] & \text{, if } A_\alpha[\mu_1,\ldots,\mu_n] > -\infty \\ & \text{ and } B_\alpha[\mu_1,\ldots,\mu_n] < +\infty \\ [A_\alpha[\mu_1,\ldots,\mu_n], +\infty) & \text{, if } A_\alpha[\mu_1,\ldots,\mu_n] > -\infty \\ & \text{ and } B_\alpha[\mu_1,\ldots,\mu_n] = +\infty \\ (-\infty, B_\alpha[\mu_1,\ldots,\mu_n]] & \text{, if } A_\alpha[\mu_1,\ldots,\mu_n] = -\infty \\ & \text{ and } B_\alpha[\mu_1,\ldots,\mu_n] < +\infty \\ \mathbb{R} & \text{, if } A_\alpha[\mu_1,\ldots,\mu_n] = -\infty \\ & \text{ and } B_\alpha[\mu_1,\ldots,\mu_n] = +\infty. \end{cases}$$

$$K_n[\mu_1,\ldots,\mu_n](t) \stackrel{\mathrm{d}}{=} \sup\left\{\alpha \mathrm{I}_{E_{\alpha}[\mu_1,\ldots,\mu_n]}(t) \mid \alpha \in (0,1)\right\}.$$

Then $K_n : [F(\mathbb{R})]^n \to U(\mathbb{R})$ is a $100 * (1-\delta)\%$fuzzy confidence interval for Y and Γ_Y.

Proof

(a) Let $(\mu_1,\ldots,\mu_n) \in [F(\mathbb{R})]^n$. Obviously $0 \le \beta \le \alpha < 1$ implies

$$A_\beta[\mu_1,\ldots,\mu_n] \le A_\alpha[\mu_1,\ldots,\mu_n] \le B_\alpha[\mu_1,\ldots,\mu_n] \le B_\beta[\mu_1,\ldots,\mu_n].$$

As the intersection of a non-increasing sequence of closed non-empty sets is not empty ,

$$\bigcap_{\alpha\in(0,1)} E_\alpha[\mu_1,\ldots,\mu_n]$$

is not empty. With the remark after Definition 3.2, it follows: $K_n[\mu_1,\ldots,\mu_n] \in F(\mathbb{R})$.
As $\{E_\alpha[\mu_1,\ldots,\mu_n] \mid \alpha \in (0,1)\}$ is a set representation of $K_n[\mu_1,\ldots,\mu_n]$, $K_n[\mu_1,\ldots,\mu_n] \in U(\mathbb{R})$ is valid.

(b) Let $(\mu_1,\ldots,\mu_n) \in [F(\mathbb{R})]^n$ and $\alpha \in [0,1)$. We want to show that

$$\inf(K_n[\mu_1,\ldots,\mu_n])_\alpha = A_\alpha[\mu_1,\ldots,\mu_n] \text{ and}$$
$$\sup(K_n[\mu_1,\ldots,\mu_n])_\alpha = B_\alpha[\mu_1,\ldots,\mu_n]$$

is valid.
Define $\alpha_r \stackrel{\mathrm{d}}{=} \alpha + 1/2r(1-\alpha)$ for $r \in I\!N$. Because of Theorem 3.6, we know that

$$\{A_{\alpha_r}[\mu_1, \ldots, \mu_n]\}_{r \in \boldsymbol{N}}$$

is monotonously non-increasing against

$$\inf\left(K_n[\mu_1, \ldots, \mu_n]\right)_\alpha$$

and

$$\{B_{\alpha_r}[\mu_1, \ldots, \mu_n]\}_{r \in \boldsymbol{N}}$$

is monotonously non-decreasing against

$$\sup\left(K_n[\mu_1, \ldots, \mu_n]\right)_\alpha.$$

Obviously

$$\Psi_\alpha \stackrel{\mathrm{d}}{=} A_\alpha[\mu_1, \ldots, \mu_n] \leq A_{\alpha_r}[\mu_1, \ldots, \mu_n]$$

is valid for all $r \in I\!N$.
For $R \in I\!N$ arbitrary choose

$$\left(x_1^{(R)}, \ldots, x_n^{(R)}\right) \in (\mu_1)_\alpha \times \ldots (\mu_n)_\alpha$$

such that

$$T_n\left(x_1^{(R)}, \ldots, x_n^{(R)}\right) \leq -R \qquad , \text{if } \Psi_\alpha = -\infty$$
$$T_n\left(x_1^{(R)}, \ldots, x_n^{(R)}\right) \leq \Psi_\alpha + 1/R \quad , \text{if } \Psi_\alpha > -\infty$$

is valid. Because of Theorem 3.4, it is possible to find an $s = s(Rk) \in I\!N$ such that

$$\left(x_1^{(R)}, \ldots, x_n^{(R)}\right) \in (\mu_1)_{\alpha_s} \times \ldots \times (\mu_n)_{\alpha_s}$$

is valid. It follows:

$$\lim_{r \to \infty} A_{\alpha_r}[\mu_1, \ldots, \mu_n] \leq A_{\alpha_s}[\mu_1, \ldots, \mu_n] \leq T_n\left(x_1^{(R)}, \ldots, x_n^{(R)}\right).$$

We can conclude:

$$\lim_{r \to \infty} A_{\alpha_r}[\mu_1, \ldots, \mu_n] \leq \Psi_\alpha.$$

All together:

$$\Psi_\alpha = \lim_{r \to \infty} A_{\alpha_r}[\mu_1, \ldots, \mu_n] = \inf\left(K_n[\mu_1, \ldots, \mu_n]\right)_\alpha.$$

In the same way, the other assertion can be shown.

(c) Let

$$(X_1, \ldots, X_n) : \Omega \to [Q(\mathbb{R})]^n$$

be a fuzzy random vector. Let $\alpha \in [0,1)$. We want to find an $A \in \mathfrak{A} \otimes \mathfrak{A}'$ with $(P \otimes P')(A) \geq 1 - \delta$ such that $(\omega, \omega') \in A$ implies

$$(\Gamma_Y[X_1, \ldots, X_n])_\alpha \subseteq \text{cl}\ (K_n[(X_1)_\omega, \ldots, (X_n)_\omega])_\alpha.$$

If $(\Gamma_Y[X_1, \ldots, X_n])_\alpha$ is empty, nothing is to be shown.
If not, we define for abbreviation:

$$\Psi_\alpha \stackrel{\text{d}}{=} \inf (\Gamma_Y[X_1, \ldots, X_n])_\alpha \text{ and}$$
$$\Phi_\alpha \stackrel{\text{d}}{=} \sup (\Gamma_Y[X_1, \ldots, X_n])_\alpha.$$

Let $N \in \mathbb{N}$ arbitrary. Choose

$$t_N \in (\Gamma_Y[X_1, \ldots, X_n])_\alpha \text{ and } u_N \in (\Gamma_Y[X_1, \ldots, X_n])_\alpha$$

such that
$t_N \leq \Psi_\alpha + \frac{1}{N}$, if $\Psi_\alpha > -\infty$
$t_N \leq -N$, if $\Psi_\alpha = -\infty$
and
$u_N \geq \Phi_\alpha - \frac{1}{N}$, if $\Phi_\alpha < +\infty$
$u_N \geq N$, if $\Phi_\alpha = +\infty$
is valid.

Because of the lemma after Definition 11.1, we can find two i.i.d. random vectors $\left(R_1^{(N)}, \ldots, R_n^{(N)}\right) \in \chi_Y^n$ and $\left(S_1^{(N)}, \ldots, S_n^{(N)}\right) \in \chi_Y^n$ with $\Gamma_Y\left(D_{R_1^{(N)}}\right) = t_N$ and $\Gamma_Y\left(D_{S_1^{(N)}}\right) = u_N$ such that

$$R_i^{(N)}(\omega, \omega') \in ((X_i)_\omega)_\alpha \text{ and } S_i^{(N)}(\omega, \omega') \in ((X_i)_\omega)_\alpha$$

is valid for $i \in \{1, \ldots, n\}$, $\omega \in \Omega$, $\omega' \in \Omega'$.
We know that

$$(P \otimes P')\left(\left\{(\omega, \omega') \in \Omega \times \Omega' \;\middle|\; \begin{array}{l} T_n\left[R_1^{(N)}(\omega, \omega'), \ldots, R_n^{(N)}(\omega, \omega')\right] \\ \leq t_N \end{array}\right\}\right)$$
$$\geq 1 - \delta_1$$

and

$$(P \otimes P')\left(\left\{(\omega,\omega') \in \Omega \times \Omega' \,\middle|\, \begin{array}{l} U_n\left[S_1^{(N)}(\omega,\omega'),\ldots,S_n^{(N)}(\omega,\omega')\right] \\ \geq u_N \end{array}\right\}\right)$$

$$\geq 1 - \delta_2$$

are valid. Define for $N \in I\!N$:

$$A_N \stackrel{\mathrm{d}}{=}$$

$$\left\{(\omega,\omega') \in \Omega \times \Omega' \,\middle|\, \begin{array}{l} T_n\left[R_1^{(N)}(\omega,\omega'),\ldots,R_n^{(N)}(\omega,\omega')\right] \leq t_N \text{ and} \\ U_n\left[S_1^{(N)}(\omega,\omega'),\ldots,S_n^{(N)}(\omega,\omega')\right] \geq u_N \end{array}\right\}$$

Then $A_N \in \mathfrak{A} \otimes \mathfrak{A}'$ and $(P \otimes P')(A_N) \geq 1 - \delta$ holds for $N \in I\!N$. Define

$$A \stackrel{\mathrm{d}}{=} \liminf_{N\to\infty} A_N = \bigcap_{N=1}^{\infty} \bigcup_{K=N}^{\infty} A_K \in \mathfrak{A} \otimes \mathfrak{A}'$$

(compare E. Henze [84], e.g.). We have

$$A = \{\omega \in \ \Omega \,|\, \omega \in A_N \ \text{ for an infinite number of } N\text{'s } \}.$$

As

$$(P \otimes P')\left(\bigcup_{K=N}^{\infty} A_K\right) \geq 1 - \delta$$

is valid for $N \in I\!N$,

$$(P \otimes P')(A) = \lim_{N\to\infty} (P \otimes P')\left(\bigcup_{K=N}^{\infty} A_K\right) \geq 1 - \delta$$

holds. By definition of A_α and B_α, we know that

$$T_n\left[R_1^{(N)}(\omega,\omega'),\ldots,R_n^{(N)}(\omega,\omega')\right] \geq A_\alpha[(X_1)_\omega,\ldots,(X_n)_\omega] \text{ and}$$
$$U_n\left[S_1^{(N)}(\omega,\omega'),\ldots,S_n^{(N)}(\omega,\omega')\right] \geq B_\alpha[(X_1)_\omega,\ldots,(X_n)_\omega]$$

is valid.
We follow that $(\omega,\omega') \in A$ implies:

$$\begin{aligned}&[t_n, u_N]\\ &\subseteq \Big[T_n\Big[R_1^{(N)}(\omega,\omega'),\ldots,R_n^{(N)}(\omega,\omega')\Big],\\ &\quad U_n\Big[S_1^{(N)}(\omega,\omega'),\ldots,S_n^{(N)}(\omega,\omega')\Big]\Big]\\ &\subseteq E_\alpha[(X_1)_\omega,\ldots,(X_n)_\omega]\end{aligned}$$

is valid for an infinite number of N's. We conclude:

$$(\Psi_\alpha,\Phi_\alpha) \subseteq E_\alpha[(X_1)_\omega,\ldots,(X_n)_\omega]$$

for $(\omega,\omega') \in A$, this implies:

$$\begin{aligned}&\left(\tilde{\Gamma}_Y[X_1,\ldots,X_n]\right)_\alpha\\ &\subseteq \text{cl } E_\alpha[(X_1)_\omega,\ldots,(X_n)_\omega] = \text{cl } (K_n[(X_1)_\omega,\ldots,(X_n)_\omega])_\alpha\end{aligned}$$

for all $(\omega,\omega') \in A$. It follows:

$$\begin{aligned}&P\left(\{\omega \in \Omega \mid (\Gamma_Y[X_1,\ldots,X_n])_\alpha\}\right)\\ &\subseteq \text{cl } (K_n[(X_1)_\omega,\ldots,(X_n)_\omega])_\alpha \geq 1-\delta.\end{aligned}$$

This completes the proof. •

It is also possible to derive one-sided fuzzy confidence intervals.

Theorem 11.11 *Let $\delta \in (0,1)$ and $n \in I\!N$. Let Y be a class of distribution functions, and $\Gamma_Y : Y \to I\!R$ be a mapping.*

(i) Let $[T_n, +\infty)$ be a $100*(1-\delta)$%confidence interval for Y and Γ_Y. Define for $(\mu_1,\ldots,\mu_n) \in [F(I\!R)]^n$, $\alpha \in [0,1)$, and $t \in I\!R$:

$$E_\alpha[\mu_1,\ldots,\mu_n](t) \stackrel{\text{d}}{=} \begin{cases}[A_\alpha[\mu_1,\ldots,\mu_n],+\infty) & ,\text{ if } A_\alpha[\mu_1,\ldots,\mu_n] > -\infty\\ I\!R & ,\text{ if } A_\alpha[\mu_1,\ldots,\mu_n] = -\infty,\end{cases}$$

where $A_\alpha[\mu_1,\ldots,\mu_n]$ is defined as in Theorem 11.10, and

$$K_n[\mu_1,\ldots,\mu_n](t) \stackrel{\text{d}}{=} \sup\left\{\alpha \mathrm{I}_{E_\alpha[\mu_1,\ldots,\mu_n]}(t) \mid \alpha \in (0,1)\right\}.$$

Then

$$K_N : [F(I\!R)]^n \to U(I\!R)$$

is a $100 * (1-\delta)\%$*fuzzy confidence interval for* Y *and* Γ_Y.

(ii) Let $(-\infty, U_n]$ *be a* $100 * (1-\delta)\%$*confidence interval for* Y *and* Γ_Y. *Define for* $(\mu_1,\ldots,\mu_n) \in [F(\mathbb{R})]^n$, $\alpha \in [0,1)$, *and* $t \in \mathbb{R}$:

$$E_\alpha[\mu_1,\ldots,\mu_n](t) \overset{\mathrm{d}}{=} \begin{cases} (-\infty, B_\alpha[\mu_1,\ldots,\mu_n]] & , \text{if } B_\alpha[\mu_1,\ldots,\mu_n] < +\infty \\ \mathbb{R} & , \text{if } B_\alpha[\mu_1,\ldots,\mu_n] = +\infty, \end{cases}$$

where $B_\alpha[\mu_1,\ldots,\mu_n]$ *is defined as in Theorem 11.10, and*

$$K_n[\mu_1,\ldots,\mu_n](t) \overset{\mathrm{d}}{=} \sup\left\{\alpha \mathrm{I}_{E_\alpha[\mu_1,\ldots,\mu_n]}(t) \mid \alpha \in (0,1)\right\}.$$

Then

$$K_n : [F(\mathbb{R})]^n \to U(\mathbb{R})$$

is a $100 * (1-\delta)\%$*fuzzy confidence interval for* Y *and* Γ_Y.

Many confidence intervals can be obtained by using monotonous functions T_n and U_n. For this important case fuzzy confidence intervals can be easily calculated.

Theorem 11.12 *Let* $\delta \in (0,1)$ *and* $n \in \mathbb{N}$. *Let* Y *be a class of distribution functions and* $\Gamma_Y : Y \to \mathbb{R}$ *be a mapping.*
Let $T_n : \mathbb{R}^n \to \mathbb{R}$ *and* $U_n : \mathbb{R}^n \to \mathbb{R}$ *be two continuous and, in each component, monotonously non-decreasing mappings such that* $T_n \leq U_n$ *holds and* $[T_n, +\infty)$ *and* $(-\infty, U_n]$ *are two* $100 * (1-\delta_1)\%$ *and* $100 * (1-\delta_2)\%$ *confidence intervals for* Y *and* Γ_Y *with* $\delta_1 + \delta_2 = \delta$.
It is possible to extend T_n *and* U_n *uniquely to arguments in* $\overline{\mathbb{R}}^n \overset{\mathrm{d}}{=} [\mathbb{R} \cup \{-\infty\} \cup \{+\infty\}]^n$.
Define for $(\mu_1,\ldots,\mu_n) \in [F(\mathbb{R})]^n$, $\alpha \in [0,1)$,$t \in \mathbb{R}$:

$$E_\alpha[\mu_1,\ldots,\mu_n] \overset{\mathrm{d}}{=}$$

$$\begin{cases} [T_n[\inf(\mu_1)_\alpha, \ldots \inf(\mu_n)_\alpha], U_n[\sup(\mu_1)_\alpha, \ldots, \sup(\mu_n)_\alpha]] \\ \qquad , \text{if } T_n[\inf(\mu_1)_\alpha, \ldots, \inf(\mu_n)_\alpha] > -\infty \\ \qquad \text{and } U_n[\sup(\mu_1)_\alpha, \ldots, \sup(\mu_n)_\alpha] < \infty \\ [T_n[\inf(\mu_1)_\alpha, \ldots \inf(\mu_n)_\alpha], +\infty) \\ \qquad , \text{if } T_n[\inf(\mu_1)_\alpha, \ldots, \inf(\mu_n)_\alpha] > -\infty \\ \qquad \text{and } U_n[\sup(\mu_1)_\alpha, \ldots, \sup(\mu_n)_\alpha] = +\infty \\ (-\infty, U_n[\sup(\mu_1)_\alpha, \ldots, \sup(\mu_n)_\alpha]] \\ \qquad , \text{if } T_n[\inf(\mu_1)_\alpha, \ldots, \inf(\mu_n)_\alpha] = -\infty \\ \qquad \text{and } U_n[\sup(\mu_1)_\alpha, \ldots, \sup(\mu_n)_\alpha] < \infty \\ \mathbb{R} \qquad , \text{if } T_n[\inf(\mu_1)_\alpha, \ldots, \inf(\mu_n)_\alpha] = -\infty \\ \qquad \text{and } U_n[\sup(\mu_1)_\alpha, \ldots, \sup(\mu_n)_\alpha] = +\infty \end{cases}$$

and

$$K_n[\mu_1,\ldots,\mu_n](t) \stackrel{\mathrm{d}}{=} \sup\left\{\alpha \mathrm{I}_{E_\alpha[\mu_1,\ldots,\mu_n]}(t) \mid \alpha \in (0,1)\right\}.$$

Then $K_n : [F(I\!R)]^n \to U(I\!R)$ *is a* $100 * (1-\delta)\%$ *fuzzy confidence interval for* Y *and* Γ_Y.

Proof Let $(\mu_1,\ldots,\mu_n) \in [F(I\!R)]^n$ and $\alpha \in [0,1)$ be arbitrary. Let $A_\alpha[\mu_1,\ldots,\mu_n]$ and $B_\alpha[\mu_1,\ldots,\mu_n]$ be defined as in Theorem 11.10.
It is sufficient to show that

$$T_n[\inf(\mu_1)_\alpha,\ldots,\inf(\mu_n)_\alpha] = A_\alpha[\mu_1,\ldots,\mu_n] \text{ and}$$

$$U_n[\sup(\mu_1)_\alpha,\ldots,\sup(\mu_n)_\alpha] = B_\alpha[\mu_1,\ldots,\mu_n] \text{ holds.}$$

We only show the first equation.

Let $T_n[\inf(\mu_1)_\alpha,\ldots,\inf(\mu_n)_\alpha] > -\infty$ be valid.
Let $(x_1,\ldots,x_n) \in (\mu_1)_\alpha \times \ldots \times (\mu_n)_\alpha$. Then $x_i \geq \inf(\mu_i)_\alpha$ is valid for $i \in \{1,\ldots,n\}$. It follows:

$$T_n(x_1,\ldots,x_n) \geq T_n[\inf(\mu_1)_\alpha,\ldots,\inf(\mu_n)_\alpha],$$

and therefore

$$A_\alpha[\mu_1,\ldots,\mu_n] =$$

$$\inf\{u \in I\!R \mid \exists(x_1,\ldots,x_n) \in (\mu_1)_\alpha \times \ldots \times (\mu_n)_\alpha \text{ with } T_n(x_1,\ldots,x_n) \leq u\}$$

$$\geq T_n[\inf(\mu_1)_\alpha,\ldots,\inf(\mu_n)_\alpha]$$

is valid.
On the other hand, let $\inf(\mu_i)_\alpha > -\infty$ be valid and let $\epsilon > 0$ be arbitrary. Choose $\rho = \rho(\epsilon) > 0$ such that

$$|x_i - \inf(\mu_i)_\alpha| \leq \rho$$

for $i \in \{1,\ldots,n\}$ implies:

$$|T_n(x_1,\ldots,x_n) - T_n[\inf(\mu_1)_\alpha,\ldots,\inf(\mu_n)_\alpha]| \leq \epsilon.$$

For $i \in \{1,\ldots,n\}$ choose $x_i \in (\mu_i)_\alpha$ such that $x_i \leq \inf(\mu_i)_\alpha + \rho$ holds. It follows:

$$A_\alpha[\mu_1,\ldots,\mu_n] \le T_n(x_1,\ldots,x_n) \le T_n[\inf(\mu_1)_\alpha,\ldots,\inf(\mu_n)_\alpha] + \epsilon.$$

This demonstrates the inequality

$$A_\alpha[\mu_1,\ldots,\mu_n] \le T_n[\inf(\mu_1)_\alpha,\ldots,\inf(\mu_n)_\alpha].$$

If $\inf(\mu_i)_\alpha = -\infty$ is vald for an $i \in \{1,\ldots,n\}$, we can conclude similarly.

If $T_n[\inf(\mu_1)_\alpha,\ldots,\inf(\mu_n)_\alpha] = -\infty$, there exists an $i_0 \in \{1,\ldots,n\}$ with $\inf(\mu_i)_\alpha = -\infty$, and it holds:

$$\lim_{t\to-\infty} T_n(x_1,\ldots,x_{i_0-1},t,x_{i_0+1},\ldots,x_n) = -\infty$$

where

$$(x_1,\ldots,x_{i_0-1},x_{i_0+1},\ldots,x_n) \in I\!R^{n-1}.$$

We conclude: $A_\alpha[\mu_1,\ldots,\mu_n] = -\infty$. •

A very important class of distribution functions is that of all normal distributions. We have the two parameters μ and σ^2. Let $\mathcal{N}$ denote the class of all normal distributions. E assigns to an $N(\mu,\sigma^2)$ distribution its expected value μ and Var its variance σ^2.

Let us first derive confidence intervals for μ. If σ^2 is known, we can solve the problem with the help of Theorem 11.10. So let us consider the case of σ^2 being unknown.

Let $n \in I\!N$ and $\delta \in (0,1)$. Let $t_{1-\delta}^{(n-1)}$ be the $(1-\delta)$-quantile of the student distribution with $(n-1)$ degrees of freedom. It is well known (compare [10,62,83,183,187]) that

$$\left[\frac{1}{n}\sum_{i=1}^n Z_i - t_{1-\delta}^{(n-1)}\frac{1}{\sqrt{n(n-1)}}\cdot\sqrt{\sum_{i=1}^n\left(Z_i-\frac{1}{n}\sum_{j=1}^n Z_j\right)^2}, +\infty\right) \quad \text{and}$$

$$\left(-\infty, \frac{1}{n}\sum_{i=1}^n Z_i + t_{1-\delta}^{(n-1)}\frac{1}{\sqrt{n(n-1)}}\text{dot}\sqrt{\sum_{i=1}^n\left(Z_i-\frac{1}{n}\sum_{j=1}^n Z_j\right)^2}\right]$$

are two usual $100*(1-\delta)$%confidence intervals for μ where $(Z_1,\ldots,Z_n)$ is an i.i.d. random vector such that Z_1 is $N(\mu,\sigma^2)$ distributed.

Lemma 11.12 *Let $n \in I\!N$ and $(\mu_1, \dots, \mu_n) \in [F(I\!R)]^n$.*
Define $S_n^2 : I\!R^n \to I\!R$ by

$$S_n^2(x_1, \dots, x_n) \overset{\mathrm{d}}{=} \sum_{i=1}^{n} \left(x_i - \frac{1}{n} \sum_{j=1}^{n} x_j \right)^2$$

for $(x_1, \dots, x_n) \in I\!R^n$.
With the help of L.A. Zadeh's extension principle (compare Definition 4.4), we can define the fuzzy set $S_n^2[\mu_1, \dots, \mu_n]$.
Define for $\alpha \in [0,1)$

$$S_\alpha[\mu_1, \dots, \mu_n] \overset{\mathrm{d}}{=}$$

$$\begin{cases} \max \left\{ \begin{array}{l} \sum_{i \in A} [\sup(\mu_i)_\alpha]^2 + \sum_{i \in \{1,\dots,n\} \setminus A} [\inf(\mu_i)_\alpha]^2 - \\ \frac{1}{n} \left[\sum_{i \in A} \sup(\mu_i)_\alpha + \sum_{i \in \{1,\dots,n\} \setminus A} \inf(\mu_i)_\alpha \right]^2 \end{array} \;\middle|\; A \subseteq \{1, \dots, n\} \right\} \\ \quad , \quad \text{if } \inf(\mu_i)_\alpha > -\infty \text{ and } \sup(\mu_i)_\alpha < +\infty \text{ holds for } i \in \{1, \dots, n\} \\ +\infty \quad , \quad \text{otherwise} . \end{cases}$$

Then

$$\sup \left(S_n^2[\mu_1, \dots, \mu_n] \right)_\alpha = S_\alpha[\mu_1, \dots, \mu_n]$$

is valid for all $\alpha \in [0,1)$.

Proof Let $\alpha \in [0,1)$.
Lemma 4.5(ii) shows that

$$\left(S_n^2[\mu_1, \dots, \mu_n] \right)_\alpha = S_n^2[(\mu_1)_\alpha, \dots, (\mu_n)_\alpha]$$
$$= \left\{ t \in I\!R \;\middle|\; \exists\, (x_1, \dots, x_n) \in (\mu_1)_\alpha \times \dots \times (\mu_n)_\alpha \text{ with } S_n^2(x_1, \dots, x_n) = t \right\}$$

is valid. Obviously

$$S_n^2(x_1, \dots, x_n) = \sum_{i=1}^{n} x_i^2 - \frac{1}{n} \left[\sum_{i=1}^{n} x_i \right]^2$$

holds for $(x_1, \ldots, x_n) \in I\!R^n$.

(a) Let $\sup(\mu_{i_0}) = +\infty$ be valid for an $i_0 \in \{1, \ldots, n\}$.
For all $N \in I\!N$ a $t_{i_0}^{(N)} \in (\mu_{i_0})_\alpha$ with $t_{i_0}^{(N)} \geq N$ can be found. There exists functions

$$B_{i_0}, C_{i_0} : I\!R^{n-1} \to I\!R$$

such that

$$S_n^2(x_1, \ldots, x_n) =$$
$$\left(1 - \frac{1}{n}\right) \cdot x_{i_0}^2 + B_{i_0}(x_1, \ldots, x_{i_0-1}, x_{i_0+1}, \ldots, x_n)\, x_{i_0} +$$
$$+ C_{i_0}(x_1, \ldots, x_{i_0-1}, x_{i_0+1}, \ldots, x_n)$$

holds for $(x_1, \ldots, x_n) \in I\!R^n$. Choose $x_i \in (\mu_i)_\alpha$ arbitrary for $i \in \{1, \ldots, n\} \backslash \{i_0\}$. Then

$$\lim_{N \to \infty} S_n^2\left(x_1, \ldots, x_{i_0-1}, x_{i_0}^{(N)}, x_{i_0+1}, \ldots, x_n\right) = +\infty$$

is valid. It follows:

$$\sup\left(S_n^2[\mu_1, \ldots, \mu_n]\right)_\alpha = +\infty.$$

(b) Let $\inf(\mu_i)_\alpha > -\infty$ and $\sup(\mu_i)_\alpha < +\infty$ be valid for all $i \in \{1, \ldots, n\}$.
(b.1) Let $(x_1, \ldots, x_n) \in (\mu_1)_\alpha \times \ldots \times (\mu_n)_\alpha$ be arbitrary. We want to apply Lemma 8.9. Define

$$\Omega \stackrel{\mathrm{d}}{=} \{1, \ldots, n\},\ \mathfrak{A} \stackrel{\mathrm{d}}{=} \mathfrak{P}(\Omega),\ P(A) \stackrel{\mathrm{d}}{=} \frac{|A|}{n}$$

for $A \subseteq \Omega$. Define

$$\Omega' \stackrel{\mathrm{d}}{=} [0,1],\ \mathfrak{A}' \stackrel{\mathrm{d}}{=} \mathfrak{B}_1 \cap [0,1],\ P' \stackrel{\mathrm{d}}{=} \mu_L$$

(the Lebesgue measure). Define

$$X(i) \stackrel{\mathrm{d}}{=} \mu_i \text{ for } i \in \{1, \ldots, n\},$$

X is a finite discrete f.r.v. Due to Lemma 8.7, we define

$$\tilde{Y}_\alpha \stackrel{\mathrm{d}}{=} \{U \in \tilde{\chi} \,|\, \forall i \in \{1, \ldots, n\}, \forall \omega' \in \Omega' : U(i, \omega') \in \mathrm{cl}\,(\mu_i)_\alpha\}.$$

Define $U \in \tilde{Y}_\alpha$ by

$$U(i,\omega') \stackrel{\mathrm{d}}{=} x_i \text{ for } i \in \{1,\ldots,n\},\ \omega' \in \Omega'.$$

Then

$$\operatorname{Var} U \leq \operatorname{Var} U_{[U \geq E\, U]}$$

(compare Lemma 8.9(i)) as well as

$$[U \geq E\, U] = A \times [0,1]$$

is valid where

$$A \stackrel{\mathrm{d}}{=} \{i \in \{1,\ldots,n\} \mid x_i \geq E\, U\}$$

$$\left\{ i \in \{1,\ldots,n\} \;\middle|\; x_i \geq \frac{1}{n} \sum_{j=1}^{n} x_j \right\}.$$

It follows:

$$\begin{aligned}
& S_n^2(x_1,\ldots,x_n) \\
=& n \operatorname{Var} U \leq n \operatorname{Var} U_{[U \geq E\, U]} \\
=& \sum_{i \in A} [\sup{(\mu_i)_\alpha}]^2 + \sum_{i \in \{1,\ldots,n\} \setminus A} [\inf{(\mu_i)_\alpha}]^2 \\
& - \left[\sum_{i \in A} \sup{(\mu_i)_\alpha} + \sum_{i \in \{1,\ldots,n\} \setminus A} \inf{(\mu_i)_\alpha} \right]^2 \\
\leq& S_\alpha[\mu_1,\ldots,\mu_n]
\end{aligned}$$

We conclude

$$\sup{\left(S_n^2[\mu_1,\ldots,\mu_n]\right)_\alpha} \leq S_\alpha[\mu_1,\ldots,\mu_n].$$

(b.2) Let $A \subseteq \{1,\ldots,n\}$ and $\epsilon > 0$ be arbitrary. Define $(x_1,\ldots,x_n) \in \mathbb{R}^n$ by

$$x_i \stackrel{\mathrm{d}}{=} \begin{cases} \sup{(\mu_i)_\alpha} & \text{, if } i \in A \\ \inf{(\mu_i)_\alpha} & \text{, if } i \in \{1,\ldots,n\} \setminus A. \end{cases}$$

S_n^2 is continuous in $(x_1, \ldots, x_n)$. Choose $\rho = \rho(x_1, \ldots, x_n) > 0$ such that $(\tilde{x}_1, \ldots, \tilde{x}_n) \in \mathbb{R}^n$ and $|x_i - \tilde{x}_i| \leq \rho$ for $i \in \{1, \ldots, n\}$ implies:

$$\left|S_n^2(x_1, \ldots, x_n) - S_n^2(\tilde{x}_1, \ldots, \tilde{x}_n)\right| \leq \epsilon.$$

Choose $(y_1, \ldots, y_n) \in (\mu_i)_\alpha \times \cdots \times (\mu_n)_\alpha$ such that

$$y_i \geq \sup{(\mu_i)_\alpha} - \rho \text{, if } i \in A, \text{ and}$$
$$y_i \leq \inf{(\mu_i)_\alpha} + \rho \text{, if } i \in \{1, \ldots, n\}.$$

It follows:

$$\sum_{i \in A} [\sup{(\mu_i)_\alpha}]^2 + \sum_{i \in \{1, \ldots, n\} \setminus A} [\inf{(\mu_i)_\alpha}]^2 -$$
$$- \frac{1}{n} \left[\sum_{i \in A} (\mu_i)_\alpha + \sum_{i \in \{1, \ldots, n\} \setminus A} \inf{(\mu_i)_\alpha} \right]^2$$
$$= S_n^2(x_1, \ldots, x_n) \leq\leq S_n^2(y_1, \ldots, y_n) + \epsilon$$
$$\leq \sup{\left(S_n^2[\mu_1, \ldots, \mu_n]\right)_\alpha} + \epsilon.$$

It follows

$$S_\alpha[\mu_1, \ldots, \mu_n] \leq \sup{\left(S_n^2[\mu_1, \ldots, \mu_n]\right)_\alpha}. \bullet$$

Theorem 11.14 *Let $n \in \mathbb{N}$ and $\delta \in (0,1)$.*
Define for $(\mu_1, \ldots, \mu_n) \in [F(\mathbb{R})]^n$ and $\alpha \in [0,1)$ $S_\alpha[\mu_1, \ldots, \mu_n]$ as in Lemma 11.12 and

$$E_\alpha[\mu_1, \ldots, \mu_n] \stackrel{\mathrm{d}}{=}$$
$$\begin{cases} \left[\frac{1}{n} \sum_{i=1}^{n} \inf{(\mu_i)_\alpha} - \frac{t_{1-\frac{\delta}{2}}^{(n-1)}}{\sqrt{n(n-1)}} \sqrt{S_\alpha[\mu_1, \ldots, \mu_n]}, \right. \\ \left. \quad \frac{1}{n} \sum_{i=1}^{n} \sup{(\mu_i)_\alpha} + \frac{t_{1-\frac{\delta}{2}}^{(n-1)}}{\sqrt{n(n-1)}} \sqrt{S_\alpha[\mu_1, \ldots, \mu_n]} \right], \\ \qquad \text{, if } \inf{(\mu_i)_\alpha} > -\infty \text{ and } \sup{(\mu_i)_\alpha} < +\infty \\ \qquad \text{holds for all } i \in \{1, \ldots, n\} \\ \mathbb{R} \qquad \text{, otherwise .} \end{cases}$$

Define

$$K_n[\mu_1,\ldots,\mu_n](t) \stackrel{\mathrm{d}}{=} \sup\left\{\alpha \cdot \mathrm{I}_{E_\alpha[\mu_1,\ldots,\mu_n]}(t) \mid \alpha \in (0,1)\right\}$$

for $t \in \mathbb{R}$. *Then*

$$K_n : [F(\mathbb{R})]^n \to U(\mathbb{R})$$

is a $100 * (1-\delta)\%$*fuzzy confidence interval for* $\mathcal{N}$ *and* E.

Proof Define for $(x_1,\ldots,x_n) \in \mathbb{R}^n$.

(i) $S_n^2(x_1,\ldots,x_n) \stackrel{\mathrm{d}}{=} \sum_{i=1}^{n}\left(x_i - \frac{1}{n}\sum_{j=1}^{n} x_j\right)^2$,

(ii) $T_n(x_1,\ldots,x_n) \stackrel{\mathrm{d}}{=} \frac{1}{n}\sum_{i=1}^{n} x_i - \frac{t_{1-\frac{\delta}{2}}^{(n-1)}}{\sqrt{n(n-1)}}\sqrt{S_n^2(x_1,\ldots,x_n)}$, and

(iii) $U_n(x_1,\ldots,x_n) \stackrel{\mathrm{d}}{=} \frac{1}{n}\sum_{i=1}^{n} x_i + \frac{t_{1-\frac{\delta}{2}}^{(n-1)}}{\sqrt{n(n-1)}}\sqrt{S_n^2(x_1,\ldots,x_n)}$.

Let $(\mu_1,\ldots,\mu_n) \in [F(\mathbb{R})]^n$ and $\alpha \in [0,1)$ be arbitrary.
If $\inf(\mu_i)_\alpha > -\infty$ and $\sup(\mu_i)_\alpha < +\infty$ holds for all $i \in \{1,\ldots,n\}$, then Theorem 11.10 demonstrates that it is sufficient to show that

$$[A_\alpha[\mu_1,\ldots,\mu_n], B_\alpha[\mu_1,\ldots,\mu_n]] \subseteq E_\alpha[\mu_1,\ldots,\mu_n]$$

holds.
Let $u \in \mathbb{R}$ be such that there exists an $(x_1,\ldots,x_n) \in (\mu_1)_\alpha \times \ldots \times (\mu_n)_\alpha$ with

$$T_n(x_1,\ldots,x_n) = \frac{1}{n}\sum_{i=1}^{n} x_i - \frac{t_{1-\frac{\delta}{2}}^{(n-1)}}{\sqrt{n(n-1)}} \cdot \sqrt{S_n^2(\mu_1,\ldots,\mu_n)} \leq u.$$

On the one hand, $x_i \geq \inf(\mu_i)_\alpha$ holds for $i \in \{1,\ldots,n\}$.
On the other hand, Lemma 11.13 and Lemma 4.5(ii) show that

$$\begin{aligned}
0 &\leq S_n^2(x_1,\ldots,x_n) \\
&\leq \sup\left\{t \in \mathbb{R} \mid \exists (t_1,\ldots,t_n) \subset (\mu_1)_\alpha,\ldots,(\mu_n)_\alpha \text{ with } S_n^2(t_1,\ldots,t_n) = t\right\} \\
&= \sup\left(S_n^2[(\mu_1)_\alpha,\ldots,(\mu_n)_\alpha]\right) = \sup\left(S_n^2[\mu_1,\ldots,\mu_n]\right)_\alpha \\
&= S_\alpha[\mu_1,\ldots,\mu_n].
\end{aligned}$$

It follows:

$$u \geq T_n(x_1, \ldots, x_n) \geq \frac{1}{n} \sum_{i=1}^{n} \inf\,(\mu_i)_\alpha - \sqrt{S_\alpha(x_1, \ldots, x_n)}.$$

Therefore

$$A_\alpha[\mu_1, \ldots, \mu_n] \geq \frac{1}{n} \sum_{i=1}^{n} \inf\,(\mu_i)_\alpha - \sqrt{S_\alpha(x_1, \ldots, x_n)}$$

is valid. In a similar way, we can show:

$$B_\alpha[\mu_1, \ldots, \mu_n] \leq \frac{1}{n} \sum_{i=1}^{n} \sup\,(\mu_i)_\alpha + \sqrt{S_\alpha(x_1, \ldots, x_n)}.$$

If $\inf\,(\mu_{i_0})_\alpha = -\infty$ or $\sup\,(\mu_{i_0})_\alpha = +\infty$ for an $i_0 \in \{1, \ldots, n\}$, then $(K_n[\mu_1, \ldots, \mu_n])_\alpha = I\!R$ is valid.
Therefore the assertion is shown. •

We can characterize $K_n[\mu_1, \ldots, \mu_n]$ as the sum of two fuzzy set if $K_n[\mu_1, \ldots, \mu_n]$ is defined as in Theorem 11.17. It holds:

$$K_n[\mu_1, \ldots, \mu_n] = \mathrm{co}\left(\frac{1}{n}\sum_{i=1}^{n} \mu_i + \frac{t^{(n-1)}_{1-\frac{\delta}{2}}}{\sqrt{n(n-1)}} \cdot \sqrt{S_n^2(\mu_1, \ldots, \mu_n)}\right)$$

$$= \frac{1}{n}\sum_{i=1}^{n} \mathrm{co}\ \mu_i + \frac{t^{(n-1)}_{1-\frac{\delta}{2}}}{\sqrt{n(n-1)}} \sqrt{S_\alpha^2(\mathrm{co}\ \mu_1, \ldots, \mathrm{co}\ \mu_n)}$$

Let $\delta \in (0,1)$ and $n \in I\!N$. Let $q_\delta^{(n-1)}$ denote the δ-quantile of the chi-square distribution with $(n-1)$ degrees of freedom.

If $Z_1, \ldots, Z_n$ is i.i.d. and Z_i is $N(\mu, \sigma^2)$ for all $i \in \{1, \ldots, n\}$, then it is well known that

$$\frac{1}{\sigma^2} \sum_{i=1}^{n} \left(Z_i - \frac{1}{n}\sum_{j=1}^{n} Z_j\right)^2$$

is chi-square distributed with $(n-1)$ degrees of freedom. It follows:

$$\left[0, \frac{1}{q_{1-\delta}^{(n-1)}} \sum_{i=1}^{n} \left(Z_i - \frac{1}{n} \sum_{j=1}^{n} Z_j \right)^2 \right]$$

is a $100 * (1-\delta)\%$confidence interval for $\mathcal{N}$ and Var .

It is easy to derive a one-sided fuzzy confidence interval for $\mathcal{N}$ and Var .

Theorem 11.15 *Let $\delta \in (0,1)$ and $n \in I\!N$.*
Define for $(\mu_1, \ldots, \mu_n) \in [F(I\!R)]^n$, $\alpha \in [0,1)$, $t \in I\!R$

$$E_\alpha[\mu_1, \ldots, \mu_n] \stackrel{\mathrm{d}}{=}$$

$$\begin{cases} \left[0, \frac{1}{q_\delta^{(n-1)}} S_\alpha[\mu_1, \ldots, \mu_n]\right], & \\ & \text{,if } \inf(\mu_i)_\alpha > -\infty \\ & \text{and } \sup(\mu_i)_\alpha < -\infty \text{ holds for } i \in \{1, \ldots, n\} \\ [0, +\infty) & \text{,otherwise} \end{cases}$$

$$K_n[\mu_1, \ldots, \mu_n](t) \stackrel{\mathrm{d}}{=} \sup\left\{\alpha \mathrm{I}_{E_\alpha[\mu_1, \ldots, \mu_n]}(t) \mid \alpha \in (0,1)\right\}.$$

Then

$$K_n : [F(I\!R)]^n \to U(I\!R)$$

is a $100 * (1-\delta)\%$fuzzy confidence interval for $\mathcal{N}$ and Var .

Now we want to derive a two-sided confidence interval for $\mathcal{N}$ and Var . In order to do so we have to calculate

$$\inf\left(S_n^2[\mu_1, \ldots, \mu_n]\right)_\alpha$$

where $n \in I\!N$, $(\mu_1, \ldots, \mu_n) \in [F(I\!R)]^n$, and $\alpha \in [0,1)$.
We assume that for all $\alpha \in [0,1)$ and $i \in \{1, \ldots, n\}$ there exist an integer $(N_i)_\alpha \geq 1$ and a system of bounded and pairwise disjoint intervals

$$\left\{ (I_j)_\alpha{}^{(i)} \mid j \in \{1, \ldots, (N_i)_\alpha\} \right\}$$

such that

$$(\mu_i)_\alpha = \bigcup_{j=1}^{(N_i)_\alpha} (I_j)_\alpha{}^{(i)}$$

holds. Define

$$(a_j)_\alpha{}^{(i)} \stackrel{\mathrm{d}}{=} \inf (I_j)_\alpha{}^{(i)} \quad \text{and} \quad (b_j)_\alpha{}^{(i)} \stackrel{\mathrm{d}}{=} \sup (I_j)_\alpha{}^{(i)}$$

for $i \in \{1, \ldots, n\}$, $\alpha \in [0,1)$, and $j \in \{1, \ldots, (N_i)_\alpha\}$. Define for $\alpha \in (0,1)$

$$\left\{c_0^{(\alpha)}, \cdots, c_M^{(\alpha)}\right\} \stackrel{\mathrm{d}}{=}$$

$$\bigcup_{i=1}^{n} (\{\inf (\mu_i)_\alpha, \sup (\mu_i)_\alpha\}) \cup \bigcup_{j=2}^{(N_i)_\alpha} \left\{(a_j)_\alpha{}^{(i)}, (b_{j-1})_\alpha{}^{(i)}, \frac{1}{2}\left[(a_j)_\alpha{}^{(i)} + (b_i)_\alpha{}^{(j)}\right]\right\}$$

such that $c_{k-1}^{(\alpha)} < c_k^{(\alpha)}$ holds for $k \in \{1, \ldots, M_\alpha\}$. Let

$$d_k^{(\alpha)} \stackrel{\mathrm{d}}{=} \frac{1}{2}\left[c_{k-1}^{(\alpha)} + c_k^{(\alpha)}\right] \quad \text{for} \quad K \in \{1, \ldots, M^{(\alpha)}\}.$$

For $t \in \mathbb{R}$ define

$$C_\alpha[t] \stackrel{\mathrm{d}}{=} \{i \in \{1, \ldots, n\} \mid t \in \mathrm{cl}\ (\mu_i)_\alpha\}$$

and a vector

$$(V_1[t]_\alpha, \cdots, V_n[t]_\alpha) \in \mathbb{R}^n \quad \text{by}$$

$$V_i[t]_\alpha \stackrel{\mathrm{d}}{=}$$

$$\begin{cases} \inf (\mu_i)_\alpha & \text{, if } \inf (\mu_i)_\alpha > t \\ \sup (\mu_i)_\alpha & \text{, if } \sup (\mu_i)_\alpha < t \\ t & \text{, if } t \in \mathrm{cl}\ (\mu_i)_\alpha \\ (b_j)_\alpha{}^{(i)} & \text{, if } t \in \left((b_j)_\alpha{}^{(i)}, \frac{1}{2}\left[(b_j)_\alpha{}^{(i)} + (a_{j+1})_\alpha{}^{(i)}\right]\right] \\ & \text{, for a } j \in \{1, \ldots, (N_i)_\alpha - 1\} \\ (a_j)_\alpha{}^{(i)} & \text{, if } t \in \left(\frac{1}{2}\left[(a_j)_\alpha{}^{(i)} + (b_{j-1})_\alpha{}^{(i)}\right], (a_j)_\alpha{}^{(i)}\right] \\ & \text{, for a } j \in \{2, \ldots, (N_i)_\alpha\} \end{cases}$$

For $t \in \mathbb{R}$ define

$$D_{\alpha,n}^2[t] \stackrel{\mathrm{d}}{=}$$

$$\begin{cases} \frac{1}{n} \sum_{i\in\{1,\ldots,n\}\backslash C_\alpha[t]} ((V_i[t])_\alpha)^2 - \frac{n}{n - \text{card } C_\alpha[t]} \left\{ \frac{1}{n} \sum_{i\in\{1,\ldots,n\}\backslash C_\alpha[t]} V_i[t]_\alpha \right\}^2, \\ \qquad , \text{if card } C_\alpha[t] < n \\ 0 \qquad , \text{if card } C_\alpha[t] = n \end{cases}$$

Define

$$T_\alpha[\mu_1, \ldots, \mu_n] =$$

$$\min \left[\begin{array}{l} \min \left\{ D^2_{\alpha,n}[(c_k)^\alpha] \;\middle|\; \begin{array}{l} k \in \{1, \ldots, M-1\} \text{ and} \\ \sum_{i\in\{1,\ldots,n\}\backslash C[(c_k)^\alpha]} V_i[(c_k)^\alpha]_\alpha \\ = (c_k)^\alpha \cdot (n - \text{card } C_\alpha[(c_k)^\alpha]_\alpha) \end{array} \right\}, \\ \min \left\{ D^2_{\alpha,n}[(d_k)^\alpha] \;\middle|\; \begin{array}{l} k \in \{1, \ldots, M-1\} \text{ and} \\ (c_{k-1})^\alpha \cdot (n - \text{card } C_\alpha[(d_k)^\alpha]) < \\ \sum_{i\in\{1,\ldots,n\}\backslash C[(c_k)^\alpha]} V_i[(C_k)^\alpha] < \\ (c_k)^\alpha \cdot (n - \text{card } C_\alpha[(c_k)^\alpha]_\alpha). \end{array} \right\} \end{array} \right]$$

Then

$$T_\alpha[\mu_1, \ldots, \mu_n] = \inf \left(S^2_n[\mu_1, \ldots, \mu_n] \right)_\alpha$$

holds for all $\alpha \in [0, 1)$.

Theorem 11.16 *Let $\delta \in (0, 1)$ and $n \in \mathbb{N}$.*
Define for $(\mu_1, \ldots, \mu_n) \in [F(\mathbb{R})]^n$, $\alpha \in [0, 1)$, $t \in \mathbb{R}$:

$$E_\alpha[\mu_1, \ldots, \mu_n] \stackrel{\text{d}}{=}$$

$$\begin{cases} \left[\frac{1}{q^{(n-1)}_{1-\frac{\delta}{2}}} \inf \left(S^2_\alpha[\mu_1, \ldots, \mu_n] \right)_\alpha , \frac{1}{q^{(n-1)}_{1-\frac{\delta}{2}}} \inf \left(S^2_\alpha[\mu_1, \ldots, \mu_n] \right)_\alpha \right], \\ \qquad \text{,if } \inf (\mu_i)_\alpha > -\infty \text{ and } \sup (\mu_i)_\alpha < +\infty \text{ holds for } i \in \{1, \ldots, n\} \\ [0, +\infty) \qquad \text{,otherwise} \end{cases}$$

$$K_n[\mu_1,\ldots,\mu_n](t) \stackrel{\mathrm{d}}{=} \sup\left\{\alpha \mathrm{I}_{E_\alpha[\mu_1,\ldots,\mu_n]}(t) \mid \alpha \in (0,1)\right\}.$$

Then

$$K_n : [F(\mathbb{R})]^n \to U(\mathbb{R})$$

is a $100 * (1-\delta)\%$*fuzzy confidence interval for* $\mathcal{N}$ *and* Var .

11.3 THE TESTING OF HYPOTHESES

Let $U_1, U_2, \ldots, U_n$ be a random sample from a population distribution $D_U \in Y$, where the functional form of the distribution function is known except for the parameter γ. Thus, f.e., the U_i's may be a random sample from $N(\gamma, 1)$, where $\gamma \in \mathbb{R}$ is not known. In many practical situations, the experimenter is interested in testing the validity of an assertion about the unknown parameter γ. So he might want to check the claim of an automobile manufacture with respect to the average fuel consumption achieved of a particular model. A problem of this type is usually referred to as a problem of testing hypotheses.

In contrast to the approach of M. Delgado et al. [35], we do not make Bayesian assumptions.

In this chapter we consider the problem of how to reject or accept hypotheses about the parameter Γ_Y in the presence of vague data. If

$$(X_1, \ldots, X_n) : \Omega \to [F(\mathbb{R})]^n$$

is a fuzzy random vector (with an original $(U_1, \ldots, U_n) \in \chi_Y^{(n)}$), $\Gamma_Y[X_1, \ldots, X_n]$ can be considered a the fuzzy perception of $\Gamma_Y(D_{U_1})$.

A hypothesis about $\Gamma_Y[X_1, \ldots, X_n]$ is called a hypothesis about the fuzzy perception of Y and Γ_Y, and we want to test such hypotheses. We restrict ourselves to the considerations hypotheses with convex fuzzy sets.

Definition 11.17 *Let* $n \in \mathbb{N}$ *and* $\delta \in (0,1)$. *Let* H_0 *be a hypothesis about the fuzzy perception of* Y *and* Γ_Y, *and* H_1 *be an alternative.*
A function

$$\phi : [F(\mathbb{R})]^n \longrightarrow \{0,1\}$$

is called a (non randomized) test for (δ, H_0, H_1), *if*

$$\tilde{P}\left(\{\omega \in \Omega \mid \phi[(X_1)_\omega, \ldots, (X_n)_\omega] = 1\}\right) \leq \delta$$

holds where

$$(X_1, \ldots, X_n) : \Omega \to [F(\mathbb{R})]^n$$

is a fuzzy random vector and $\Gamma_Y[X_1, \ldots, X_n]$ fulfills the null hypothesis. For a set $A \subseteq \Omega$, we define :

$$\tilde{P}(A) \leq \delta \overset{\text{d}}{\Leftrightarrow} \exists B \in \mathfrak{A} \otimes \mathfrak{A}' \ \text{ with } \ (P \otimes P')(B) \leq \delta$$

such that $A \times \Omega' \subseteq B$.

We want to compare $\Gamma_Y[X_1, \ldots, X_n]$ with a fixed convex fuzzy set μ_0. We distinguish the test "=" against "≠", "≤" against ">", and "≥" against "<". We test on N levels $\alpha_1, \ldots, \alpha_N$ and reject the null hypothesis, if at least K levels lead to rejection.

Theorem 11.18 *Let $n \in \mathbb{N}$, $\delta \in (0,1)$, and $N \in \mathbb{N}$.
Let $\{\alpha_1, \ldots, \alpha_N\} \subseteq [0,1)$, $K \in \{1, \ldots, N\}$, and $\mu_0 \in U(\mathbb{R})$.
Let $[T_n, +\infty)$ and $(-\infty, U_n]$ be two one-sided $100 * (1-\delta_1)\%$ and $100 * (1-\delta_2)\%$ (usual) confidence intervals for Y and Γ_Y with $\delta_1 + \delta_2 = \delta \frac{K}{N}$ and $T_n \leq U_n$.
Define for $(\mu_1, \ldots, \mu_n) \in [F(\mathbb{R})]^n$ and $\alpha \in [0,1)$ $A_\alpha[\mu_1, \ldots, \mu_n]$ and $B_\alpha[\mu_1, \ldots, \mu_n]$ as in Theorem 11.10.
Define for $(\mu_1, \ldots, \mu_n) \in [F(\mathbb{R})]^n$*

$$\phi_i(\mu_1, \ldots, \mu_n) \overset{\text{d}}{=} \begin{cases} 1 & , \text{if} \inf(\mu_0)_{\alpha_i} < A_{\alpha_i}[\mu_1, \ldots, \mu_n] \text{ or} \\ & \quad \sup(\mu_0)_{\alpha_i} > B_{\alpha_i}[\mu_1, \ldots, \mu_n] \\ 0 & , \text{otherwise} \end{cases}$$

for $i \in \{1, \ldots, N\}$, and

$$\phi(\mu_1, \ldots, \mu_n) \overset{\text{d}}{=} \begin{cases} 1 & , \text{if} \sum_{i=1}^{N} \phi_i(\mu_1, \ldots, \mu_n) \geq K \\ 0 & , \text{otherwise.} \end{cases}$$

Then

$$\phi : [F(\mathbb{R})]^n \longrightarrow \{0,1\}$$

is a test for

H_0: "The convex hull of the fuzzy perception of Y and Γ_Y is equal to μ_0 "

against

H_1: "It is not equal to μ_0"

on the significance level δ.

Proof Let $K_n[\mu_1,\ldots,\mu_n]$ be defined as in Theorem 11.10 for $(\mu_1,\ldots,\mu_n) \in [F(\mathbb{R})]^n$. We know that

$$\inf \mathrm{cl}\ (K_n[\mu_1,\ldots,\mu_n])_\alpha = \inf (K_n[\mu_1,\ldots,\mu_n])_\alpha = A_\alpha[\mu_1,\ldots,\mu_n]$$

and

$$\sup \mathrm{cl}\ (K_n[\mu_1,\ldots,\mu_n])_\alpha = \sup (K_n[\mu_1,\ldots,\mu_n])_\alpha = B_\alpha[\mu_1,\ldots,\mu_n]$$

is valid for all $\alpha \in [0,1)$.
Theorem 11.10 shows that K_n is a $100 * (1 - \frac{K}{N} \cdot \delta)\%$ fuzzy confidence interval for Y and Γ_Y.
Let $(X_1,\ldots,X_n)$ be a fuzzy random vector. Then

$$\tilde{P}\left(\{\omega \in \Omega \mid (\Gamma_Y[X_1,\ldots,X_n])_\alpha \not\subseteq \mathrm{cl}\ (K_n[(X_1)_\omega,\ldots,(X_n)_\omega])_\alpha\}\right) \leq \delta \cdot \frac{K}{N}$$

is valid for all $\alpha \in [0,1)$. It follows:

$$\tilde{P}\left(\left\{\omega \in \Omega \left| \begin{array}{l} \inf (\Gamma_Y[X_1,\ldots,X_n])_\alpha < A_\alpha[(X_1)_\omega,\ldots,(X_n)_\omega] \text{ or} \\ \sup (\Gamma_Y[X_1,\ldots,X_n])_\alpha > B_\alpha[(X_1)_\omega,\ldots,(X_n)_\omega] \end{array} \right.\right\}\right)$$

$$\leq \delta \cdot \frac{K}{N}$$

for all $\alpha \in [0,1)$.
Let the null hypothesis be fulfilled, i.e. $\mathrm{co}\ (\Gamma_Y[X_1,\ldots,X_n]) = \mu_0$. For abbreviation define

$$A \stackrel{\mathrm{d}}{=} \{\omega \in \Omega \mid \phi[(X_1)_\omega,\ldots,(X_n)_\omega] = 1\}.$$

We want to show that $\tilde{P}(A) \leq \delta$ holds.
We know that there exist sets $B_i \in \mathfrak{A} \otimes \mathfrak{A}'$ with $(P \otimes P')(B_i) \leq \delta \cdot \frac{K}{N}$ such that

$$\begin{aligned} &\{\omega \in \Omega \mid \phi_i[(X_1)_\omega,\ldots,(X_n)_\omega] = 1\} \times \Omega' \\ &= \left\{\omega \in \Omega \left| \begin{array}{l} \inf (\mu_0)_\alpha < A_{\alpha_i}[(X_1)_\omega,\ldots,(X_n)_\omega] \text{ or} \\ \sup (\mu_0)_\alpha > B_{\alpha_i}[(X_1)_\omega,\ldots,(X_n)_\omega] \end{array} \right.\right\} \times \Omega' \subseteq B_i \end{aligned}$$

holds for $i \in \{1, \ldots, N\}$.
Define a usual random variable Y by setting

$$Y \stackrel{\mathrm{d}}{=} \sum_{i=1}^{N} \mathrm{I}_{B_i}.$$

Obviously $E\,Y$ exists, so Markov's inequality (compare [31,52,183]) yields:

$$P\,(Y \geq \lambda E\,Y) \leq \frac{1}{\lambda}$$

for all $\lambda > 1$. If $\lambda \stackrel{\mathrm{d}}{=} 1/\delta$, then we obtain, as $E\,Y \leq \delta \cdot K$ holds:

$$P\left(\sum_{i=1}^{N} \mathrm{I}_{B_i} \geq K\right) \leq \delta.$$

Obviously $(\omega, \omega') \in A \times \Omega'$ implies:

$$\sum_{i=1}^{N} \mathrm{I}_{B_i}(\omega, \omega') \geq K.$$

So

$$\tilde{P}(A) \leq \delta$$

follows. •

We can also test one-sided hypotheses.

Theorem 11.19 *Let $n \in \mathbb{N}$, $\delta \in (0,1)$, and $N \in \mathbb{N}$.
Let $\{\alpha_1, \ldots, \alpha_N\} \subseteq [0,1)$, $K \in \{1, \ldots, N\}$, and $\mu_0 \in U(\mathbb{R})$.*

*(i) If $(-\infty, U_n]$ is a (usual) one-sided $100 * (1 - \delta \cdot \frac{K}{N})\%$ confidence interval for U and Γ_Y, let $B_\alpha[\mu_1, \ldots, \mu_n]$ be defined as in Theorem 11.10 for $(\mu_1, \ldots, \mu_n) \in [F(\mathbb{R})]^n$ and $\alpha \in [0,1)$.
Define for $(\mu_1, \ldots, \mu_n) \in [F(\mathbb{R})]^n$:*

$$\phi_i(\mu_1, \ldots, \mu_n) \stackrel{\mathrm{d}}{=} \begin{cases} 1, & \text{if } \sup{(\mu_0)_{\alpha_i}} > B_{\alpha_i}[\mu_1, \ldots, \mu_n] \\ 0, & \text{otherwise} \end{cases}$$

for $i \in \{1, \ldots, N\}$, and

$$\phi(\mu_1, \ldots, \mu_n) \stackrel{\mathrm{d}}{=} \begin{cases} 1, & \text{if } \sum_{i=1}^{N} \phi_i(\mu_1, \ldots, \mu_n) \geq K \\ 0, & \text{otherwise.} \end{cases}$$

Then $\phi : [F(\mathbb{R})]^n \to \{0,1\}$ *is a test for*

H_0*: "The convex hull of the fuzzy perception of* Y *and* Γ_Y *is greater or equal* μ_o*"*

against

H_1*: "It is less than* μ_0 *"*

on the significance level δ*.*

(ii) *If* $[T_n, +\infty)$ *is a (usual) one-sided* $100 * (1 - \delta \cdot \frac{K}{N})\%$ *confidence interval for* Y *and* Γ_Y*, let* $A_\alpha[\mu_1, \ldots, \mu_n]$ *be defined as in Theorem 11.10 for* $(\mu_1, \ldots, \mu_n) \in [F(\mathbb{R})]^n$ *and* $\alpha \in [0,1)$ *.*
Define for $(\mu_1, \ldots, \mu_n) \in [F(\mathbb{R})]^n$*:*

$$\phi_i(\mu_1, \ldots, \mu_n) \stackrel{\mathrm{d}}{=} \begin{cases} 1, & \text{if } \inf(\mu_0)_{\alpha_i} < A_{\alpha_i}[\mu_1, \ldots, \mu_n] \\ 0, & \text{otherwise} \end{cases}$$

for $i \in \{1, \ldots, N\}$ *, and*

$$\phi(\mu_1, \ldots, \mu_n) \stackrel{\mathrm{d}}{=} \begin{cases} 1, & \text{if } \sum_{i=1}^{N} \phi_i(\mu_1, \ldots, \mu_n) \geq K \\ 0, & \text{otherwise.} \end{cases}$$

Then $\phi : [F(\mathbb{R})]^n \to \{0,1\}$ *is a test for*

H_0*: "The convex hull of the fuzzy perception of* U *and* Γ_Y *is less or equal* μ_o*"*

against

H_1*: "It is greater than* μ_0 *"*

on the significance level δ*.*

Proof

(i) Let $K_n[\mu_1, \ldots, \mu_n]$ be defined as in Theorem 11.11(ii) for $(\mu_1, \ldots, \mu_n) \in [F(\mathbb{R})]^n$. We know that

$$\sup \mathrm{cl}\ (K_n[\mu_1, \ldots, \mu_n])_\alpha = \sup (K_n[\mu_1, \ldots, \mu_n])_\alpha = B_\alpha[\mu_1, \ldots, \mu_n]$$

is valid for all $\alpha \in [0,1)$.
$K_n : [F(\mathbb{R})]^n \longrightarrow U(\mathbb{R})$ is a $100 * (1 - \frac{K}{N} \cdot \delta)\%$ fuzzy confidence interval for Y and Γ_Y.
We conclude:
If $(X_1, \ldots, X_n)$ is a fuzzy random vector, then

$$\tilde{P}(\{\omega \in \Omega \mid \sup(\Gamma_Y[X_1, \ldots, X_n])_\alpha > B_\alpha[(X_1)_\omega, \ldots, (X_n)_\omega]\}) \leq \delta \cdot \frac{K}{N}$$

is valid for all $\alpha \in [0,1)$.
The assertion can be followed similarly to the proof of Theorem 11.18.

(ii) can be shown by analogy.

Now we want to apply our results to the important class $\mathcal{N}$ of all normal distributions and the parameter mapping E assigning to each $N(\mu, \sigma^2)$ distributions its parameter μ. We have to combine Theorem 4.1 and the last two theorems.

Theorem 11.20 *Let* $n \in \mathbb{N}$, $\delta \in (0,1)$, *and* $N \in \mathbb{N}$.
Let $\{\alpha_1, \ldots, \alpha_N\} \subseteq [0,1)$, $K \in \{1, \ldots, N\}$, *and* $\mu_0 \in U(\mathbb{R})$.
Let $t_{1-\delta}^{(n-1)}$ *denote the* $(1-\delta)$*-quantile of the student distribution with* $(n-1)$ *degrees of freedom.*

(i) Define for $(\mu_1, \ldots, \mu_n) \in [F(\mathbb{R})]^n$:

$$\phi_i(\mu_1, \ldots, \mu_n) \stackrel{\mathrm{d}}{=}$$

$$\begin{cases} 1, & \text{if } \inf(\mu_0)_{\alpha_i} < \frac{1}{n} \sum\limits_{j=1}^{n} \inf(\mu_j)_{\alpha_i} - \frac{t^{(n-1)}_{1-\frac{\delta \cdot K}{2 \cdot N}}}{\sqrt{n(n-1)}} \sqrt{S_{\alpha_i}(\mu_1, \ldots, \mu_n)} \\ & \text{or } \sup(\mu_0)_{\alpha_i} > \frac{1}{n} \sum\limits_{j=1}^{n} \sup(\mu_j)_{\alpha_i} + \frac{t^{(n-1)}_{1-\frac{\delta \cdot K}{2 \cdot N}}}{\sqrt{n(n-1)}} \sqrt{S_{\alpha_i}(\mu_1, \ldots, \mu_n)} \\ 0, & \text{otherwise} \end{cases}$$

for $i \in \{1, \ldots, N\}$, *and*

$$\phi(\mu_1, \ldots, \mu_n) \stackrel{\mathrm{d}}{=} \begin{cases} 1, & \text{if } \sum\limits_{i=1}^{N} \phi_i(\mu_1, \ldots, \mu_n) \geq K \\ 0, & \text{otherwise.} \end{cases}$$

Then $\phi : [F(\mathbb{R})]^n \to \{0,1\}$ *is a test for*

H_0: *"The convex hull of the fuzzy perception of* $\mathcal{N}$ *and* E *is equal to* μ_0*"*
against
H_1: *"It is not equal to* μ_0 *"*
on the significance level δ.

(ii) Define for $(\mu_1, \ldots, \mu_n) \in [F(\mathbb{R})]^n$:

$$\phi_i(\mu_1, \ldots, \mu_n) \stackrel{\mathrm{d}}{=}$$

$$\begin{cases} 1, & \text{if } \sup(\mu_0)_{\alpha_i} > \frac{1}{n} \sum\limits_{j=1}^{n} \sup(\mu_j)_{\alpha_i} + \frac{t^{(n-1)}_{1-\frac{\delta \cdot K}{N}}}{\sqrt{n(n-1)}} \sqrt{S_{\alpha_i}(\mu_1, \ldots, \mu_n)} \\ 0, & \text{otherwise} \end{cases}$$

for $i \in \{1,\ldots,N\}$, *and*

$$\phi(\mu_1,\ldots,\mu_n) \stackrel{\mathrm{d}}{=} \begin{cases} 1, & \text{if } \sum_{i=1}^{N} \phi_i(\mu_1,\ldots,\mu_n) \geq K \\ 0, & \text{otherwise.} \end{cases}$$

Then $\phi : [F(\mathbb{R})]^n \to \{0,1\}$ *is a test for*

H_0*: "The convex hull of the fuzzy perception of* N *and* E *is greater or equal to* μ_0*"*

against

H_1*: "It is less than* μ_0 *"*

on the significance level δ*.*

(iii) Define for $(\mu_1,\ldots,\mu_n) \in [F(\mathbb{R})]^n$*:*

$$\phi_i(\mu_1,\ldots,\mu_n) \stackrel{\mathrm{d}}{=}$$

$$\begin{cases} 1, & \text{if } \inf(\mu_0)_{\alpha_i} < \frac{1}{n}\sum_{j=1}^{n} \inf(\mu_j)_{\alpha_i} - \frac{t^{(n-1)}_{1-\frac{\delta \cdot K}{N}}}{\sqrt{n(n-1)}} \sqrt{S_{\alpha_i}(\mu_1,\ldots,\mu_n)} \\ 0, & \text{otherwise} \end{cases}$$

for $i \in \{1,\ldots,N\}$, *and*

$$\phi(\mu_1,\ldots,\mu_n) \stackrel{\mathrm{d}}{=} \begin{cases} 1, & \text{if } \sum_{i=1}^{N} \phi_i(\mu_1,\ldots,\mu_n) \geq K \\ 0, & \text{otherwise.} \end{cases}$$

Then $\phi : [F(\mathbb{R})]^n \to \{0,1\}$ *is a test for*

H_0*: "The convex hull of the fuzzy perception of* $\mathcal{N}$ *and* E *is less or equal to* μ_0*"*

against

H_1*: "It is greater than* μ_0 *"*

on the significance level δ*.*

In part 1 of this chapter, we assumed that we deal with an i.i.d. fuzzy random vector $(X_1,\ldots,X_n)$ such that $D_{(\underline{Y}_1)_\alpha}$ and $D_{(\overline{Y}_1)_\alpha}$ belong to Y and $D_{(\underline{X}_1)_\alpha}$ and $D_{(\overline{X}_1)_\alpha}$ belong to Y. With the help of a chi-squared test of goodness of fit, we can test whether this assertion is fulfilled or not.

Let $n \in I\!N$ and $\delta \in (0,1)$. Let $(Z_1,\ldots,Z_n)$ be i.i.d. It is well known that $\Phi : I\!R^n \to I\!R$ is called a chi-squared test of goodness of fit for the null hypothesis "$D_{Z_1} \in Y$" against "$D_{Z_1} \notin Y$" on the significance level δ if $D_{Z_1} \in Y$ implies

$$P[\Phi(Z_1,\ldots,Z_n) = 1] \leq \delta.$$

In [52], [83], [128] a test function Φ can be found for this problem with the help of the maximum-likelihood-estimators and the chi-square distribution. With the help of Markov's inequality it is easy to derive a test for the problem mentioned above.

Theorem 11.21 *Let $n \in I\!N$, $\delta \in (0,1)$, and $N_1, N_2 \in I\!N$.
Let $\{\alpha_1,\ldots,\alpha_{N_1}\} \subseteq [0,1)$, $\{\beta_1,\ldots,\beta_{N_2}\} \subseteq (0,1]$, and $K \in \{1,\ldots,N_1+N_2\}$.
Let Y be a class of distribution functions, let $\Phi : I\!R^n \to \{0,1\}$ be a (non randomized) chi-square test of goodness of fit for Y on the significance level $\delta \cdot K/2\,(N_1+N_2)$. Define for $(\mu_1,\ldots,\mu_n) \in [F\,(I\!R)]^n$*

$$\phi_1(\mu_1,\ldots,\mu_n) \stackrel{\mathrm{d}}{=}$$
$$\sum_{i=1}^{N_1} \Phi\left[\inf(\mu_1)_{\alpha_i},\ldots,\inf(\mu_n)_{\alpha_i}\right] + \sum_{i=1}^{N_1} \Phi\left[\sup(\mu_1)_{\alpha_i},\ldots,\sup(\mu_n)_{\alpha_i}\right] +$$
$$\sum_{i=1}^{N_2} \Phi\left[\inf(\mu_1)_{\overline{\beta}_i},\ldots,\inf(\mu_n)_{\overline{\beta}_i}\right] + \sum_{i=1}^{N_2} \Phi\left[\sup(\mu_1)_{\overline{\beta}_i},\ldots,\sup(\mu_n)_{\overline{\beta}_i}\right]$$

and

$$\phi[\mu_1,\ldots,\mu_n] \stackrel{\mathrm{d}}{=} \begin{cases} 1, & \text{if } \phi_1(\mu_1,\ldots,\mu_n) \geq K \\ 0, & \text{if } \phi_1(\mu_1,\ldots,\mu_n) < K. \end{cases}$$

Let $(X_1,\ldots,X_n)$ be an i.i.d. fuzzy random vector. Then

$$\Phi : [F\,(I\!R)]^n \longrightarrow \{0,1\}$$

is a test for

*H_0: "$\forall \alpha \in [0,1): D_{(\underline{Y}_1)_\alpha} \in Y \wedge D_{(\overline{Y}_1)_\alpha} \in Y$ and
$\forall \alpha \in (0,1]: D_{(\underline{X}_1)_\alpha} \in Y \wedge D_{(\overline{X}_1)_\alpha} \in Y$"
against the negation of H_0
on a significance level δ.*

In a similar way, other tests can be constructed. With the help of a Kolmogorov-Smirnov-test, we can test the null hypothesis "$F_X = F$" where $(X_1, \ldots, X_n)$ is an i.i.d. fuzzy random vector with the distribution function $F_X : I\!R \to U_C([0,1])$, and $F : I\!R \to U_C([0,1])$ is a given mapping.

With the help of Wilkoxons's rank sum-test (compare [20]), it is possible to test
"$\forall x \in I\!R : F_X(x) = F_Y(x)$" against "$\exists x \in I\!R : F_X(x) \neq F_Y(x)$"
or
"$\forall x \in I\!R : F_X(x) \leq F_Y(x)$" against "$\exists x \in I\!R : F_X(x) \geq F_Y(x)$"
on a given significance level where $(X_1, \ldots, X_{n_1}, Y_1, \ldots, Y_{n_2})$ is completely independent such that

$$F_{X_i} = F_X \quad (i = 1, \ldots, n_1) \text{ and } F_{X_j} = F_Y \quad (j = 1, \ldots, n_2).$$

We must assert that $D_{(\underline{Y}_1)_\alpha}$ and $D_{(\overline{Y}_1)_\alpha}$ are continuous for $\alpha \in [0,1)$ and converge against 0 resp. 1, as x converges against $-\infty$ resp. $+\infty$.

12 On a software tool for statistics with vague data

In the preceding chapters, we demonstrated how to perform statistical analyses in the presence of vague data. The problem with these methods is that the practical calculations turn out to be intricate even if the samples are small.

We therefore present a software tool by which we can support users who want to apply our statistical methods. The main aim is to develop efficient algorithms for the calculation of statistical values. This requires a suitable data structure for the storage of vague data.

Another purpose is to define a user-friendly surface for our system in order to increase the acceptance of this tool. We distinguish between two different types of "employers" of our system known from knowledge based systems:

- First there is a user who only has access to vague data. This type of employer wants to make statistical evaluations with the given data. It is not necessary for him to understand the internal representation of vague data. For the user the results of the analyses are also vague data.
- Second we consider an expert who defines languages of vague data and describes the vague data by fuzzy sets. He chooses the statistical methods and finally decides how to transform results of the fuzzy set theory into vague data.

The expert prepares the environment for the work with the system, whereas the user makes the statistical evaluations without knowing the details of the internal representation. A user and an expert may be the same person. Thus we obtain this diagram:

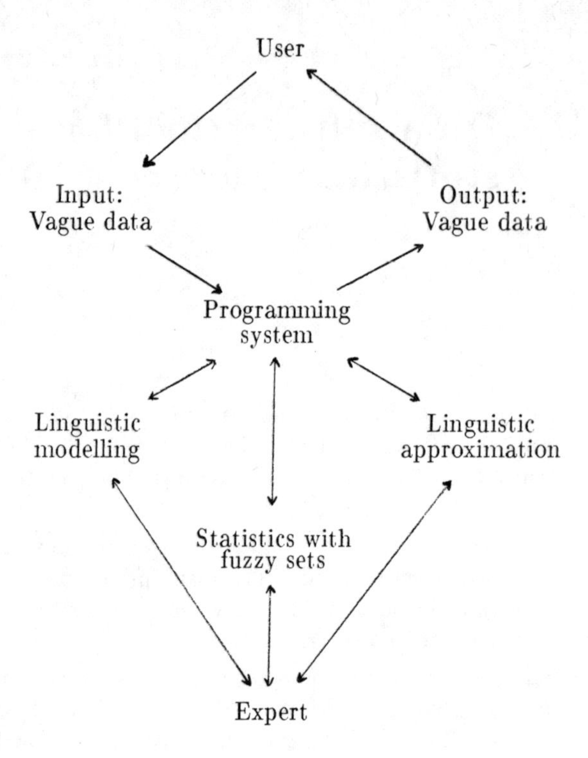

Fig. 12.1. The system components of our software tool.

Throughout this chapter we will use a small tutorial example concerning life time considerations in order to demonstrate some facilities of the programming system.

12.1 LINGUISTIC MODELLING

During statistical evaluations, a user must describe the data material. Since we are interested in vague data, we have to consider languages of vague data. In order to make this notion precise, we define formal languages.

Definition 12.1 *Let* $T = \{t_1, \ldots, t_k\}$, $k \in I\!N$, *and let*

$$O = \left\{o_1^1, \ldots, o_{m_1}^1, \ldots, o_1^n, \ldots, o_{m_n}^n\right\},$$

$n \in I\!N$, $m_1, \ldots, m_n \in I\!N$, *be finite sets. We say that* L *is generated by* T *and* O, *if the following properties hold:*

(i) $T \subseteq L$,

(ii) if $\{l_1, \ldots, l_n\} \subseteq L$, *then*

$$\left\{o_1^1(l_1), \ldots, o_{m_1}^1(l_1), \ldots, o_1^n(l_1, \ldots, l_n), \ldots, o_{m_n}^n(l_1, \ldots, l_n)\right\} \subseteq L$$

(iii) L *is the smallest language, for which (i) and (ii) hold.*

T is called a term set of vague data, O is called the set of operators. If an expression l is given, then we can decide by a finite algorithm whether l is in L or not.

Example 12.2 As a typical example, we consider evaluations concerning life times of mayflies. A very simple language consists of the term set

$$\begin{aligned} T = &\{\text{between } s \text{ and } t \text{ hours} \mid s, t \in I\!R, 0 \leq s < t \leq 24\} \\ &\cup \{\text{exactly } s \text{ hours} \mid s \in I\!R, 0 \leq s \leq 24\} \\ &\cup \{\text{approximately } s \text{ hours} \mid s \in I\!R, 0 \leq s \leq 24\} \\ &\cup \{\text{young}\} \end{aligned}$$

and the operators

$$O = \{o_1^1, o_2^1, o_1^2, o_2^2\},$$

where o_1^1 = *very*, o_2^1 = not, o_1^2 = and, o_2^2 = *or*.

Using these common notions, we see that it is possible to express rather complicated expressions such as

"between 1 and 2 hours *or* not *very* young

with this simple language L.

Elements of L are sometimes called values of a linguistic variable [214].

With our definition of a formal language, the syntactical properties of linguistic values are described. In the following we assign to each expression in a language its meaning.

In Chapter 5 we recognized that $C_n(I\!R)$ is a suitable class for representing fuzzy sets, so we give the following definition of a semantic of L.

Definition 12.3 *Each mapping M from L to $C_n(\mathbb{R})$ is called a semantic of L.*

We propose a practical method for defining semantics.
Let $N : T \to C_n(\mathbb{R})$ be a mapping that assigns to each element of the term set its meaning, and let N^* be a mapping that assigns to operators its meaning, i.e.

$$\left\{o_1^1, \ldots, o_{m_1}^1\right\} \xrightarrow{N^*} \{p^1 \mid p^1 : C_n(\mathbb{R}) \to C_n(\mathbb{R})\}$$
$$\left\{o_2^1, \ldots, o_{m_2}^2\right\} \xrightarrow{N^*} \left\{p^2 \,\middle|\, p^2 : [C_n(\mathbb{R})]^2 \to C_n(\mathbb{R})\right\}$$
$$\ldots$$
$$\{o_n^1, \ldots, o_{m_n}^n\} \xrightarrow{N^*} \{p^n \mid p^n : [C_n(\mathbb{R})]^n \to C_n(\mathbb{R})\}$$

Then there is a unique extension M of N from T to L such that M is structure preserving. By structure preserving we mean that M is a homomorphism with the property of

$$M\left(o_i^k(l)\right) = \left(N^*\left(o_i^k\right)\right)(l),$$

for $k \in \{1, \ldots, n\}$, $i \in \{1, \ldots, m_k\}$, $l = (l_1, \ldots, l_k) \in L^k$.

The uniqueness follows by induction about the structure of the elements in L.

These considerations show that we have to specify the map N^* in order to obtain a semantic M. Let us continue with our example concerning mayflies.

Example 12.2 (continuation)
From Chapter 4 we know that it is reasonable to define

$$N^*(\underline{and}) = \wedge$$
$$\text{, where } \wedge : C_n^2(\mathbb{R}) \to C_n(\mathbb{R}),\ (\mu \wedge \mu')(x) = \min(\mu(x), \mu'(x))$$
$$N^*(\underline{or}) = \vee$$
$$\text{, where } \vee : C_n^2(\mathbb{R}) \to C_n(\mathbb{R}),\ (\mu \wedge \mu')(x) = \max(\mu(x), \mu'(x))$$
$$N^*(\underline{not}) = \overline{}$$
$$\text{, where } \overline{} : C_n(\mathbb{R}) \to C_n(\mathbb{R}),\ \overline{\mu}(x) = 1 - \mu(x)$$

The quantity $\underline{very}$ is a so called linguistic hedge [214]. The representation of this operator depends of course on its context. In our context we choose subjectively

$$N^*\left(\underline{very}\right) = \sim \text{ , where } \sim : C_n(\mathbb{R}) \to C_n(\mathbb{R}),\ \tilde{\mu} = \left[\mu^2\right]_n.$$

Notice that μ^2 is not defined with the help of the extension principle (see p.30) but pointwise, i.e.

$$\mu^2[t] = (\mu[t])^2$$

holds for $t \in \mathbb{R}$.
If, for example, $n = 10$ and $\mu \stackrel{\text{d}}{=} \frac{1}{2}\mathrm{I}_{[0,24]} \in C_{10}(\mathbb{R})$ holds, then $\mu^2 = \frac{1}{4}\mathrm{I}_{[0,24]}$ is not in $C_{10}(\mathbb{R})$, but $\left[\mu^2\right]_{10} = 0.2 \cdot \mathrm{I}_{[0,24]}$ is. This is the reason for choosing the fuzzy sets $\left[\mu^2\right]_n$.

The next problem is to model the vague data in T by fuzzy set of $C_n(\mathbb{R})$. Of course we set

$$N(\text{exactly } s \text{ hours }) = \mathrm{I}_{\{s\}} \in C_n(\mathbb{R}) \text{ , and}$$
$$N(\text{between } s \text{ and } t \text{ hours }) = \mathrm{I}_{[s,t]} \in C_n(\mathbb{R}).$$

How can we model arbitrary vague data in T? It depends again on the context which fuzzy set may be chosen. We consider first an example concerning particle shape analysis.

Example 12.4 In mining sciences, engineers are interested in approximately characterizing the shape of a three dimensional particle by two-dimensional regular geometric figures. This is interesting with respect to the behaviour of the particle when sinking in water. If we have rather round particles, then the shape should be characterized by figures of the form

$$E_a \stackrel{\text{d}}{=} \left\{ (x,y) \in \mathbb{R}^2 \,\middle|\, \left(\frac{x}{a}\right)^2 + y^2 = r^2 \right\} \quad ,a > 0$$

where $2 \cdot r$ denotes the largest diameter of the particle .

For each $a > 0$, we want to determine the "degree of acceptability" $\mu_A(a)$ with respect to the shape E_a of the particle. We obtain a mapping $\mu_A : \mathbb{R}^+ \to [0,1]$.

H. Bandemer [4 to 8] has derived a "fuzzy shape analysis" method for similar problems. As we are not interested in the exact shape but only in a characterization of it, the fourier shape analysis (compare [10,9], e.g.) seems to be unappropriate here.

We examine the particle in a fixed position, and by a vertical projection, we obtain a two-dimensional grey-tone picture. It can be described by a function $g : \mathbb{R}^2 \to [0,1]$. $g(x,y) = 1$ indicates that the particle has its greatest vertical extension in the point (x,y). If the vertical extension of the particle in (x, y) is zero, then $g(x, y) = 0$. The fuzzy contour h of the particle is a function $h : I\!R^2 \rightarrow [0,1]$ given by the pointwise fuzzy entropy (compare A.De Luca and S. Termini [33] and R. Kruse [119])

$$h(x,y) \stackrel{\mathrm{d}}{=} 2\min\{g(x,y), 1 - g(x,y)\} \quad \text{for} \quad (x,y) \in I\!R^2.$$

Then we calculate the fuzzy set μ_A where

$$\mu_A(a) \stackrel{\mathrm{d}}{=} \begin{cases} \dfrac{\int_{E\alpha} h(x,y)d(x,y)}{\int_{E\alpha} d(x,y)} & \text{, if } a > 0 \\ 0 & \text{, if } a \leq 0. \end{cases}$$

In practical application we approximate $\mu_A(a)$ with the help of a finite set $\{(x_r, y_r) \mid r = 1, 2, \ldots, N\}$.

Let us return to our example concerning mayflies. How should we model approximately s hours <u>and</u> young ? Answers to this problem are of three types.
The <u>descriptive approach</u> studies how people actually assign numerical degrees of acceptability when required to do so. This provides sociological/psychological information but only a weak basis for a formal structure. We refer to A.M. Norwich and I.B. Turksen [162] and T. Saaty [185] for a survey of these techniques.
The <u>normative approach</u> assumes nothing about the nature of the degrees of acceptability, but asks how meaning can, in principle , be assigned to vague data such that rational behaviour can be deduced. This method provides a language which has a well-defined interpretation and which simultaneously provides a basis for a formal system of reasoning. Unfortunately there are a lot of descriptive approaches but only very few normative approaches (S. French [55]).
In our book we take an <u>axiomatic approach</u> , i.e. we restrict ourselves to support the user in defining fuzzy sets.

Our software tool enables the expert to define the language of vague data and to describe the vague data by giving several internal given classes of fuzzy sets, function tables, and operations on fuzzy sets. We can use for example the statement

young := fexpo (0, 0.5)

to model the vague data "young" by a fuzzy set of the fexpo-class with the membership function

$$\text{fexpo}\,(a,b) = \begin{cases} \exp\,(-b(x-a)) & ,\text{ if } x \geq a \\ 0 & ,\text{ else} \end{cases} \quad , \quad \text{where } \; b > 0.$$

We have

$$N(\text{ young }) = [\text{fexpo}\,(0, 0.5)]_n.$$

In order to describe the data " approximately s hours" we consider again the class $\{g^{(a,b)} \mid a \in I\!R,\ b \geq 0\}$ (p. 12). $g^{(a,b)}$ will here be denoted by gauss (a, b). With the command

$$\text{approximately}\,(s) := \text{gauss}\,(s, 1.5)$$

the assignment

$$N(\text{ approximately } s \text{ hours }) \overset{\text{d}}{=} [\text{gauss}\,(s, 1.5)]_n.$$

is fulfilled.

The expert works in a dialogue with the computer using plot facilities. He is able to express statement like

"the insect was young <u>and</u> <u>not</u> between 1 and 2 hours old."

He can check the internal representation as a fuzzy set, and he is able to modify the mapping N until the vague data are described to his satisfaction.

12.2 LINGUISTIC APPROXIMATION

In the preceding considerations we described how, for any given problem context, a data base composed of a language L is generated. Obviously this language will be problem dependent, but in general L should be large enough such that any possible vague data of the problem should be described.

However, in most practical cases, L does not have to be infinite, since only an approximate description of each particular situation is required. Moreover, L must be easily understandable. Thus, complex syntactical structures should be avoided if possible.

In our context of statistical evaluations, the output produced by our software tool should have the same level as the input, i.e. vague data. Some problems appear, if we insist on this requirement. Suppose we want to add two vague data. The natural way is to represent these data by the associated fuzzy sets in $C_n(\mathbb{R})$, and then to add these fuzzy sets. The result μ is in $C_n(\mathbb{R})$, but there may exist no $l \in L$, such that $M(l) = \mu$, if the language L is not rich enough.

In practical application, we can restrict ourselves to find a vague datum $l \in L$ for a given fuzzy set $\mu \in C_n(\mathbb{R})$ which satisfies the property $\mu \subseteq M(l)$ (see page 8) and which is "close" to μ. This problem is referred to as linguistic approximation (P.P. Bonissone [13]).

We give two kinds of solution to this problem. The first is an existence theorem which provides a complete solution of the problem, but it has less practical relevance. The second gives some hints for implementing a heuristic approach which gives sufficiently good approximations.

Consider the class

$$\tau = \left\{ t^{(a,b,c,d)} \mid a \leq b \leq c \leq d, \;\; a,b,c,d \in \mathbb{R} \right\} \subseteq M(L),$$

the class of all trapezoidal fuzzy sets (compare page 56). We define for $n \in \mathbb{N}$ the set

$$\tau_n \stackrel{\mathrm{d}}{=} \{[t]_n \mid t \in \tau\}.$$

Elements of this class may be viewed as a kind of flou sets (page 6.) with sure region $[b, c]$ and maximum region $[a, d]$. The classes τ_n obviously play a central role in modelling vague data.
We assume in the following that the language L is rather rich; i.e. we require that $M(L) \supseteq \tau_n$ holds. A further assumption is that the language contains the operators <u>and</u> and <u>or</u>. We want to prove the following property:

(∗) For every $\mu \in C_n(\mathbb{R})$ there is an $l \in L$ such that $M(l) = \mu$.

In order to begin the proof, we define the set cl (τ_n) as the smallest class of fuzzy set in $C_n(\mathbb{R})$ such that

(i) $t \in \tau_n \;\Rightarrow\; t \in \mathrm{cl}\,(\tau_n)$,
(ii) $t, t' \in \mathrm{cl}\,(\tau_n) \;\Rightarrow\; t \wedge t' \in \mathrm{cl}\,(\tau_n)$, and
(iii) $t, t' \in \mathrm{cl}\,(\tau_n) \;\Rightarrow\; t \vee t' \in \mathrm{cl}\,(\tau_n)$

hold.

Theorem 12.5 cl $(\tau_n) = C_n(\mathbb{R})$ *is valid for* $n \in \mathbb{N}$.

The proof of this theorem is very simple:

Proof Since each set of a set representation of $\mu \in C_n(\mathbb{R})$ is a finite union of intervals, μ is a finite union of fuzzy sets of the class

$$\left\{ \frac{i}{n} \cdot \left[t^{(a,a,b,b)} \right]_n \mid a, b \in \mathbb{R}, a \leq b, \; i = 1, \ldots, n \right\}$$

(notice that for $\mu \in F(\mathbb{R})$ and $\lambda \in [0,1]$ $\lambda \cdot \mu$ is defined pointwise, i.e. $(\lambda \cdot \mu)(t) = \lambda \cdot \mu(t)$ holds for $t \in \mathbb{R}$).
On the other hand, we have

$$\frac{i}{n} \left[t^{(a,a,b,b)} \right]_n =$$

$$\left[t^{(a,a,b,b)} \right]_n \wedge \left[t^{a-4i(b-a),(4n-4i)(b-a)+a,(4n-4i)(b+a)+a,4n-4i(b-a)+a} \right]_n \wedge$$

$$\wedge \left[t^{a-4i(b-a),a-4i(b-a),a-4i(b-a),4n-4i(b-a)+a} \right]_n .$$

Therefore each $\mu \in C_n(\mathbb{R})$ is in cl (τ_n); the other inclusion is satisfied by definition. •

From this theorem follows $(*)$.

The proof of the theorem is constructive, but it does not give a reasonable procedure for the linguistic approximation problem. In our software tool, we use methods of heuristic searches [163]. The details of this search are described elsewhere. In this book we describe the notion "closeness" of an approximation.

We measure closeness by a generalized Hausdorff metric. The linguistic approximation problem for $\mu \in C_n(\mathbb{R})$ such that $\exists x \in \mathbb{R} : \mu(x) = 1$ is the following:

Find an $l \in L$ such that $\mu \subseteq M(l)$, i.e. $\mu(t) \leq M(l)(t)$ is valid and such that

$$d(M(l), \mu) = \max \left\{ d_H \left[M(l)_{\overline{(i/n)}}, \mu_{\overline{(i/n)}} \right] \mid i = 1, \ldots, n \right\}$$

is less than a given bound $\epsilon > 0$.

Since the level sets are finite unions of intervals, and since $\mu \subseteq M(l)$ implies

$$A \stackrel{\mathrm{d}}{=} M(l)_{\overline{(i/n)}} \subseteq \mu_{\overline{(i/n)}} \stackrel{\mathrm{d}}{=} B,$$

and therefore

$$\sup_{a \in A} \inf_{b \in B} |a - b| = d_H[A, B]$$

holds, the calculation of the Hausdorff metric is rather simple (compare Example 6.3). The Hausdorff metric serves as a value for the hill-climbing.

12.3 EXAMPLES

Example 12.2 (continuation)
We assume that the user examines only four insects. The ages of this (unrealistic) sample are:

$$\text{age}\,(1) \stackrel{d}{=} \text{between 4 and 5 hours} \quad \underline{\text{or}} \quad \text{approximately 7 hours},$$

$$\text{age}\,(2) \stackrel{d}{=} \text{between 1 and 2 hours},$$

$$\text{age}\,(3) \stackrel{d}{=} \text{approximately 1 hour}, \quad \text{and}$$

$$\text{age}\,(4) \stackrel{d}{=} \text{young}.$$

(a) If life time is to be considered, it is often reasonable to assume that the underlying random mechanism is exponentially distributed with the parameter $\lambda > 0$. This means, if the (usual) random variable U describes the life time, then it has the probability density function

$$f_U(t) = \begin{cases} \frac{1}{\lambda} e^{-t/\lambda} & , \text{if } t \geq 0 \\ 0 & , \text{if } t < 0. \end{cases}$$

As $E\,U = \lambda$ is valid, Theorem 11.8 shows that the arithmetic mean is an unbiased and strong consistent estimator for the fuzzy perception of λ; i.e. $\frac{1}{4} \sum_{i=1}^{4} \text{co } M(\text{age } (i))$ is a "sensible " estimator for Y and E where Y is the class of all exponential distributions. We obtain:

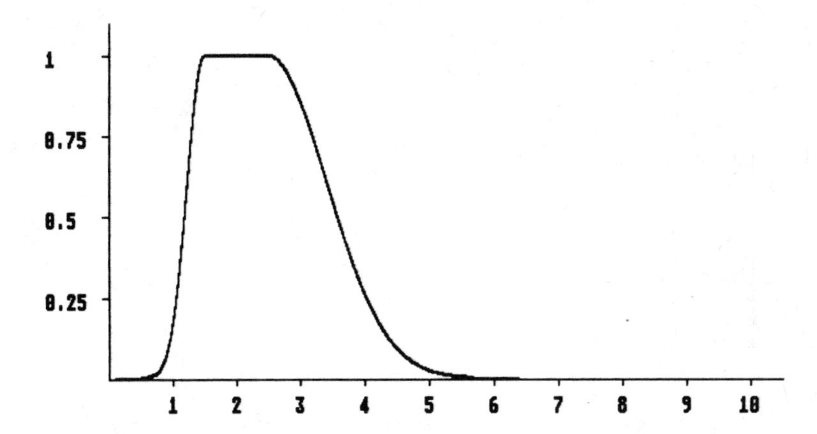

Fig. 12.2. Estimation of the characteristic "age".

The module "linguistic approximation" enables the user to get an appropriate linguistic value for this set. We obtain, f.e., the vague datum

"between 2 and 3 hours <u>or</u> approximately 1 hour " .

(b) Let $\Gamma_\delta(n)$ denote the δ-quantile of the gamma distribution with $p = n$ and $b = 1$; i.e. the δ-quantile of the distribution of U where U has the probability density

$$f_U(t) = \begin{cases} \frac{1}{(n-1)\cdot\ldots\cdot 2\cdot 1} \cdot t^{n-1} \cdot e^{-t} & \text{, if } t \geq 0 \\ 0 & \text{, if } t < 0. \end{cases}$$

It is easy to demonstrate that

$$\left[0, \frac{\sum_{i=1}^{n} U_i}{\Gamma_\delta(n)}\right] \quad \text{and} \quad \left[\frac{\sum_{i=1}^{n} U_i}{\Gamma_{1-\delta}(n)}, +\infty\right)$$

are two (usual) one-sided $100 * (1 - \delta)\%$confidence intervals for λ if $(U_1, \ldots, U_n)$ is i.i.d. and U_1 is exponentially distributed with the parameter λ. Let Y be the class of all exponential distributions and E assign the parameter λ.

Define for $n \in \mathbb{N}$, $(\mu_1, \ldots, \mu_n) \in [F(\mathbb{R})]^n$, $\alpha \in [0,1)$ and $t \in \mathbb{R}$ according to Theorem 11.12

$$E_\alpha[\mu_1, \ldots, \mu_n] \stackrel{\mathrm{d}}{=}$$

$$\begin{cases} \left[\dfrac{\sum_{i=1}^{n} \inf(\mu_i)_\alpha}{\Gamma_{1-\delta/2}(n)}, \dfrac{\sum_{i=1}^{n} \sup(\mu_i)_\alpha}{\Gamma_{\delta/2}(n)} \right] & \text{, if } \begin{array}{l} \inf(\mu_i)_\alpha \geq 0 \text{ and} \\ \sup(\mu_i)_\alpha < +\infty \\ \text{for all } i \in \{1, \ldots, n\} \end{array} \\ [0, \infty) & \text{, if } \begin{array}{l} \inf(\mu_i)_\alpha < 0 \text{ or} \\ \sup(\mu_i)_\alpha = +\infty \\ \text{for an } i \in \{1, \ldots, n\} \end{array} \end{cases} .$$

$$K_n[\mu_1, \ldots, \mu_n](t) \stackrel{\mathrm{d}}{=} \sup\left\{ \alpha \mathrm{I}_{E_\alpha[\mu_1, \ldots, \mu_n]}(t) \mid \alpha \in (0,1) \right\}.$$

Then $K_n : [F(\mathbb{R})]^n \rightarrow U(\mathbb{R})$ is a $100 * (1-\delta)\%$ fuzzy confidence interval for Y and E. We set $\delta = 10\%$ and have $n = 4$. In our case, we obtain the following result:

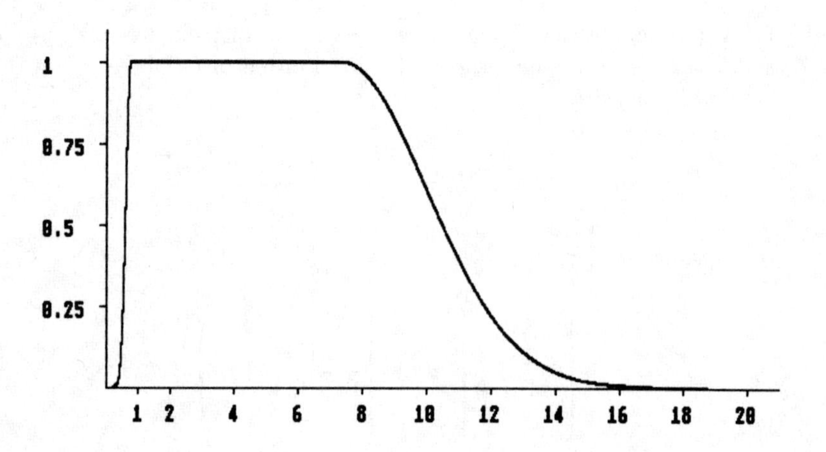

Fig. 12.3. Fuzzy confidence interval for Y and E.

(c) With help of the lemma of Gliwenko-Cantelli, we can estimate $F_X(3)$. We obtain:

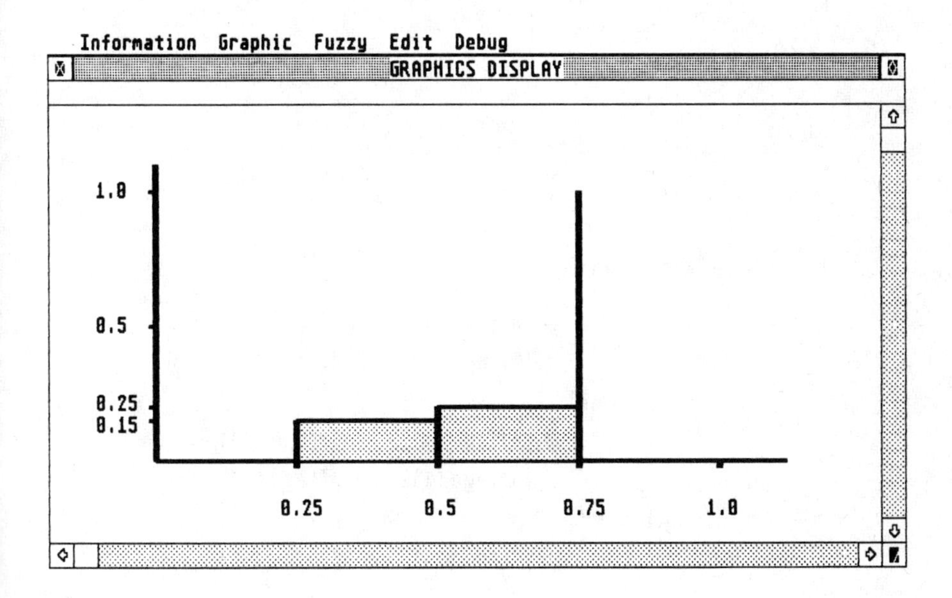

Fig. 12.4. Estimation of $\tilde{F}_X(3)$.

(d) Now we define $T_0 \stackrel{\mathrm{d}}{=}$ approximately 9 hours .
We want to test the null hypothesis

H_0: "The convex hull of the fuzzy perception of Y and E is greater or equal to $M(T_0)$"

against

H_1: " It is less than $M(T_0)$ "

on the significance level $\delta = 10\%$.

We choose $N = 6$, $\{\alpha_1, \ldots, \alpha_6\} \stackrel{\mathrm{d}}{=} \{0.1, 0.3, 0.5, 0.7, 0.8, 0.9\} \subseteq [0, 1)$, and $K = 3$. According to Theorem 11.19, we define for $i \in \{1, \ldots, 6\}$, $n \in \mathbb{N}$, $(\mu_1, \ldots, \mu_n) \in [F(\mathbb{R})]^n$:

$$\phi_i(\mu_1, \ldots, \mu_n) = \begin{cases} 1, & \text{if } \sup(\mu_0)_{\alpha_i} > \dfrac{\sum_{j=1}^{n} \sup(\mu_j)_{\alpha_i}}{\Gamma^{(n)}_{\delta/2}} \\ 0, & \text{otherwise} \end{cases}$$

and

$$\phi(\mu_1, \ldots, \mu_n) = \begin{cases} 1, & \text{if } \sum_{i=1}^{6} \phi(\mu_1, \ldots, \mu_n) \geq 3 \\ 0, & \text{otherwise .} \end{cases}$$

In our case we obtain

$$\begin{aligned} &\phi_1[M(\text{age}(1)), \ldots, M(\text{age}(4))] = \ldots \\ &= \phi_3[M(\text{age}(1)), \ldots, M(\text{age}(4))] = 0 \end{aligned}$$

and

$$\begin{aligned} &\phi_4[M(\text{age}(1)), \ldots, M(\text{age}(4))] = \ldots \\ &= \phi_6[M(\text{age}(1)), \ldots, M(\text{age}(4))] = 1 \end{aligned}$$

which yields

$$\phi[M(\text{age}(1)), \ldots, M(\text{age}(4))] = 1;$$

i.e. the null hypothesis has to be rejected.

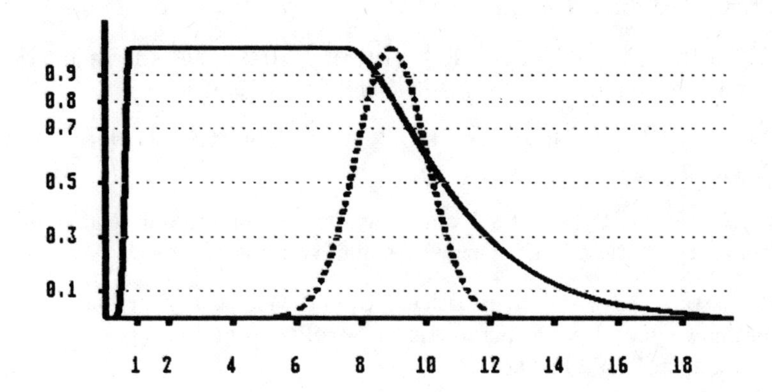

Fig. 12.5. $M(T_0)$ and the confidence interval of part (b).

REFERENCES

[1] ARROW,K.J. and HAHN,F.H. (1971). General competitive analysis. Holden Day Inc., San Francisco.

[2] ARTSTEIN,Z. and VITALE,R.A. (1975). A strong law of large numbers of random compact sets. Ann. Probability 3, 879-882.

[3] AUMANN,J. (1965). Integral of set valued functions. J. Math. Anal. Appl. 12, 1-22.

[4] BANDEMER,H. (1985). Evaluating explicit functional relationships from fuzzy observations. Fuzzy Sets and Systems 16, 41-52.

[5] BANDEMER,H. and SCHMERLING,S.(1984). Construction of estimation regions for the parameters of explicit functional relationships. Biometr. J. 26, 123-130.

[6] BANDEMER,H. and SCHMERLING,S. (1985). Evaluating an explicit functional relationship by fuzzyfying the statement of its satisfying. Biometr. J. 27, 149-157.

[7] BANDEMER,H. and SCHMERLING,S. (1985). Methods to estimate parameters in explicit functional relationships. Freiberger Forschungshefte D 170, 69-90.

[8] BANDEMER, H. and GERLACH, W. (1985). Evaluating implicit functional relationships from fuzzy observations. Freiberger Forschungshefte D 170, 101-118.

[9] BANDEMER, H., ALBRECHT, M. and KRAUT, A. (1985). On using Fourier series in characterizing particle shape. Particle characterizing 2, 98-103.

[10] BEDDOW, J.K. (1984). Analysis. Particle Characterizing in Technology. Vol II: Morphological CRC Press Inc., Boca Raton.

[11] BELLMAN,R.E. and GIERTZ,M. (1973). On the analytic formalism of the theory of fuzzy sets. Inform. Sci. 5, 149-156.

[12] BILLINGSLEY,P. (1968). Convergence of probability measures. Wiley, New York.

[13] BONISSONE,P.P. (1982). A fuzzy sets based linguistic approach: Theory and applications. In: GUPTA,M.M. and SANCHEZ,E. Approximate reasoning in decision analysis. North Holland Publ., Amsterdam.

[14] BREIMAN,L. (1968). Probability. Addison Wesley, Reading, Massachusetts.

[15] BUCK-EMDEN,R., CORDES,R. and KRUSE,R. (1987). Processor power considerations - An application for fuzzy markov chains . Fuzzy Sets and Systems, 21, 289-301.

[16] BUTANARIU,D. (1985). A note on the measurability of the fuzzy mappings. Preprint.

[17] CHAPIN,E.W. (1974). Set-valued theory.
Part I: Notre Dame J. Formal Logic 4, 619-634.
Part II: Notre Dame J. Formal Logic 5, 255-267.

[18] CHOQUET,G. (1954). A theory of capacities. Ann. Inst. Fourier (Grenoble) V, 131-295.

[19] CHOW,Y.S. and TEICHER,H. (1978). Probability theory. Springer Verlag, New York.

[20] CRAMER,H. (1961). Mathematical methods of statistics. Princeton University Press.

[21] CRESSIE,N. (1979). A central limit theorem for random sets. Z. Wahrscheinlichkeitstheorie verw. Gebiete 49, 33-47.

[22] CZOGALA,E. (1982). Probabilistic sets under various operations. BUSEFAL 11, 5-13.

[23] CZOGALA,E. (1983). A generalized concept of a fuzzy probabilistic controller. Fuzzy Sets and Systems 11, 287-297.

[24] CZOGALA,E. (1983). An idea of generalized fuzzy probabilistic control algorithm. BUSEFAL 13, 55-65.

[25] CZOGALA,E. (1983). On distribution function description of probabilistic sets and its application in decision making. Fuzzy Sets and Systems 10, 21-29.

[26] CZOGALA,E. (1984). Probabilistic sets in decision making and control. Verlag TÜV Rheinland, Köln.

[27] CZOGALA,E., GOTTWALD,S. and PEDRYCZ,W. (1982). Contribution to application of energy measure of fuzziness. Fuzzy Sets and Systems 8, 205-214.

[28] CZOGALA,E., GOTTWALD,S. and PEDRYCZ,W. (1983). Logical connectives of probabilistic sets. Fuzzy Sets and Systems 10, 299-308.

[29] CZOGALA,E. and HIROTA,K. (1986). Probabilistic sets: Fuzzy and stochastic approach to decision, control and recognition processes. Verlag TÜV Rheinland, Köln.

[30] CZOGALA,E. (1984). An introduction to probabilistic L-valued logic Fuzzy Sets and Systems 13, 179-185.

[31] DE FINETTI,D. (1974). Theory of probability, Vol I,II. Wiley, New York.

[32] DE LUCA,A. and TERMINI,S. (1972). Algebraic properties of fuzzy sets. J. Math. Anal. Appl. 40, 373-386.

[33] DE LUCA,A. and TERMINI,S. (1972). Definition of a non probabilistic entropy in the setting of fuzzy set theory. Information and Control 20, 301-312.

[34] DEBREU,G. (1965). Integration of correspondences. Proceed. Fifth Berkeley Symp. Math. Statist. and Probability 2, 351-372. Univ. California Press.

[35] DELGADO,M., VERDEGAY,J.L. and VILA,M.A. (1985). Testing fuzzy hypotheses. A Bayesian approach. In: GUPTA,M.M., KANDEL,A., BANDLER,W. and KISZKA,J.B. Approximate reasoning in expert systems, 307-316. North Holland Publ., Amsterdam.

[36] DEMPSTER, A.P. (1967). Upper and lower probabilities induced by a multivalued mapping. Ann. Math. Statistics 38, 325-339.

[37] DUBOIS,D. and PRADE,H. (1980). Fuzzy Sets and Systems: Theory and appliations. Academic Press, New York.

[38] DUBOIS,D. and PRADE,H. (1983). Fuzzy sets and statistical data. Ensembles Flous-82, Rapport du LIST 174.

[39] DUDA,R.O., HART,P.E., KONOLIGE,P. and REBOH,R. (1979). A computer based consultant for mineral exploration. Final Tech. Report, SRI International, Menlo Park, California.

[40] DUDEWICZ,E. (1976). Introduction to statistics and probability. Holt, Rinehard and Winston, New York.

[41] EIKE,M. and FRECKMANN,J. (1985). Entwurf einer Sprache und Realisierung eines Interpreters zur Durchführung von Rechenoperationen und statistischen Untersuchungen mit unscharfen Daten. Diplomarbeit Braunschweig.

[42] ELKER,J., POLLARD, D. and STUTE,W. (1979). Gliwenko-Cantelli theorems for classes of convex sets. Adv. Appl. Prob. 11, 820-833.

[43] ETEMADI,N. (1981). An elementary proof of the strong law of large numbers. Z. Wahrscheinlichkeitstheorie verw. Gebiete 55, 119-122.

[44] FERON,R. (1972). Characteristiques de positions et de dispersion des aleatoires et des multialeatoires. Publ. Econometriques 5, fasc.2, 99-125.

[45] FERON,R. (1976). Economie d'echange aleatoire floue. Comptes Rendus Acad. Sciences, Serie A, 1379-1384.

[46] FERON,R. (1976). Ensembles flous, ensembles aleatoires flous et economie aleatoire floue. Publ. Econometriques 9, fasc.1, 25-64.

[47] FERON,R. (1976). Ensembles aleatoires flous. C.R. Acad. Sci. Paris, Ser. A. 282, 903-906.

[48] FERON,R. (1976). Ensembles flous attaches a un ensemble aleatoire flou. Publ. Econometriques 9, fasc.2, 51-66.

[49] FERON,R. (1979). Sur les ensembles aleatoires flous dont la fonction d'aparentance prend ses valeurs dans un treillis distributif ferme.

[50] FERON,R. (1979). Sur les notions de distance et d'ecart dans une structure floue et leur application aux ensembles aleatoires flous. Compte Rendus Acad. Sciences Serie A, 35-38.

[51] FERON,R. (1981). Probabilistic and statistical study of random fuzzy sets whose referencial is R**n. In: LASKER, G.E. Applied systems and cybernetics, Vol VI, 2831-283. Pergamon Press, New York.

[52] FISZ,M. (1980). Wahrscheinlichkeitstheorie und mathematische Statistik, 10 . Auflage. VEB Deutscher Verlag der Wissenschaften, Berlin(Ost).

[53] FORTET,R. and KAMBOUIZA,M. (1976). Ensembles aleatoires et ensembles flous . Publ. Econometriques 9, No. 1.

[54] FREELING, A.N.S. (1981). Fuzzy probabilities and the value of coherence. In: LASKER,G.E. Applied systems and cybernetics, Vol VI, 2991-2996 Pergamon Press, New York.

[55] FRENCH,S. (1984). Fuzzy decision analysis, some criticism. Management Science 20, 29-44.

[56] FU,K.S., ISHIZUKA,M. and YAO,J.T.P. (1982). Application of fuzzy sets in earthquake engineering, In: YAGER,R.R. Recent developments in fuzzy set and possibility theory, 504-518. Pergamon Press, New York.

[57] FUNG,L.W. and FU,K.S. (1974). The k'th optimal policy algorithm for decision-making in a fuzzy environment. In: EYKHOFF,P. Identification and system parameter estimation, 102. North Holland Publ., Amsterdam.

[58] FUNG,L.W. and FU,K.S. (1975). An axiomatic approach to rational decision-making in a fuzzy environment. In: ZADEH,L.A., FU,K.S., TANAKA,K. and SHIMURA,M. Fuzzy sets and their applications to cognitive and decision processes, 227-256. Academic Press, New York.

[59] GAENSSLER,P. and STUTE,W. (1979). Empirical processes: A survey of results for independent and identically distributed variables. Ann. probability 7, 193-243.

[60] GAINES, B.R. (1978). Fuzzy and probability uncertainty logic. Information and control 38, 154-169.

[61] GENTILHOMME,Y. (1968). Les ensembles flous en linguistique. Cahiers de Ling. Theor. et Appl., 47-65.

[62] GILES,R.G. (1985). The concept of grade of membership. 1.IFSA Congress, Palma de Mallorca.

[63] GINE,E., HAHN,M.G. and ZINN,J. (1982). Limit theorems for random sets: An application of probability in Banach space results. Proceed. Fourth Int. Conf. on Probability in Banach Space Results, to appear.

[64] GIRI,N. (1975). Introduction to probability and statistics.
Part II:. Statistics: Marcel Dekker Inc., New York. Textbooks and monographs, Vol 7.

[65] GOGUEN,J.A. (1967). L-fuzzy sets. J. Math. Anal. Appl. 18, 145-174.

[66] GOGUEN,J. A. (1969). The logic of inexact concepts. Synthese 19, 325-373.

[67] GOGUEN,J. A. (1974). Concept representation in natural and artificial languages: Axioms, extensions and applications for fuzzy sets. Int. J. Man-Machine Stud. 6, 513-561.

[68] GOODMAN,I.R. (1980). Identification of fuzzy sets with class of canonical induced random sets. Proceed. 19th IEEE Conf. Decision and Control, 352-357.

[69] GOODMAN,I.R. (1981). Application of a combined probabilistic and fuzzy set technique to the attribute problem in ocean surveillance. Proceed. 20th IEEE Conf. Decision and Control, 1409-1412.

[70] GOODMAN,I.R. (1981). Fuzzy sets as random level sets: Implications and extensions of the basic results. In: LASKER,G.E. Applied systems and cybernetics, Vol VI, 2757-2766. Pergamon Press, New York.

[71] GOODMAN,I.R. (1982). Characterizations of n-ary fuzzy set operations which induce homomorphic random set operations. In: GUPTA,M.M. and SANCHEZ,E. Fuzzy information and decision processes, 203-212. North Holland Publ., New York.

[72] GOODMAN,I.R. (1982). Fuzzy sets as equivalence classes of random sets. In: YAGER,R.R. Recent developments in fuzzy sets and possibility theory, 327-432. Pergamon Press, New York.

[73] GOODMAN,I.R. (1982). Fuzzy sets as equivalence classes of random sets. In: YAGER,R.R. Recent developments in fuzzy set and possibility theory. Pergamon Press, New York.

[74] GOODMAN,I.R. (1982). PACT: Possibility approach to correlation and tracking. Proceed. 16th Asilomar Conf. Circuits, Syst., Comp. (IEEE), 359-363.

[75] GOODMAN,I.R. (1984). An approach to the target data association problem using subjective and statistical information. Proceed. Am. Contr. Conf., 587-592.

[76] GOODMAN,I.R. (1984). Some new results concerning random sets and fuzzy sets. Inform. Sci. 34, 93-113.

[77] GOODMAN,I.R. and NGUYEN,H.T. (1986). Uncertainty models for knowledge-based systems. North Holland Publ. New York.

[78] GRENADER,U. (1976). Pattern synthesis, lectures in pattern theory, Vol I. Springer Verlag, New York.

[79] GRENADER,U. (1978). On mathmatical semantics: A pattern theoretic view. Division of applied mathematics, Brown University.

[80] GRENADER,U. (1978). Pattern analysis, lectures in pattern theory, Vol II. Springer Verlag, New York.

[81] HALMOS,P.R. (1950). Measure theory. Van Nostrand, New York.

[82] HALMOS,P.R. (1974). Measure theory. Springer Verlag, New York.

[83] HARTUNG,J. (1985). Statistik. R. Oldenbourg, München.

[84] HENZE,E. (1985). Einführung in die Masstheorie. 2.Auflage, Bibliographisches Institut, Mannheim.

[85] HILDENBRANDT,W. (1974). Core and equilibria of a large economy. Princeton University Press.

[86] HIROTA,K. (1977). Concepts of probabilistic sets. Proceed. IEEE Conf. on Decision and Control, 1361-1366, New Orleans.

[87] HIROTA,K. and IIJIMA,T. (1978). A decision making model- A new approached based on the concepts of probabilistic sets. Int. Conf. on Cybernetics and Society, Tokyo, 1348-1353.

[88] HIROTA,K. and IIJIMA,T. (1979). Analysis of a single probabilistic set based on the concept of subjective entropy. Systems Computers Controls 10, 55-64.

[89] HIROTA,K. and IIJIMA,T. (1979). Logical basis in probabilistic set theory. Systems Computers Controls 10, 45-54.

[90] HIROTA,K. (1979). Extended fuzzy expressions of probabilistic sets. In: GUPTA,M.M., RAGADE,R.K. and YAGER,R.R. Advances in fuzzy set theory and applications, 201-214. North Holland Publ., Amsterdam.

[91] HIROTA,K. (1981). Concepts of probabilistic sets. Fuzzy Sets and Systems 5, 31-46.

[92] HIROTA,K., CZOGALA,E. and PEDRYCZ,W. (1981). Probabilistic sets in identification of fuzzy systems. Internat. Conf. Systems Sciences 7, Wroclaw, 84-86.

[93] HIROTA,K. and PEDRYCZ,W. (1981). Fuzzy-probabilistic algorithms in identification of fuzzy systems. IFAC 8th Triennial World Congress, Vol V, Kyoto, 52-56.

[94] HIROTA,K. and PEDRYCZ,W. (1982). Fuzzy systems identification via probabilistic sets. Inform. Sci. 27.

[95] HIROTA,K. and PEDRYCZ,W. (1982). On identification of fuzzy systems under existence of vagueness. IFAC, Theory and Application of Digital Control, Vol 1, New Delhi, 11-15.

[96] HIROTA,K. and PEDRYCZ,W. (1983). Analysis and synthesis of fuzzy systems by the use of probabilistic sets. Fuzzy Sets and Systems 10, 1-13.

[97] ISHIZUKA,M., FU,K.S. and YAO,J.T.P. (1981). Inference procedure with uncertainty for problem reduction method. Technical Report TREE, 81-83.

[98] ISHIZUKA,M. ,FU,K.S. and YAO,J.T.P. (1982). A rule-based inference with fuzzy set for structural damage assessment. In: GUPTA,M.M.

and SANCHEZ,E. Approximate reasoning in decision analysis, 261-268. North Holland Publ., Amsterdam.

[99] JAHN,K.U. (1975). Intervall-wertige Mengen. Math. Nachr. 68, 115-132.

[100] JAIN,N.C. and MARCUS,M.B. (1975). Central limit theorem for C(S) valued random variables. J. Funct. Anal. 19, 216-231.

[101] KAMPE DE FERIET,J. (1982). Interpretation of membership functions of fuzzy sets in terms of plausibility and belief. In: GUPTA,M. and SANCHEZ,E. Fuzzy information and decision process, 93-98. North Holland Publ., Amsterdam.

[102] KANDEL,A. (1979). On fuzzy statistics. In: GUPTA,M.M., RAGADE,R.K. and YAGER,R.R. Advances in fuzzy set theory and applications, 181-200. North Holland Publ., Amsterdam.

[103] KAUFMANN,A. and GUPTA,M.M. (1985). Introduction to fuzzy arithmetic. Van Nostrand, New York.

[104] KENDALL,D.G. (1974). Foundations of a theory of random sets. In: HARDING,E.F. and KENDALL, D.G. Stochastic geometry, 322-376. Wiley, New York.

[105] KENDALL,D.G. (1974). Random sets and integral geometry. In: HARDING,E.F. and KENDALL, D.G. Stochastic geometry, 322-376. Wiley, New York.

[106] KENDALL,D.G. and MORAN,P.A. (1963). Geometrical probability. Griffin, London.

[107] KLEMENT,E.P. (1982). Construction of fuzzy sigma-algebras using triangular norms. J. Math. Anal. Appl. 85, 543-565.

[108] KLEMENT,E.P., PURI,M.L. and RALESCU, D.A. (1983). Central limit theorem for fuzzy random variables. to appear.

[109] KLEMENT,E.P. PURI,M.L. and RALESCU, D.A. (1983). Law of large numbers for fuzzy random variables. Submitted for publication.

[110] KLEMENT,E.P., SCHWYHLA,W. and LOWEN,R. (1981). Fuzzy probability measures. Fuzzy Sets and Systems 5, 21-30.

[111] KRANTZ,D.H., LUCE,R.D., SUPPERS,P. and TVERSKY,A. (1971). Foundations of measurement, Vol I. Academic Press, New York.

[112] KRUSE,R. (1982). The strong law of large numbers for fuzzy random variables. Inform. Sci. 28, 233-241.

[113] KRUSE,R. (1983). Schätzfunktionen für Parameter von unscharfen Zufallsvariablen. Habilitationsschrift, Technische Universität Braunschweig.

[114] KRUSE,R. (1983). On the entropy of fuzzy events. Kybernetes 12, 53-57.

[115] KRUSE,R. (1984). Probabilistische Mengen. Wiss. Abt. der Braunschw. Wiss. Gesellschaft 36, 7-13.

[116] KRUSE,R. (1984). Statistical estimation with linguistic data. Inform. Sci. 33, 197-207.

[117] KRUSE,R. (1985). On a language and an interpreter for calculation and statistics on linguistical data. Proceed. First IFSA Congress, Palma de Mallorca.

[118] KRUSE,R. (1986). Characteristics of linguistic random variables. In: DI NOLA,A. and VENTRE,A.G.S. Topics in the mathematics of fuzzy systems, 219-230. Verlag TÜV Rheinland, Köln.

[119] KRUSE,R. (1986). On the entropy of lambda-additive fuzzy measures. J. Math. Anal. Appl. 117, to appear.

[120] KRUSE.R. (1987). On the variance of random sets. J. Math. Anal. Appl. 121, to appear.

[121] KRUSE,R. (1988). On a software tool for statistics with linguistic data. Fuzzy Sets and Systems, 25.

[122] KRUSE,R. and MEYER,K. D. (1986). Statistics with fuzzy data. In: BANDLER,W. and KANDEL,A. Recent developments in the theory and applications of fuzzy sets. Proceed. North American Fuzzy Inform. Processing Soc. '86 (NAFIPS '86), New Orleans.

[123] KRUSE,R. and MEYER,K. D. (1987). Fuzzy Markov chains and their applications to processor power considerations. Proceed. of the 11th Polish Conference of the Theory of Machines and Mechanism, Zakopane, Poland.

[124] KRUSE,R. and MEYER,K. D. (1987). Confidence areas for the parameters of linguistic random variables. In: FEDRIZZI,M. and KACPRZYK,J. Combining fuzzy imprecision with probabilistic uncertainty in decision making. Springer Verlag, Berlin(West).

[125] KRUSE,R. and MEYER,K. D. (1987). Confidence intervals for the parameter of the normal distribution in the presence of vague data. Proceed. 2nd Polish Symposium on Interval and Fuzzy Mathematics, Poznan.

[126] KURATOWSKI,K. and RYLL-NARDZEWSKI,C. (1965). A general theory of selectors. Bull. Polish Acad. Sci. 13, 397-403.

[127] KWAKERNAAK,H. (1978). Fuzzy random variables.
Part I : Definitions and theorems. Inform. Sci. 15, 1-15.
Part II: Algorithms and examples for the discreet case. Inform. Sci. 17, 253-278.

[128] LEHMANN,E.L. (1959). Testing statistical hypotheses. Wiley, New York.

[129] LEHMANN,E.L. (1975). Nonparametrics: Statistical methods based on ranks. Holden Day Inc., San Francisco.

[130] LOWEN,R. (1978). On fuzzy complements. Inform. Sci. 14, 107-113.

[131] LOWEN,R. (1980). Convex fuzzy sets. Fuzzy Sets and Systems 3, 291-310.

[132] LYASHENKO,N.N. (1979). On limit theorems for sums of independent compact random subsets in the Euclidean space. Zap. Nauch. Sem. Leningrad Otd. Mat. Inst. 85, 113-128 (in russian).

[133] LYASHENKO,N.N. (1983). Statistics of random compacts in Euclidean space. J. Soviet Math. 21, 76-92.

[134] LYASHENKO,N.N. (1986). Random sets and image analysis. Theor. Prob. 30(1), 212-219, Meating abstract.

[135] MATHERON,G. (1975). Random sets and integral geometry. Wiley,New York.

[136] MEYER,K.D. and KRUSE,R. (1985). Fuzzy statistics: Limit theorems. Preprint. Braunschweig.

[137] MEYER,K.D. and KRUSE,R. (1985). On the variance of a fuzzy random variable. Preprint. TU Braunschweig.

[138] MEYER,K.D. and KRUSE,R. (1985). Statistics with imprecise data.
Part I : Distribution functions and i.i.d. sequences.
Part II: On the theorem of Gliwenko.
Preprint. Braunschweig.

[139] MEYER,P.A. (1966). Probabilites et potentiel. Hermann, Paris.

[140] MICHAEL,E. (1951). Topologies on spaces of subsets. Trans. Amer. Math. Soc. 71, 152-182.

[141] MIYAKOSHI,M. and SHIMBO,M. ( ). Set representations of a fuzzy set and its applications to Jensen's inequality. Fuzzy Sets and Systems, to appear.

[142] MIYAKOSHI,M. and SHIMBO,M. (1984). A strong law of large numbers for fuzzy random variables. Fuzzy Sets and Systems 12, 133-142.

[143] MIYAKOSHI,M. and SHIMBO,M. (1984). An individual ergodic theorem for fuzzy random variables . Fuzzy Sets and Systems 13, 285-290.

[144] MIZUMOTO,M. and TANAKA,K. (1979). Some properties of fuzzy numbers. In: GUPTA,M.M., RAGADE,R.K. and YAGER,R.R. Advances in fuzzy set theory and applications, 153-164. North Holland Publ., Amsterdam.

[145] MOORE,R.E. (1966). Interval analysis. Prentince Hall, Englewood Cliffs, New Jersey.

[146] MUELLER,P.H. (1975). Wahrscheinlichkeitsrechnung und mathematische Statistik - Lexikon der Stochastik, 2.Auflage. Akademie Verlag, Berlin(Ost).

[147] NAHMIAS,S. (1978). Fuzzy variables. Fuzzy Sets and Systems 1, 97-110.

[148] NAHMIAS,S. (1979). Fuzzy variables in a random environment. In: GUPTA,M.M., RAGADE,R.K. and YAGER,R.R. Advances in fuzzy set theory and applications, 165-180. North Holland Publ., Amsterdam.

[149] NAN-IUN,Z. (1982). A preliminary study of the theoretical basis of fuzzy sets. In: WANG,P.P. Advances in fuzzy set theory and applications. Plenum Press, New York.

[150] NEGOITA,C.V. and RALESCU, D.A. (1975). Applications of fuzzy sets to system analysis. Birkhäuser Verlag, Basel.

[151] NEGOITA,C.V. and RALESCU, D.A. (1975). Representation theorems for fuzzy concepts. Kybernetes 4, 169-174.

[152] NGUYEN,H.T. (1977). On fuzziness and linguistic probabilities. J. Math. Anal. Appl. 61, 658-671.

[153] NGUYEN,H.T. (1978). On conditional possibility distributions. Fuzzy Sets and Systems 1, 299-309.

[154] NGUYEN,H.T. (1978). On random sets and belief functions. J. Math. Anal. Appl. 65, 531-542.

[155] NGUYEN,H.T. (1979). Some mathematical tools for linguistic probabilities. Fuzzy Sets and Systems 2, 53-65.

[156] NGUYEN,H.T. (1984). On entropy of random sets and possibility distributions. In: BEZDEK,J. The analysis of fuzzy information. CRC Press, to appear.

[157] NGUYEN,H.T. (1984). On modelling of linguistic information using random sets. Inform. Sci. 34, 265-274.

[158] NICKEL,K.L. (1980). Interval mathmatics 1980. Academic Press, New York.

[159] NORWICH,A.M. and TURKSEN,I.B. (1982). The construction of membership functions. In: YAGER,R.R. Fuzzy set and possibility theory. Pergamon Press, New York.

[160] NORWICH,A.M. and TURKSEN,I.B. (1982). Meaningfulness in fuzzy set theory. In: YAGER,R.R. Fuzzy set and possibility theory. Pergamon Press, New York.

[161] NORWICH,A.M. and TURKSEN,I.B. (1982). The fundamental measurement of fuzziness. In: YAGER,R.R. Fuzzy set and possibility theory. Pergamon Press, New York.

[162] NORWICH,A.M. and TURKSEN,I.B. (1984). A model for the measurement of membership and the consequences of its empirical implementation. Fuzzy Sets and Systems 12, 1-25.

[163] PEARL,J. (1984). Heuristics. Addison Wesley, Reading.

[164] PEIZHUANG,W. (1982). From fuzzy statistics to random subsets and a preliminary study of the theoretical basis of fuzzy sets. In: WANG,P.P. Advances in fuzzy set theory and applications. Plenum Press, New York.

[165] PEIZHUANG,W. and NAN-IUN,Z. (1983). Falling space, the probabilistic description of fuzzy subsets. J. Math. Res. and Exposition (China) 3, 163-178.

[166] PEIZHUANG,W. and SANCHEZ,E. (1982). Treating a fuzzy subset as a fallable random subset. In: GUPTA,M.M. and SANCHEZ,E. Fuzzy information and decision processes, 213-220. North Holland Publ., New York.

[167] PEIZHUANG,W. and SANCHEZ,E. (1984). Hyperfields and random sets. In: SANCHEZ,E. Proceeding of the IFAC symp. on fuzzy information, knowledge representation and decision analysis. Pergamon Press, New York, 335-339.

[168] PEIZHUANG,W. and XIHUI,L. (1984). Set valued statistics. J. Engin. Math. 1, 43-54.

[169] PEIZHUANG,W., XIHUI,L. and SANCHEZ,E. (1986). Set valued statistics and its application to earthquake engineering. Fuzzy Sets and Systems 86, 201-364.

[170] PFANZAGL,J. (1966). Allgemeine Methodenlehre der Statistik I,II, 2. Auflage. Goeschen-de Gruyter, Berlin(West).

[171] PURI,M.L. and RALESCU, D.A. (1983). Differentials of fuzzy functions. J. Math. Anal. Appl. 91, 552-558.

[172] PURI,M.L. and RALESCU, D.A. (1983). Strong law of large numbers for Banach space valued random sets. Ann. probability 11, 222-224.

[173] PURI,M.L. and RALESCU, D.A. (1985). The concept of normality. Ann. Probability 13, 1373-1379.

[174] PURI,M.L. and RALESCU, D.A. (1985). Limit theorems for random compact sets in Banach space. Math. Proceed. Camb. Phil. Soc. 1985, Vol. 97, 151-158.

[175] PURI,M.L. and RALESCU, D.A. (1986). Fuzzy random variables. J. Math. Anal. Appl. 114, 409-422.

[176] RADSTROEM,H. (1952). An embedding theorem for spaces of convex sets. Proceed. Amer. Math. Soc. 3, 165-169.

[177] RALESCU,A.L. and RALESCU,D.A. (1984). Probability and fuzziness. Inform. Sci. 34, 85-92.

[178] RALESCU,D.A. (1979). A survey of the representation of fuzzy concepts and applications. In: GUPTA,M.M. RAGADE,R.K. and YAGER,R.R. Advances in fuzzy set theory and applications, 77-91. North Holland Publ.

[179] RALESCU,D.A. (1982). Fuzzy logic and statistical estimation. Proceed. 2nd World Conference on Mathematics at the Service of Man

[180] RIPLEY,B.D. (1976). Locally finite random sets: Foundations for point process theory. Ann. Prob. 4, 983-994.

[181] RIPLEY,B.D. (1981). Spatial statistics. Wiley, New York.

[182] ROBBINS,H.E. (1944). On the measure of a random set. Ann. Math. Statist. 15, 70-74.

[183] ROHATGI,V.K. (1976). An introduction to probability theory and mathematical statistics. Wiley, New York.

[184] SAATY,T.L. (1974). Measuring the fuzziness of sets. J. of Cybernetics 4, 53-61.

[185] SAATY,T.L. (1986). Scaling the membership function. Europ. J. Operations Research 25, 320-329.

[186] SAVAGE,L.J. (1971). Elicitation of personal probabilities and expectations. J. Am. Statist. Assoc. 66, 783-801.

[187] SAVAGE,L.T. (1972). The foundations of statistics (2nd ed.). Dover Publ. New York.

[188] SCHEFE,P. (1980). On foundations of reasoning with uncertain facts and vague concepts. Inter. J. Man-Machine Studies 12, 35-62.

[189] SCHWEIZER,B. and SKLAR,A. (1961). Associative functions and statistical triangle inequalities. Publ. Math. Debrecen 8, 169-186.

[190] SCHWEIZER,B. and SKLAR,A. (1983). Probability metric spaces. North Holland Publ., New York.

[191] SHAFER,G. (1976). A mathematical theory of evidence. Princeton Univ. Press.

[192] SHORTLIFFE,E.H. (1976). Computerbased medical consultations: MYCIN. American Elsevier, New York.

[193] SERRA,J. (1982). Image Analysis and mathematical morphology. Academic Press, London.

[194] SKALA,H.T. (1977). On many valued-logics, fuzzy sets, fuzzy logics and their applications. Fuzzy Sets and Systems 1, 129-149.

[195] SPIEGELHALTER,D.J. and KNILL-JONES,R.P. (1984). Statistical and knowledge based approaches to clinical decision-support systems with an application in gastroeterology. J. Roy. Statist. Soc. Ser. A 147, 35-77.

[196] STEIN,W.E. and TALATI,K. (1981). Convex fuzzy random variables. Fuzzy Sets and Systems 6, 271-283.

[197] SUGENO,M. (1974). Theory of fuzzy integrals and its applications, Ph. D. Thesis. Institute of Technology, Tokyo.

[198] TAYLOR,R.L. (1978). Stochastic convergence of weighted sums of random elements in linear spaces, lecture notes in mathematics, 672. Springer Verlag, Berlin(West).

[199] THOLE,U., ZIMMERMANN,H.-J. and ZYSNO,P. (1979). On the suitability of minimum and product operators for the intersection of fuzzy sets. Fuzzy Sets and Systems 2, 167-180.

[200] TORGESON,W.S. (1958). Theory and methods of scaling. Wiley, New York.

[201] TRADER, D.A. and EDDY,W.F. (1981). A central limit theorem for the Minkowski sum of random sets. Preprint, Carnegie-Mellon University, Pittsburgh.

[202] VITALE,R.A. (1981). A central limit theorem for random convex sets. Preprint, Claremont Graduate School, Claremont, California.

[203] WALD,A. (1966). Statistical decision functions. Wiley, New York.

[204] WALLEY,P. and FINE,T.L. (1982). Towards a frequentiest theory of upper and lower probability. Ann. Statist. 10, 741-761.

[205] WATADA, J., FU,K.S. and YAO,J.T.P. (1984). Linguistic assessment of structural damage. Pergamon Press, New York, 335-339.

[206] WEBER,S. (1983). A general concept of fuzzy connectives, negations and implications based on t-norms and t-conorms. Fuzzy Sets and Systems 11, 115-134.

[207] WEIL,W. (1982). An application of the central limit theorem for Banach space-valued random variables to the theory of random sets. Z. Wahrscheinlichkeitstheorie verw. Gebiete 60, 203-208.

[208] XIHUI,L., MENGMEI,W. and PEIZHUANG,W. (1984). Application of fuzzy set theory to earthquake damage forecast. J. of Building Structure 5, 26-43.

[209] YAGER,R.R. (1980). On a general class of fuzzy connectives. Fuzzy Sets and Systems 4, 235-242.

[210] YAGER,R.R. (1983). Probabilities from fuzzy observations. Tech. Report MII-207, Iona College, New Rochelle, New York.

[211] YAGER,R.R. (1986). A characterization of the extension principle. Fuzzy Sets and Systems 4, 235-242.

[212] ZADEH,L.A. (1965). Fuzzy sets. Information and Control 8, 338-353.

[213] ZADEH,L.A. (1968). Probability measures of fuzzy events. J. Math. Anal. Appl. 23, 421-427.

[214] ZADEH,L.A. (1975). The concept of a linguistic variable and its application to approximate reasoning.

Part I : Inform. Sci. 8, 199-249.
Part II : Inform. Sci. 8, 301-357.
Part III: Inform. Sci. 9, 43-80.

[215] ZADEH,L.A. (1978). Fuzzy sets as a basis for a theory of possibility. Fuzzy Sets and Systems 1, 3-28.

[216] ZADEH,L.A. (1980). Fuzzy sets versus probability, Proceed. IEEE 68, No 3, 421.

[217] ZADEH,L.A. (1981). Possibility theory and soft data analysis. In: COBB,L. and THRALL,R.M. Mathematical frontiers of the social and policy scences. Westview Press, Boulder, Co, 69-129.

[218] ZADEH,L.A. (1981). Test-score semantics for natural languages and meaning representation via PRUF. In: RIEGER, B.B. Empirical semantics. Brockmeyer, Bochum.

[219] ZADEH,L.A. (1983). The role of fuzzy logic in management of uncertainty in expert systems. Fuzzy Sets and Systems 11, 199-227.

[220] ZIMMERMANN,H.J. (1985). Fuzzy set theory - and its applications. Kluwer - Nijhoff Publ., Boston and Dordrecht.

LIST OF SYMBOLS

Symbol	Page	Meaning
$B_\omega^{(N)}$	174	variance estimator
$B[t]$	102	
b_U^X	66	
$\mathfrak{C}$	198, 201	
$\{c_0, \ldots, c_M\}$	102	
$C_\alpha[p_1, \ldots, p_n]$	90	
$c_{i,A}^{(i_0)}(p_1, \ldots, p_n)$	95	
cl A	18	closure of $A \subseteq \mathbb{R}$
$cl\,(\tau_n)$	238	
$C_n\,(\mathbb{R})$	43	
co A	17	convex hull of $A \subseteq \mathbb{R}$
co μ	17	convex hull of $\mu \in F\,(\mathbb{R})$
co X	76	convex hull of a f.r.v. X
$C_\omega^{(N)}$	176	
$\mathfrak{C}\,(\Sigma)$	188	class of all nonempty compact subsets of a nonempty set Σ
$\mathfrak{C}\,(\mathbb{R})$	64	class of all nonempty compact subsets of $\mathbb{R}$
$C[t]$	102	
D	81	
d_1	53	metric on $F_C\,(\mathbb{R})$
$\xrightarrow{d_1}$	58	Hausdorff convergence with respect to d_1
$\{d_1, \ldots, d_M\}$	103	
d_∞	53	pseudometric on $F\,(\mathbb{R})$
$\xrightarrow{d_\infty}$	58	Hausdorff convergence with respect to d_∞
$D^2_{\alpha}, n[t]$	221	
$D^2_{\alpha}, p_1, \ldots, p_n(c_1, \ldots, c_n)$	87	
$D^2_{\alpha}(c_1, \ldots, c_n)$	99	

$D^2_{\alpha,p_1,\ldots,p_n}[t]$	108	
d_H	50	Hausdorff pseudometric on $\mathfrak{P}(I\!R)$
$D^2_{\alpha,p_1,\ldots,p_n}(A)$	100	
$\xrightarrow{d_H,Q}$	60, 61	Hausdorff convergence with respect to Q
$D_j[t]$	102	
$D^k X$	72	central moment of order k of a f.r.v. X with respect to χ
$\tilde{D}^k X$	72	central moment of order k of a f.r.v. X with respect to $\tilde{\chi}$
D_U	193	distribution function of a r.v. U
E	212	mapping assigning an $N(\mu,\sigma^2)$-distribution the parameter μ
$E(A)$	10, 22	class of all fuzzy subsets of a non-empty set A
$E_\alpha[\mu_1,\ldots,\mu_n]$	205	
$E_{\alpha,p_1,\ldots,p_n}(c_1,\ldots,c_n)$	87	
$E_{\alpha,p_1,\ldots,p_n}[t]$	104	
$E^2_{\alpha,p_1,\ldots,p_n}(c_1,\ldots,c_n)$	87	
$E^2_{\alpha,p_1,\ldots,p_n}[t]$	104	
$E_j[t]$	102	
$\emptyset$	11	empty set
$E_\omega^{(N)}$	166	estimator for the expected value
$=$	18	equality
$E\,X$	72	expected value of the f.r.v. X with respect to χ
$\tilde{E}\,X$	72	expected value of the f.r.v. X with respect to $\tilde{\chi}$
$E\,X^k$	72	moment of order k of the f.r.v. X with respect to χ
$\tilde{E}\,X^k$	72	moment of order k of the f.r.v. X with respect to $\tilde{\chi}$

$F_C(K)$	23	class of all upper semicontinuous fuzzy sets of $K \subseteq \mathbb{R}$
$F_C(\mathbb{R})$	20	class of all upper semicontinuous fuzzy sets of $\mathbb{R}$
$f_{i_0,A}(p_1,\ldots,p_n)$	96	
$F(K)$	23	class of all normal fuzzy sets of $K \subseteq \mathbb{R}$
$Fl(\mathbb{R})$	5	class of all flou sets of $\mathbb{R}$
$Fl_L(\mathbb{R})$	6	class of all L-flou sets of $\mathbb{R}$
$F_L(\mathbb{R})$	7	class of all L-sets of $\mathbb{R}$
$F_{\text{Lip}}(\mathbb{R})$	189	class of all Lipschitz fuzzy setsof $\mathbb{R}$
$F(\mathbb{R})$	10	class of all normal fuzzy sets of $\mathbb{R}$
f.r.v.	64	fuzzy random variable
f.r.v. *	189	fuzzy random variable in the sense of Klement, Puri, Ralescu
f.r.vector	68	fuzzy random vector
$\tilde{F}_X$	131	distribution function of the f.r.v. X
$\tilde{F}_{(X_1,\ldots,X_n)}$	132	distribution function of the f.r.vector $(X_1,\ldots,X_n)$
$g^{(a,b)}$	12, 42	Gaųssian fuzzy set with the parameters a and b
$\Gamma_\delta^{(n)}$	241	δ-quantile of the Γ-distribution with $p = n$ and $b = 1$
Γ_i	80	
$\geq$	17	bigger or equal (for fuzzy sets)
Γ_Y	193	parameter valued mapping
$\Gamma_Y[X_1,\ldots,X_n]$	194	fuzzy parameter of the f.r.vector $(X_1,\ldots,X_n)$
$h_{a,b}$	15	fuzzy set with the parameters a and b

$(H, \| \|\ \|)$	50	normed Banach space
I_A	10	indicator function of a set A
$I_A[p_1, \ldots, p_n]$	90	
i.i.d.	141	independent and identically distributed
K_n	204	fuzzy confidence interval
L	233	formal language
$\|\|L\|\|$	77, 188	norm of the random set L
$\leq$	19	less or equal
$a \vee b$	7	maximum of two real numbers a and b
$\bigvee_{i \in I} \mu_i$	24	union of the family $\{\mu_i \mid i \in I\}$ of fuzzy sets
$a \wedge b$	7	minimum of two real numbers numbers a and b
$\bigwedge_{i \in I} \mu_i$	24	disjunction of the family $\{\mu_i \mid i \in I\}$ of fuzzy sets
$M(l)$	234, 238	
μ	10	fuzzy set (of the real line)
μ_α	11	strong α-cut of the fuzzy set μ
$\mu_{\overline{\alpha}}$	11	α-level-set of the fuzzy set μ
$\overline{\mu}$	24	complement of the fuzzy set μ
$\|\mu\|$	33	absolute value of the fuzzy set μ
μ^k	33	k-th power of the fuzzy set μ
$[\mu]_n$	43	representation of the fuzzy set μ by the software tool
$\mu_X(U)$	66	degree of acceptability that $U \in \chi$ is original of the f.r.v. X

$\tilde{\mu}_X(U)$	67	degree of acceptability that $U \in \tilde{\chi}$ is original of the f.r.v. X
$\mu_X^{\mathfrak{Y}}(U)$	66	degree of acceptability that $U \in \mathfrak{Y}$ is original of the f.r.v. X
$\mathbb{N}$		the set of all positive integers
N	234	mapping assigning to a term a fuzzy set
$N \in \mathbb{N}$	212	class of all normal distributions
N^*	234	mapping assigning to an operator an operation on $F(\mathbb{R})$
N_α	19	
$(N_i)_\alpha$	80	(for an interval set representation of μ_i)
$N(K)$	23	class of all fuzzy sets of $K \subseteq \mathbb{R}$ with an interval set representation
$N(\mathbb{R})$	19	class of all fuzzy sets of $\mathbb{R}$ with an interval set representation
O	233	set of operators
$(\Omega, \mathfrak{A}, P)$	63	
$(\tilde{\Omega}, \tilde{\mathfrak{A}}, \tilde{P})$	66	
$(\Omega', \mathfrak{A}', P')$	67, 194	
$(\hat{\Omega}, \hat{\mathfrak{A}}, \hat{P})$	66	
$(\Omega \times \Omega', \mathfrak{A} \otimes \mathfrak{A}', P \otimes P')$	67	product probability space
$o_k^i(l)$	234	
P	63, 194	probability measure
$\xrightarrow{P}$	198	convergence in probability (of f.r.v.'s)
$\tilde{P}$	66	
P'	67, 194	

$\hat{P}$	66	
$P \otimes P'$	67	product probability measure
$\tilde{P}(A) \leq \delta$	223	
$P^{(a,b)}$	12	fuzzy set with the parameters a and b
Φ	57, 189	linear isometry
$\Phi(A_1, \ldots, A_n)$	30	image of n subsets $A_1, \ldots, A_n \subseteq \mathbb{R}$ under $\Phi : \mathbb{R}^n \to \mathbb{R}$
$\Phi_\alpha[p_1, \ldots, p_n]$	104	
$\Phi(\mu_1, \ldots, \mu_n)$	30	the image of $(\mu_1, \ldots, \mu_n) \in [F(\mathbb{R})]^n$ under $\Phi : \mathbb{R}^n \to \mathbb{R}$
$\phi[X_1, \ldots, X_n]$	68	
$\phi(\mu_1, \ldots, \mu_n)$	223	test function, depending on $(\mu_1, \ldots, \mu_n) \in [F(\mathbb{R})]^n$
$\phi_i(\mu_1, \ldots, \mu_n)$	223	test function, depending on $(\mu_1, \ldots, \mu_n) \in [F(\mathbb{R})]^n$
p.i.i.d.	141	pairwise independent and identically distributed
p.i.i.s.	142	pairwise independent and identically shaped
$P_i^{(N)}$	166	relative frequency of μ_i
P_n	86	
$\mathfrak{P}(\mathbb{R})$	4, 10	set of all subsets of $\mathbb{R}$
$\Psi_\alpha[p_1, \ldots, p_n]$	87	
$\mathbb{Q}$	54, 55	the set of all rational numbers
$q_{a,b}$	15	fuzzy set with the parameters a and b
$q_\delta^{(n-1)}$	218	quantile of the chi-square-distribution
$Q(K)$	23	class of all fuzzy sets on $K \subseteq \mathbb{R}$ with a normal set representation

$Q(\mathbb{R})$	23	class of all fuzzy sets on $\mathbb{R}$ with a normal set representation
$\mathbb{R}$	4	set of all real numbers
$\overline{\mathbb{R}}^n$	121	closure of $\mathbb{R}^n$
r.v.		random variable
Σ	77	class of f.r.v.'s
S_α	98	
$\tilde{S}_\alpha$	84	
$S_\alpha[\mu_1,\ldots,\mu_n]$	213	
Σ_α	100	
$S_n[x_1,\ldots,x_n]$	122	empirical distribution function of $(x_1,\ldots,x_n) \in \overline{\mathbb{R}}^n$
$S_n[\mu_1,\ldots,\mu_n]$	122	empirical distribution function of $(\mu_1,\ldots,\mu_n) \in [F(\mathbb{R})]^n$
$S_n^2(x_1,\ldots,x_n)$	213, 217	
$S_n^2[\mu_1,\ldots,\mu_n]$	213	
$\subseteq$	18	subset
$\sum_{i=1}^{n} x_i A_i$	30, 32	linear combination of n subsets $A_1,\ldots,A_n$ of $\mathbb{R}$ where $(x_1,\ldots,x_n) \in \mathbb{R}^n$
$\sum_{i=1}^{n} x_i \mu_i$	30	linear combination of $(\mu_1,\ldots,\mu_n) \in [F(\mathbb{R})]^n$ where $(x_1,\ldots,x_n) \in \mathbb{R}^n$
$\sum x_i X_i$	68	
supp X	189	support of the f.r.v.* X
$s(x,y)$	26	t-conorm of x and y where $(x,y) \in \mathbb{R}^2$
T	233	term set of vague data

$\mathcal{T}$	238	class of all trapezoids in $C_n(I\!R)$
$t^{(a,b,c,d)}$	59	trapezoid with the parameters a, b, c, and d
$T_\alpha[\mu_1,\ldots,\mu_n]$	221	for calculating a fuzzy confidence
$t_\delta^{(n-1)}$	212	δ-quantile of the student distribution
T_L	64	Choquet's capacity
$T_n[\mu_1,\ldots,\mu_n]$	178, 197, 199	estimator for a fuzzy parameter where $(\mu_1,\ldots,\mu_n)\in[F(I\!R)]^n$
$t_n[\mu_1,\ldots,\mu_n]$	199	image of $(\mu_1,\ldots,\mu_n)\in[F(I\!R)]^n$ under the estimator t_n
$T_n(x_1,\ldots,x_n)$	203	value of a confidence interval $[T_n,+\infty)$ at $(x_1,\ldots,x_n)\in I\!R^n$
$\mathcal{T}_n$	238	class of all trapezoids in $C_n(I\!R)$
$\Theta_p X$	121	p-quantile of the f.r.v. X with respect to χ
$\tilde{\Theta}_p X$	121	p-quantile of the f.r.v. X with respect to $\tilde{\chi}$
U	63, 67	random variable
U_A	84, 98	r.v. depending on a set $A\in\mathfrak{A}\otimes\mathfrak{A}'$
$U_C(I\!R)$	19	class of all bounded convex fuzzy sets on $I\!R$
$U_{CL}(I\!R)$	189	class of fuzzy sets on $I\!R$
$U(K)$	22	class of all convex fuzzy sets on $K\subseteq I\!R$
$U_n(x_1,\ldots,x_n)$	203	value of a confidence interval $(-\infty,U_n]$ at $(x_1,\ldots,x_n)\in I\!R^n$
$U(I\!R)$	16	class of all convex fuzzy sets on $I\!R$
Var	212	mapping assigning to an $N(\mu,\sigma^2)$ - distribution its parameter σ^2
Var X	62	variance of the f.r.v. X with respect to χ

$\widetilde{\mathrm{Var}}\, X$	72	variance of the f.r.v. X with respect to $\tilde{\chi}$
$\mathrm{Var}\ \alpha, p_1, \ldots, p_n[t]$	104	
$(V_i)(\omega)$	143	random set depending on $\omega \in \Omega$
$(\underline{V}_i)$	143	
$(\overline{V}_i)$	143	
$V_i[t]$	102	
$V[t]$	103	
X	64	fuzzy random variable
χ	65	class of all $\mathfrak{A} \to \mathfrak{B}_1$- measurable r.v.'s
$\tilde{\chi}$	67	class of all $\mathfrak{A} \otimes \mathfrak{A}' \to \mathfrak{B}_1$-measurable r.v.'s
X_α	69	
$(X_1, \ldots, X_n)$	68	fuzzy random vector
$(\overline{X}_i)_\alpha$		
$(\underline{X}_i)_\alpha$		
χ_n^Y	194	class of all Y-distributed i.i.d. random vectors of size n
Y	193	class of (usual) distribution functions
Y_α	83	
$\tilde{Y}_\alpha$	81	
$\underline{Y}_\alpha$	69	
$\overline{Y}_\alpha$	69	
$[\underline{Y}_\alpha \leq x]$	69	
$[\overline{Y}_\alpha \leq x]$	69	
$(\underline{Y}_i)_\alpha$	77, 135	
$(\overline{Y}_i)_\alpha$	77, 135	

SUBJECT INDEX

A

B

C

D

E

F

K

L

M

N

O

W

Y

Z